Attachment of Organisms to the Gut Mucosa

Volume I

Editor

Edgar C. Boedeker, M.D.
Chief, Department of Gastroenterology
Walter Reed Army Institute of Research
Washington, D.C., and
Associate Professor of Medicine
Uniformed Services University of the Health Sciences
Bethesda, Maryland

CRC Press
Taylor & Francis Group
Boca Raton London New York

CRC Press is an imprint of the
Taylor & Francis Group, an **informa** business

First published 1984 by CRC Press
Taylor & Francis Group
6000 Broken Sound Parkway NW, Suite 300
Boca Raton, FL 33487-2742

Reissued 2018 by CRC Press

Library of Congress Cataloging in Publication Data

Main entry under title:

Attachment of organisms to the gut mucosa.

Papers presented at a research workshop held in
Reston, Va., Oct. 1981 and sponsored by the American
Gastroenterological Association.
Bibliography: p.
Includes indexes.
1. Gastric mucosa—Microbiology—Congresses.
2. Gastrointestinal system—Microbiology—Congresses.
3. Attachment mechanisms (Biology)—Congresses.
I. Boedeker, Edgar C., 1942- . II. American
Gastroenterological Association. [DNLM: 1. Bacteria—
Physiology—Congresses. 2. Cell membrane—Microbiology
—Congresses. 3. Escherichia coli—Physiology Congresses. 4. Intestinal mucosa—Microbiology—Congresses.
5. Gastrointestinal diseases—Etiology—Congresses.
6. Pili, bacterial—Physiology—Congresses. WI 100
A883 1981]
QR171.I6A88 1984 616.3'301 83-23217
ISBN 0-8493-5286-X (v. 1)
ISBN 0-8493-5287-8 (v. 2)

A Library of Congress record exists under LC control number: 83023217

ISBN 13: 978-1-315-89086-9 (hbk)
ISBN 13: 978-1-351-06996-0 (ebk)

Visit the Taylor & Francis Web site at http://www.taylorandfrancis.com and the
CRC Press Web site at http://www.crcpress.com

PREFACE

Over the past decade it has become clear that a number of important infectious diseases of the gastrointestinal tract in man and animals are mediated by the attachment of microorganisms to the gut mucosa. The attached organisms gain a selective advantage in being able to divide while resisting normal clearance mechanisms. In the gastrointestinal tract, as in other organ systems, mucosal adherence appears to depend on specific receptor-receptor interactions between surface structures elaborated by bacteria and naturally occurring components of the host mucosal surface.

The most rapid progress in this area has been made by microbiologists in defining bacterial adherence factors. These workers have localized the adhesins (or colonization factor antigens) of enteric pathogens to a class of hair-like appendages termed pili (or fimbriae). In an increasing number of cases, these structures have been isolated, purified, and characterized immunologically and biochemically, and their relations to other virulence properties investigated. As might be predicted from the pronounced species specificity of many enteric infections, the adherence pili thus far defined from various strains are antigenically distinct, although they share similar morphology, physical properties, and subunit structures.

Information has also begun to appear regarding the specific receptors on the host gastrointestinal surface which mediate adherence. In addition, a great deal of information is available concerning the structure/function relationships of the intestinal epithelium and of its overlying glycocalyx and mucus layers. This information has been developed particularly from the point of view of the intestinal mucosa as an efficient digestive-absorptive surface capable of interacting specifically with a variety of luminal components.

In October 1981, a research workshop on "The Attachment of Micro-Organisms to the Intestinal Mucosa" was held in Reston, Virginia, under the sponsorship of the American Gastroenterological Association and with the support of the National Institutes of Health (National Institute of Arthritis, Metabolic, Digestive and Kidney Diseases, National Institute of Allergy and Infectious Diseases, and the Fogarty International Center). This forum brought together two groups of researchers: microbiologists who have been studying colonization factor antigens and gastroenterologists (including physiologists, biochemists, and morphologists) who have been studying the gastrointestinal mucosal surface. The result was an exchange of basic, complementary information through a thorough and systematic review of current knowledge in a field experiencing many recent advances. The present volumes are based on the presentations made at the Reston workshop, and the editor is grateful to the contributors for providing complete reviews of each area as well as for including new data on developments from their laboratories.

Edgar C. Boedeker, M.D.

THE EDITOR

Edgar C. Boedeker, M.D. is currently chief of the Department of Gastroenterology at the Walker Reed Army Institute of Research, Washington, D.C. and Associate Professor of Medicine at the Uniformed Services University of the Health Sciences, Bethesda, Maryland. He graduated from Harvard College, Cambridge, Massachusetts in 1964 and received his M.D. degree from Washington University School of Medicine (St. Louis) in 1968. From 1968 to 1971 he completed internship and resideny training in Internal Medicine as Barnse Hospital-Washington University Medical Center in St. Louis. He then trained in gastroenterology as an NIH research fellow in medicine as Boston University School of Medicine from 1971 to 1973. He is a diplomate of the American Board of Internal Medicine, and of its subspecialty board in Gastroenterology. He returned to Barnse Hospital-Washington University to serve as Chief Resident in Internal Medicine and Assistant Director of the General Medical Service (1973 to 1974) and during this period co-edited the 21st edition of the Washington University manual of Medical Therapeutics. He entered active duty with the U.S. Army in 1974 and is currently a lieutenant colonel (promotable) in the Medical Corps. He has served as Chief of the Gastroenterology Department at the Water Reed Army Institute of Research since 1974. His research interests are in defining the determinants of the Associations between microorganisms and the intestine, and in utilizing this knowledge to develop vaccines against enteric infections. He is a Fellow of the American College of Physicians, and his memberships include the American Gastroenterological Association, the American Association for Microbiology, and the American Federation for Clinical Research.

CONTRIBUTORS

Edgar C. Boedeker, M.D.
Chief, Department of Gastroenterology
Walter Reed Army Institute of Research
Washington, D.C.; and
Associate Professor of Medicine
Uniformed Services University of the Health Sciences
Bethesda, Maryland

J. Robert Cantey, M.D.
Staff Physician
Veterans Administration Medical Center; and
Professor of Medicine
Medical University of South Carolina
Charleston, South Carolina

Christopher P. Cheney, Ph.D.
Research Biochemist
Department of Gastroenterology
Division of Medicine
Walter Reed Army Institute of Research
Washington, D.C.

K. J. Cheng, Ph.D.
Research Scientist
Agriculture Canada Research Station
Lethbridge, Alberta, Canada

Steven Clegg, Ph.D.
Assistant Professor
Department of Microbiology
University of Iowa
Iowa City, Iowa

J. W. Costerton, Ph.D.
Professor of Biology
University of Calgary
Calgary, Canada

Carl F. Deneke, Ph.D.
Special Scientific Staff
Infectious Disease Service
New England Medical Center
Boston, Massachusetts

William O. Dobbins, III, M.D.
Associate Chief of Staff for Research
Veterans Administration Medical Center; and
Professor of Internal Medicine
University of Michigan Medical School
Ann Arbor, Michigan

Dolores G. Evans, Ph.D.
Associate Professor of Clinical Microbiology
The University of Texas Medical School
Houston, Texas

Doyl J. Evans, Ph.D.
Department of Microbiology
The University of Texas Medical School
Houston, Texas

Ahmed Faris, B.S.
Department of Bacteriology and Epizootology
Biomedicum
Uppsala, Sweden

Ralph A. Giannella, M.D.
Professor of Medicine
Director, Division of Digestive Diseases
University of Cincinnati
College of Medicine
Cincinnati, Ohio

R. J. Gibbons, Ph.D.
Senior Staff Member
Forsyth Dental Center
Boston, Massachusetts

Donald Goldmann, M.D.
Hospital Epidemiologist
Director, Bacteriology Laboratory
Children's Hospital; and
Assistant Professor of Pediatrics
Harvard Medical School
Boston, Massachusetts

Sherwood L. Gorbach, M.D.
Chief, Infectious Diseases
New England Medical Center; and
Professor of Medicine and Microbiology
Tufts University School of Medicine
Boston, Massachusetts

Nicholas G. Guerina, M.D., Ph.D.
Research Associate
Department of Infectious Diseases; and
Clinical Fellow in Medicine
Children's Hospital
Boston, Massachusetts

Stellan Hjerten, Ph.D.
Professor of Biochemistry
Institute of Biochemistry
University of Uppsala
Uppsala, Sweden

Lindsey R. Inman
Graduate Student
Department of Pathology
Medical University of South Carolina
Charleston, South Carolina

Richard E. Isaacson, Ph.D.
Assistant Professor of Epidemiology
University of Michigan
Ann Arbor, Michigan

Mats Lindahl, B.Sc.
Department of Bacteriology and
Epizootiology
College of Veterinary Medicine
Biomedicum
Uppsala, Sweden

Harley W. Moon
Veterinary Pathologist
USDA-National Animal Disease Center
Ames, Iowa

Marian R. Neutra, Ph.D.
Associate Professor of Anatomy
Harvard Medical School
Boston, Massachusetts

K. R. Rozee, Ph.D.
Professor of Microbiology
Dalhousie University
Halifax, Nova Scotia, Canada

Paul L. Runnels
Veterinary Pathologist
USDA-National Animal Disease Center
Ames, Iowa

David A. Sack, Ph.D.
Assistant Professor of Medicine
Division of Geographic Medicine
Johns Hopkins University
School of Medicine
Baltimore, Maryland

Dwayne C. Savage, Ph.D.
Professor of Microbiology
University of Illinois
Urbana, Illinois

R. Sellwood, Ph.D.
Agricultural and Food Research Council
Institute for Research in Animal Diseases
Compton, Newbury, England

Nathan Sharon, Ph.D.
Head, Department of Biophysics
Dean, Faculty of Biophysics-
Biochemistry
The Weizmann Institute of Science
Rehovot, Israel

Grace M. Thorne, Ph.D.
Assistant Professor of Medicine
Tufts University School of Medicine
Boston, Massachusetts

Martin H. Ulshen, M.D.
Associate Professor of Pediatrics
University of North Carolina
School of Medicine
Chapel Hill, North Carolina

Torkel Wadström, M.D., Ph.D.
Professor
Department of Bacteriology
Biomedicum
Uppsala, Sweden

Jacqueline L. Wolf, M.D.
Assistant Professor of Medicine Harvard
Medical School; and
Associate Physician
Brigham and Women's Hospital
Boston, Massachusetts

TABLE OF CONTENTS

Volume I

Morphology and Bacterial Adherence

Volume II

Enteric Bacterial Adherence Systems in Animals and Man

Chapter 1

ADHERENCE OF THE NORMAL FLORA

Dwayne C. Savage

TABLE OF CONTENTS

I. INTRODUCTION

A. Normal Vs. Indigenous Microflora

The concept is widely held that a "normal microflora" lives on the skin and mucous membranes of the upper respiratory, genitourinary, and intestinal tracts of man and other animals. The term "normal microflora" may mislead, however, as an encompassing description of the microbial populations to be found on body surfaces and in the contents of certain areas of the intestines.[1] Findings from studies conducted primarily within the past decade support a concept that the body surfaces and cavities and the microorganisms found on or in them constitute open ecosystems obeying all natural laws governing such ecosystems.[2] Thus, the microorganisms found in the systems can be recognized to be either autochthonous species (i.e., indigenous inhabitants of communities established on or in the body), or allochthonous (transient) species only passing through the region or temporarily colonizing a habitat vacated by its autochthonous inhabitants for one reason or another. Allochthonous species isolated at any given time from a specimen from the body should not be regarded as members of a microbiota "normal" to the region from which the specimen was taken.[2]

The concept of normalcy of the microflora is complicated by yet another problem. Some microbial species that can be isolated from the human body can cause diseases in humans under certain circumstances, i.e., be pathogenic.[3,4] Both allochthonous and autochthonous species may have such capacities. Thus, a microbial isolate thought to be a member of the "normal" microflora may well be the etiological agent of a disease the person is suffering at the time the culture was taken. "Normal" as applied to the microflora of the body is an awkward concept at best. More rational is reference to the microorganisms isolated from a body surface at any given time as "the microflora" or "microflora of the ——" (area of the body at issue). Such reference does not confuse and leaves open the question (when information is not available) whether or not the isolate is indigenous or transient.

B. Goals of This Chapter

Microorganisms of many genera and species have been found associating closely with epithelial surfaces of mucous membranes in various areas of the bodies of animals of numerous species including man.[5-7] Some of these microbial types are known to adhere to substances on or in the mucosal epithelia with which they associate. In this chapter some such microorganisms are discussed. The discussion has two goals; (1) to summarize evidence that indigenous microorganisms associate with the surfaces of cells in the epithelium of mucosae in various areas of the mammalian gastrointestinal tract, and (2) to describe and evaluate some evidence from one approach to study the mechanisms by which microorganisms adhere to mucosal epithelial cells.

II. ASSOCIATION OF INDIGENOUS MICROORGANISMS WITH GASTROINTESTINAL MUCOSAL EPITHELIA

A. Diversity of Animals, Microorganisms, and Epithelial Habitats

1. Microbial Associations and Changes in Epithelial Architecture

Microorganisms of numerous species have been seen in close association with epithelial surfaces in all parts of the gastrointestinal tracts of putatively normal laboratory rodents and chickens. Likewise, they have been observed associating with surfaces in various areas of the tracts of mammals other than rodents and including man (Table 1). Most such information has been derived from studies in which scanning and transmission electron microscopy were used to detect the organisms on the surfaces. Thus, in most cases, little is known about the mechanisms by which the microbes maintain themselves in the epithelial habitats. In fact,

Table 1
MICROORGANISMS ASSOCIATING WITH EPITHELIA IN THE GASTROINTESTINAL TRACTS OF MAMMALS AND BIRDS[a]

Microbial genus[b]	Animal type	Area of tract	Type of epithelium
Microorganisms that Associate with No Major Change in Epithelial Histology			
Lactobacillus	Rats, mice, hamsters, swine	Stomach	Keratinized squamous
Lactobacillus	Chickens	Crop	Keratinized squamous
Torulopsis	Rats, mice, guinea pigs (?)[c]	Stomach	Columnar
Spirochetes	Humans, cats, dogs	Stomach	Columnar
Gram-positive and Gram-negative bacteria	Humans	Small intestine	Columnar
Rod- and coccal-shaped bacteria	Guinea pigs	Small intestine	Columnar
Fusiform-shaped bacteria (*Clostridium* and other genera)	Rats, mice	Cecum, colon	Columnar
Fusiform- and helical-shaped bacteria	Guinea pigs	Cecum, colon	Columnar
Helical-shaped bacteria; spirochetes	Rats, mice, dogs	Cecum, colon	Columnar
Gram-positive and Gram-negative bacilli	Koalas	Cecum	Pseudostratified columnar
Spirochetes	Swine	Colon	Columnar
Gram-positive and Gram-negative bacteria	Chickens, swine, humans	Large bowel	Columnar
Rod-, coccal-, and helical-shaped bacteria	Sheep, cows	Rumen	Squamous
Microorganisms that Associate with Major Change in Epithelial Histology			
Filamentous bacteria	Rats, mice, chickens, dogs (?)	Small intestine	Columnar
Rod-shaped bacteria	Rats	Colon	Columnar
Spirochetes and helical-shaped bacteria	Rhesus monkeys	Colon	Columnar
Fusiform-shaped bacteria	Guinea pigs	Cecum	Columnar

[a] Table adapted from Reference 7; see also References 10 and 11.
[b] Or morphological type.
[c] (?) evidence shows only that some individuals tested contain microbial type indicated.

in most cases, evidence is not available concerning the structure(s) to which the organisms adhere on the surface, or in many instances, even whether they "adhere" to any structure at all. A reasonable theoretical case can be made that microbes can colonize mucus on a particular surface and only appear in microscopic preparations to adhere to epithelial cells underlying the mucus.[7] In any case, in associating with epithelia, microorganisms may or may not obviously alter the architecture of underlying epithelial cells.[7]

2. *Microorganisms that Associate and Do Not Change Epithelial Architecture*

Many microbial species have been reported to associate with and not obviously alter the architecture of the gastrointestinal mucosa (Table 1). In some of these instances, the microorganisms surely adhere to the underlying mucosa. Indigenous lactic acid bacteria that associate with the keratinizing squamous epithelium in the esophagi and stomachs of rodents, the crops of chickens, and the pars oesophagia of swine undoubtedly adhere to the keratinized cells on the surface of the epithelium.[8] At this writing, however, such a statement cannot be made about microorganisms reported to associate with and not alter the architecture of mucosa covered with columnar secreting epithelium (Table 1). Some such microbial types, for instance the *Torulopsis* species that colonizes the gastric secreting epithelium in rodents,[7] may adhere by some means to structures on those surfaces.[9] In no other cases, however, does evidence from experimentation support an hypothesis that indigenous organisms reported to associate with and not alter the architecture of columnar epithelia also adhere to some structure on the surfaces.

3. Microorganisms that Associate and Change Epithelial Architecture

By contrast, indigenous microorganisms that alter the architecture of the mucosal surfaces with which they associate clearly adhere to structures on those surfaces (Table 1). For example, the segmented, filamentous bacteria observed on the epithelium of the small bowels of laboratory rodents and chickens and, under certain circumstances also dogs, undoubtedly adhere to the membranes of the columnar epithelial cells themselves (Table 1). Similarly, rod-shaped bacteria reported to be present in the colons of rats and spirochetes and spiral-shaped bacteria that colonize cecal crypts in rats and the colonic mucosa in rhesus monkeys (and possibly also man) surely adhere by one end to the microvillous membrane (or its glycocalyx) of the epithelial cells (Table 1). However, such documented cases are rare for putatively indigenous microorganisms.

III. MECHANISMS BY WHICH INDIGENOUS MICROORGANISMS ASSOCIATE WITH GASTROINTESTINAL EPITHELIAL SURFACES

A. Gastrointestinal Epithelial "Surfaces"

Epithelial surfaces in gastrointestinal tracts are complex architecturally and physiologically. Indeed in reference to their colonization by microorganisms, such surfaces can best be regarded as three-dimensional compartments. On secreting gastrointestinal epithelium, such compartments lie on dynamic microvillous membranes consisting of a lipid-bilayer and its associated intrinsic and extrinsic glyco- and lipoproteins.[12] Some of the latter molecules undoubtedly form the glycocalyx overlying the membrane, which is in turn overlain by a layer of mucinous glyco- and (probably also) lipoproteins.[13] In addition, the complex compartment thus formed contains substances secreted by epithelial cells and materials from the digesta being absorbed by such cells. In other words, from the microbiological point of view, the compartments are incredibly complex nutritionally and environmentally.

Such complexity makes quite difficult studies of the mechanisms by which microorganisms associate with secreting epithelia. Somewhat less complex, perhaps, are studies of the mechanisms by which microorganisms, such as lactobacilli, associate with the nonsecreting squamous epithelia found in the crops of birds and the stomachs of mammals of some species (Table 1). Such surfaces are less complex in architecture and physiology than secreting epithelia. Indeed some such surfaces have on them little if any mucin.[22] Microorganisms undoubtedly adhere to the keratinized squamous cells in order to remain in the habitats on such epithelia.[7,14] As described below, we have been studying one such association.

B. Adhesion of Lactobacilli to Murine Gastric Epithelia

1. Specificity

As noted above, strains of various *Lactobacillus* species associate with the surface of stratified squamous epithelium in the stomachs of mice (Table 1). Also as noted, in associating with the epithelial cells, the bacteria undoubtedly adhere to them. The adhesion involves some specificity from animal species to species. Strains of lactobacilli isolated from the crops of domestic chickens and some other birds adhere to epithelial cells taken from chicken crops and incubated in vitro, while strains isolated from mammals of several species do not.[15] Conversely, *Lactobacillus* strains isolated from rats adhere in vitro to squamous

Table 2
SPECIFICITY BY WHICH ISOLATES OF *LACTOBACILLUS* SPECIES ASSOCIATE WITH THE GASTRIC NONSECRETING EPITHELIUM IN MONOASSOCIATED GENOTOBIOTIC MICE[a]

Lactobacillus species	Isolated from	No. strains tested	Population level[b]	Stomach layers[c]
fermentum	Mice[d]	2	8—9	16/16
Unidentified	Mice[e]	2	7—8	21/22
leichmanii	Mice[e]	1	7—8	1/5
acidophilus	Human[f]	7	5—8	0/42
acidophilus	Human[f]	1	6—7	0/6
acidophilus	Chicken[f]	5	6—8	0/30
acidophilus	Swine[f]	5	6—8	1/30

[a] Adapted from Reference 5.
[b] Range of the level (Log_{10} No. bacteria/g whole wet stomach) estimated by viable count method.
[c] Layers visualized microscopically of Gram-positive bacteria on the keratinized squamous epithelium of nonsecreting part of the stomach; each bacterial strain was tested in at least six mice.
[d] Isolated in author's laboratory.
[e] Isolated in New Zealand and provided to the author by Gerald Tannock of the Department of Microbiology, University of Otago, Dunedin, N.Z.
[f] Isolated and provided to the author by Marvin L. Speck, Department of Food Science, North Carolina State University, Raleigh, N.C.

gastric cells removed from rats, while strains from other mammals and some birds do not.[16] Similarly, strains isolated from swine adhere to swine gastric cells, while, with a few exceptions, strains from chickens and some other mammals do not.[14] In my laboratory, we have tested whether or not the specificity detected with bacteria and epithelial cells exposed to each other in vitro, can be detected as well in a living animal. Our method was to monoassociate germ-free animals with strains of *Lactobacillus* species isolated from animals of several species.[5,17]

All strains of the *Lactobacillus* species tested colonized the gastrointestinal tracts of monoassociated mice (Table 2). Animals associated with each strain yielded evidence of lactobacilli in the stomachs and cecums at population levels comparable with those at which such bacteria are detected in the same regions in mice with gastrointestinal microbiotas acquired naturally. Only strains originally isolated from mice could be found, however, in deep layers on epithelial surfaces in histological sections of the stomachs of the animals. All such layers, appeared only on the keratinized stratified squamous epithelium of the stomachs and not on the columnar epithelium in the stomachs or the cecums. These findings confirm in most essentials the findings made in experiments in which the lactobacilli and tissue cells were mixed in vitro.[14-16] As a whole, the findings indicate substantial specificity of particular strains of lactobacilli for associating with and adhering to the epithelium in particular animal species.

Such data also reinforce the hypothesis that adhesion to a squamous crop or gastric surface is important for lactobacilli to be able to colonize the upper part of the gastrointestinal tract of an animal under ordinary circumstances (i.e., not in a germ-free isolator). Nevertheless, the capacity to adhere to the surface is undoubtedly insufficient by itself to ensure that the microorganisms can colonize the epithelial habitat.[7,18] A particular strain of *Lactobacillus* sp. (strain 100-16) isolated from a mouse and strain of *L. acidophilus* (100-33) isolated from

Table 3
CHARACTERISTICS OF *LACTOBACILLUS* STRAINS AND THEIR COLONIZATION OF THE MOUSE STOMACH[a]

Lactobacillus				Antibody[e]		
Species	Strain[b]	Animal origin[c]	Plasmid[d]	16^S	16^R	Epithelial layer[f]
Unidentified	100-1	Mouse(USA)	+	+	−	4+
Unidentified	100-5	Mouse(USA)	+	+	−	4+
Unidentified	100-14	Mouse(NZ)	+	+	−	4+
fermentum	$100\text{-}16^S$	Mouse(USA)	+	+	−	4+
fermentum	$100\text{-}16^R$	Mouse(USA)	−	−	+	None
acidophilus	100-12	Human	−	−	+	None
acidophilus	100-21	Human	−	−	−	±
acidophilus	100-23	Swine	−	−	−	±
acidophilus	100-28	Swine	−	−	+	±
acidophilus	100-33	Swine	−	+	−	±

[a] See Reference 17.
[b] Our number.
[c] USA, United States of America; NZ, New Zealand.
[d] Larger than 90 Mdal molecular weight.
[e] Two preparations of rabbit globulins; one from animals injected with cells of *L. fermentum* $100\text{-}16^S$, the other from animals injected with *L. fermentum* $100\text{-}16^R$.
[f] As detected by frozen-section histology of the stomachs of ex-germ-free mice monoassociated with the *Lactobacillus* strain indicated.[5]

a pig adhere in vitro in equally high numbers to the surfaces of the epithelia on both the secreting and nonsecreting portions of the stomachs of mice.[8] In monoassociated ex-germ-free mice, however, strain 100-33 does not colonize either the nonsecreting or secreting epithelium of the stomach (Table 3), while 100-16 forms a thick layer on the nonsecreting but not the secreting surface under comparable conditions. Such findings suggest that the capacity to adhere to epithelial cells (at least in vitro) does not by itself dictate that a *Lactobacillus* strain can colonize the gastric squamous surface in living animals.[7,18] The capacity to adhere to the epithelial surface is surely necessary for lactobacilli to colonize the stomachs of animals living under conventional conditions and bearing a total microflora acquired naturally. Nevertheless, the capacity of the bacteria to colonize the surface may be specified as much by nutritional and environmental conditions as by a capacity to bind to the epithelium.

2. *Mechanisms*

Finding from research conducted within the past few years by investigators in England indicate that lactobacilli adhere to squamous epithelia in the chicken crop or swine stomach via acidic polysaccharides containing binding sites for concanavalin A.[19,20] Data from work by investigators in Japan on lactobacilli adhering to squamous cells from the stomachs of rats suggest, however, that macromolecules other than polysaccharides may be involved in the adhesion.[16] The studies by the investigators in both countries were conducted with bacteria and animal cells exposed to each other in vitro. In the experiments in both cases, the investigators chemically treated in various ways either the bacterial or the animal cells in efforts to determine what types of treatment inhibited the adhesion. When they successfully inhibited adhesion, the investigators would infer from the type of chemical treatment used something about the nature of the macromolecules mediating the adhesion. We have done similar experiments using an adhesion assay where lactobacilli and mouse stomach cells are exposed to each other in vitro. We have concluded that the experimental method has a strong likelihood of giving misleading results.[7,8]

Bacterial cells grown in vitro in culture media have surface properties that may differ from those of cells of the same strain growing in a living animal. Similarly, animal cells, particularly intestinal epithelial cells, when removed from the living animal and placed in buffered fluid of some type, tend to change in character by losing macromolecules from their surfaces and changing in other ways. Indeed, intestinal epithelial cells from adult animals usually vacuolate, enlarge, and lyse within minutes after being placed in common cell culture media. (Savage, personal observations and personal communication from other investigators.) Two such changed and changing systems, when exposed and allowed to interact in vitro, may or may not interact molecularly as they would in vivo. Moreover, when either the bacterial or animal cells are treated biochemically in some way in attempts to learn the mechanisms mediating the interaction, the altered surfaces of the cells may not be affected by the treatment in the same way in vitro as they would in the living host. Thus, the information gained from the test conducted in vitro may mislead concerning the mechanisms mediating adhesion in vivo. Because of this rationale, we are studying the mechanisms mediating adhesion of lactobacilli to gastric squamous cells by attempting to use approaches involving some techniques of molecular genetics.

We believe that studies of the mechanisms of adhesion will progress most effectively if the bacterial ligand is identified from the stable basis of its genetics, rather than from results of binding studies that are designed arbitrarily. Basing our approach on the observation that genes on plasmids code for the K-antigen ligands by which *E. coli* binds to intestinal epithelium,[21] we have been screening *Lactobacillus* strains for plasmids, and attempting to relate genes on the plasmids to the macromolecules mediating *Lactobacillus* binding to squamous cells.

We have screened for plasmids numerous strains of *Lactobacillus* species isolated from animals of several species.[17] Most strains isolated from mice, and able to colonize the gastric squamous epithelium in monoassociated ex-germ-free mice, have been found to contain large plasmids of about 90 Mdal in molecular weight (Table 3). By contrast, a spontaneous variant of one of the mouse strains (100-16) and numerous strains of *Lactobacillus* species isolated from animals other than mice lack the large plasmid, and cannot colonize the gastric squamous epithelium in monoassociated animals (Table 3).

Lactobacillus strains containing the large plasmid produce in broth media and in colonies growing on agar surfaces an extracellular material that reacts antigenically with rabbit antibodies induced by cells of *L. fermentum* $100\text{-}16^S$ (plasmid-bearer) but not with antibodies induced by cells of the variant, $100\text{-}16^R$ (Table 3). Extracellular substances from strains other than $100\text{-}16^R$ lacking the plasmid also do not react with the antibody. The extracellular material from strain $100\text{-}16^S$ contains protein, carbohydrate, and possibly also lipoteichoic acid. We are purifying and chemically characterizing substances from the material and testing them for any role in adhesion of the bacteria. Once substances mediating the adhesion have been identified, then efforts will be made to localize the genes coding for the substances. Obviously, the large plasmids are particular targets of such study.

IV. SUMMARY

Indigenous microorganisms of numerous species associate with epithelial surfaces in most major regions of the gastrointestinal canals of all mammals and birds that have been examined so far. Some of the microbial species adhere to the epithelium with which they associate. In some cases, the mechanisms by which the microorganisms adhere to the surface have been studied experimentally. In aggregate, results from such studies have been confusing, perhaps because both the microbial and epithelial surfaces are complex. Described herein is an approach to studying such mechanisms, based upon analysis of the genetics of the adhering microorganism. Such an approach may lead to unambiguous definition of adhesion mechanisms.

REFERENCES

1. **Savage, D. C.,** Microbial ecology of the gastrointestinal tract, in *Proc. Symp. Nutrition and the Intestinal Flora,* Symposia of the Swedish Nutrition Foundation, Vol. 15, Hallgren, B., Ed., Swedish Nutrition Foundation, 1982, in press.
2. **Savage, D. C.,** Microbial ecology of the gastrointestinal tract, *Ann. Rev. Microbiol.,* 31, 107, 1977.
3. **Savage, D. C.,** Colonization by and survival of pathogenic bacteria on intestinal mucosal surfaces, in *Absorption of Microorganisms to Surfaces,* Bitton, G. and Marshall, K. C., Eds., Wiley-Interscience, New York, 1980, 175.
4. **Jones, G. W.,** Some aspects of the interaction of microbes with the human body, in *Contemporary Microbial Ecology,* Ellwood, D. KC., Hedger, J. N., Latham, M. J., Lynch, J. M., and Slater, J. H., Eds., Academic Press, London, 253.
5. **Savage, D. C.,** Introduction to mechanisms of association of indigenous microbes, *Am. J. Clin. Nutr.,* 32, 113, 1979.
6. **Lee, A.,** Normal flora of animal intestinal surfaces, in *Absorption of Microorganisms to Surfaces,* Bitton, G. and Marshall, K. C., Eds., Wiley-Interscience, New York, 1980, 145.
7. **Savage, D. C.,** Adherence of normal flora to mucosal surfaces, in *Bacterial Adherence,* Beachey, E. H., Ed., Chapman and Hall, London, 1980, 31.
8. **Kotarski, S. F. and Savage, D. C.,** Models for study of the specificity by which indigenous lactobacilli adhere to murine gastric epithelium, *Infect. Immun.,* 26, 966, 1979.
9. **Suegara, N., Siegel, J. E., and Savage, D. C.,** Ecological determinants in microbial colonization of the murine gastrointestinal tract: adherence of *Torulopsis pintolopesii* to epithelial surfaces, *Infect. Immun.,* 25, 139, 1979.
10. **Knehans, A. W. and O'Dell, B. L.,** Intestinal microflora in the guinea pig as observed by scanning electron microscopy. Effect of fibrous dietary supplements, *J. Nutr.,* 110, 1543, 1980.
11. **McKenzie, R. A.,** The caecum of the koala, *Phascolarctos cinereus:* light, scanning and transmission electron microscopic observations on its epithelium and flora, *Aust. J. Zool.,* 26, 249, 1978.
12. **Brasitus, T.,** Lipid dynamics and lipid-protein interaction in rat enterocyte plasma membranes, in *Attachment of Organisms to the Gut Mucosa,* Vol. 2, Boedeker, E. C., Ed., CRC Press, Boca Raton, Fla., 1984.
13. **Allen, A.,** The structure and function of gastrointestinal mucus, in *Attachment of Organisms to the Gut Mucosa,* Vol. 2, Boedeker, E. C., Ed., CRC Press, Boca Raton, Fla., 1984.
14. **Barrow, P. A., Brooker, B. E., Fuller, R., and Newport, M. J.,** The attachment of bacteria to the gastric epithelium of the pig and its importance in the microecology of the intestine, *J. Appl. Bacteriol.,* 48, 147, 1980.
15. **Fuller, R. and Turvey, A.,** Bacteria associated with the intestinal wall of the fowl *(Gallus domesticus), J. Appl. Bacteriol.,* 3, 617, 1971.
16. **Suegara, N., Morotomi, M., Watanabe, T., Kawai, Y., and Mutai, M.,** Behavior of microflora in the rat stomach: adhesion of lactobacilli to the keratinized epithelial cells of the rat stomach in vitro, *Infect. Immun.,* 12, 173, 1975.
17. **Lin, J. H-C. and Savage, D. C.,** Colonization by lactobacilli of the nonsecreting epithelium of the mouse stomach, Abstr. Annu. Meet. Am. Soc. Microbiol., 1982, 104.
18. **Artwohl, J. E. and Savage, D. C.,** Determinants in microbial colonization of the murine gastrointestinal tract: pH, temperature, and energy-yielding metabolism of *Torulopsis pintolopesii, Appl. Environ. Microbiol.,* 37, 697, 1979.
19. **Brooker, B. E. and Fuller, R.,** Adhesion of lactobacilli to the chicken crop epithelium, *J. Ultrastruct. Res.,* 52, 21, 1975.
20. **Fuller, R.,** Nature of the determinant responsible for the adhesion of lactobacilli to chicken crop epithelial cells, *J. Gen. Microbiol.,* 87, 245, 1975.
21. **Elwell, L. P. and Shipley, P. L.,** Plasmid-mediated factors associated with virulence of bacteria to animals, *Ann. Rev. Microbiol.,* 34, 465, 1980.
22. **Savage, D. C.,** personal observations.

Chapter 2

ADHERENCE OF THE ORAL FLORA

R. J. Gibbons

TABLE OF CONTENTS

I. INTRODUCTION

The oral cavity has several unique features which has made it a useful model for studying the association of bacteria with host tissues. It contains keratinized and nonkeratinized epithelial surfaces in addition to those of the teeth. Each of these surfaces harbors a distinct collection of bacteria, and they are readily accessible for study. Furthermore, oral secretions can be obtained more easily and in greater quantity than those from other areas of the alimentary canal, or from other mucosal sites. Consequently, it is possible to investigate the influence of components of salivary secretions, such as IgA antibodies, lysozyme, mucins, etc., on colonizing bacteria more conveniently than in many other body locations. In view of these features, it is not surprising that studies of the ecology of bacteria in the mouth were among the first to reveal the surprising specificity involved in bacterial adherence to host tissues, and the fundamental role that this plays in microbial colonization of mucosal surfaces.[1-3] These studies also provided a conceptual basis for understanding the innate resistance or susceptibilities of different hosts and tissues to infectious agents.[4] Because of this, studies of bacterial adherence are currently attracting the interests of increasing numbers of investigators.

The adhesion of bacteria to teeth and to other surfaces of the mouth has been the subject of several recent reviews.[5-8] Therefore, the following discussion will focus on some of the principles which have evolved from these studies which appear to be applicable to host-parasite relationships occurring throughout the gastrointestinal canal. Examples of novel bacterial adherent interactions will also be described which may be relevant to intestinal environments.

II. ADHESION OF BACTERIA TO SURFACES OF THE MOUTH

Interest in the attachment of bacteria to surfaces of the mouth developed because of the microbial accumulations which form on the teeth. These are called dental plaques, and they are responsible for initiating dental caries and various forms of periodontal diseases.[7] Plaques accumulate most rapidly in sites which are protected from normal cleansing mechanisms, i.e., along the gingival margin, in areas between the teeth, and in protected fissures.[7] Depending upon their bacterial composition, plaques developing near the gums may expose the tissues to antigens and various toxins which often results in destruction of the periodontal tissues.[9] Other plaques on the crowns of teeth may contain highly acidogenic organisms, such as *Streptococcus mutans*, and can subject the underlying tooth enamel to high concentrations of acid from the fermentation of dietary sugars. These acids dissolve the hydroxyapatite mineral of the teeth, and carious lesions result.[10,11] Last year, the American public spent over $16 billion for the treatment of dental diseases. In addition, these diseases also cause considerable pain and loss of productive time.

To understand how plaques form, models were devised to mimic bacterial attachment to teeth. Teeth are covered by a thin amorphous film, termed the acquired pellicle, which is formed by the selective adsorption of salivary glycoproteins to the hydroxyapatite mineral of enamel.[12,13] Bacteria which attach to the teeth, therefore, interact with adsorbed salivary macromolecules making up the pellicle.[7,14] To mimic teeth, powdered enamel or hydroxyapatite discs or beads have been incubated with saliva to form experimental pellicles; these are then used as adsorbents for studying bacterial attachment.[15-17] Use of such models revealed that various oral bacterial species adsorbed with a surprising degree of selectivity to pellicle surfaces. Similar observations have been made concerning the attachment of bacteria to epithelial cells scraped from the buccal surfaces of volunteers.[1] The specificity of bacterial attachment has also been demonstrated in the mouths of human volunteers by introducing mixtures of bacterial strains which were tagged or labeled by being made resistant to an antibiotic.[1,3] The proportions of one species relative to another which could be recovered from various surfaces of the mouth were determined by culturing on antibiotic-containing media. These studies substantiated the specificity of bacterial attachment in vivo and with all microorganisms studied to date, a correlation has been noted between their relative adherence to a given oral surface and the extent to which they naturally colonize that site.[5,8]

As a result of these studies, it became clear that bacteria must possess a highly developed recognition system which is capable of distinguishing between the surface components of different types of oral epithelial cells, and also between the glycoprotein pellicle present on the teeth. Secondly, the correlations noted between the adhesion of *Streptococcus, Lactobacillus, Neisseria,* and *Bacteroides* species to oral surfaces and their natural colonization suggests that attachment must be a powerful determinant of colonization of these tissues.[1,5,6,8] These experiments led to the simple realization that in environments which contain surfaces exposed to a fluid flow, such as the mouth, the nasopharyngeal area, portions of the intestinal canal, the eye, heart, bladder, etc., bacteria must attach to a surface to avoid being removed by the flow of secretions bathing the surface. The rate of bacterial growth in the mouth,[5] or in the intestinal canal[18] appears to be too slow to enable bacteria to persist at a site free in the fluids.

While bacterial colonization on mucosal surfaces becomes limited because of desquamation, colonization of the teeth is not affected in this manner, and, therefore, bacteria can accumulate on the teeth in masses several hundreds of microns thick. The vulnerability of such a nondesquamating tissue to microbial attack is well illustrated by the fact that dental caries and periodontal diseases are among the most prevalent infections which afflict humans. The same reasons also appear to account, in part, for the high susceptibility to infections of prosthetic devices placed in systemic locations.

III. MECHANISMS OF BACTERIAL ATTACHMENT TO ORAL SURFACES

The mechanisms involved in bacterial attachment to surfaces of the mouth are not fully understood. Incomplete evidence suggests that bacteria may become loosely associated with oral tissues by ionic and other physical forces of low specificity, but these do not appear to be sufficient to promote colonization.[8] Rather, more firm attachment often appears to depend upon the formation of polymeric bridges between the microorganism and the surface. In at least some instances, these bridges appear to involve lectin-like components, termed "adhesins," of the bacterium which bind to specific saccharides in receptors on epithelial cells or in the glycoprotein pellicle of the teeth. Several lines of evidence support this model. For example, electron microscopic observations of oral streptococci[19] and *Actinomyces* species[20] show that the organisms possess thin fimbriae-like fibrils on their surface which seem to mediate their attachment. In the case of *S. salivarius*[19] and *S. pyogenes,*[22] these surface fibrils can be removed by trypsin-treatment and such denuded streptococci no longer attach well to epithelial cells. In addition, the attachment of some oral bacteria is inhibited by certain sugar constituents of glycoproteins or glycolipids. Thus, adsorption of *S. mutants*[23] and *Leptotrichia*[24] species to experimental pellicles is inhibited by galactosides. Galactosides also inhibit hemagglutination,[25] and certain coaggregation[26] reactions mediated by *Actinomyces* strains. These observations suggest that the organisms interact with galactoside-containing receptors on oral tissues. Since bacteria are well known to possess sugar-binding proteins and surface-bound enzymes with specific binding properties, it is reasonable to suppose that they may also function as adhesins. It should also be mentioned that mucins and certain other nonimmune components of oral secretions can bind to bacteria and inhibit their attachment to experimental pellicles as well as to epithelial cells.[27] This suggests that these secretory glycoproteins are structurally related to the receptors on host tissues to which bacteria attach, and they can therefore serve as competitive inhibitors. This notion is supported by evidence which suggests that blood group-reactive glycoproteins or glycolipids on oral epithelial cells and in the pellicle on teeth serve as receptors for some oral organisms. The saccharide nature of the receptors is further suggested by the observation that pretreatment of experimental pellicles or epithelial cells with certain lectins can affect the attachment of *S. mutans* and other oral streptococci.[27,28]

The adsorption of several oral species to saliva-treated hydroxyapatite surfaces follows the kinetics of a Langmuir adsorption isotherm and analyses of such isotherms indicate that pellicles contain different numbers of binding sites for different organisms.[17,29] This suggests that different species of oral bacteria interact with different receptors in the pellicle. This is also supported by competition experiments which have shown that cells of similar serotypes of *S. sanguis* compete for each other for binding sites on experimental pellicles, whereas cells of other streptococci, such as *S. mutans*, *S. mitis*, and *S. salivarius*, do not compete with *S. sanguis*.[29,30] Also, the adsorption of *S. mutans* and *Leptotrichia buccalis* to experimental pellicles is inhibited by galactosides,[23] but the adsorption of *Bacteroides gingivalis*,[31] *Actinomyces viscosus*, and *A. naeslundii*[32] is not.

IV. ACCUMULATION OF BACTERIA ON THE TEETH

Once bacteria have attached to the pellicle and are proliferating on teeth, a second type of adherent interaction takes place between the bacterial cells which enables them to resist removal by oral cleansing forces, and thereby accumulate in layers hundreds of microns thick.[7,8] Substances in the oral environment which bind to bacteria and induce their aggregation are thought to promote accumulation of organisms which are proliferating on the teeth. Components of saliva, including blood group-reactive mucins,[27,33] high molecular weight glycoproteins,[34] and antibodies[35] have been shown to bind to bacteria and probably function this way.

The coherence of some organisms also can be mediated by extracellular polymeric material synthesized by the organisms themselves. This appears to be the case with *S. mutans*, *A. viscosus*, and *A. naeslundii* since they form thick accumulations on solid surfaces in vitro in broth cultures.[7]

The role of extracellular polymers in promoting bacterial accumulation has been especially well studied with *S. mutans*.[7,36] This organism is known to form thick plaques on the teeth of experimental animals which are fed diets which contain sucrose, but not diets which contain glucose. Strains of *S. mutans* have been shown to synthesize complex mixtures of extracellular glucans specifically from sucrose, and not from other common carbohydrates, and the synthesis of these polymers is associated with the ability of the organism to accumulate on solid surfaces in vivo and in vitro. One type of glucan formed consists of glucose units which are predominantly alpha-1:6 linked and is similar to classical dextrans; varying proportions of alpha-1:3 and alpha-1:4 bonds may also be present in the molecules. Other glucans synthesized are predominantly alpha-1:3 linked and tend to be quite insoluble; these polymers have been named "mutan."[7,11,36] Both types of glucans have been detected in human dental plaques.

S. mutans cells contain at least two glucosyltransferases (dextransucrases) which bind to glucan molecules and continue their synthesis from sucrose. In addition, the streptococci also contain on their surfaces a glucan-binding protein which is devoid of enzyme activity. The possession of these ligands enables the streptococci to interact with high molecular weight glucan molecules, and the organisms aggregate in the presence of preformed glucan. The sum of the interactions of these ligands with glucan molecules is thought to link the streptococci together, and thereby foster their accumulation on the teeth.[5,7]

It was initially thought that glucan synthesis was also important for the attachment of *S. mutans* cells to the pellicle on teeth, but this does not appear to be true. *S. mutans* cells can attach to and colonize the teeth of laboratory animals fed diets free of sucrose, though the streptococci do not accumulate and form thick plaques under these conditions.[37] Also, the minimum infective dose of a mutant of *S. mutans*, which forms little or no cell-associated glucosyltransferase, is similar to that of its parent in experimental animals.[38] Other studies have shown that glucan synthesis from sucrose does not foster the retention of *S. mutans* cells associated with experimental pellicles or with human teeth.[16] In view of this evidence, the importance of glucan synthesis seems to be the promotion of streptococcal coherence once the organisms have become firmly attached to teeth and are proliferating. This property is important in the caries process because it enables *S. mutans* cells to accumulate in sufficient masses on teeth to generate enough acid from dietary sugars to cause enamel demineralization.

In contrast to *S. mutans, A. viscosus*, and *A. naeslundii* cells accumulate on solid surfaces in vitro in the presence of a variety of carbohydrates.[39] Thus, the components which foster their accumulation are not related to the glucans. The fimbriae of these organisms have been demonstrated to play a role in cell attachment and aggregation,[20,21,26] and the organisms synthesize an extracellular heteropolysaccharide slime composed of *N*-acetyl-glucosamine and small amounts of glucose and galactose.[40] They also produce a fatty acid substituted amphipathic polysaccharide containing mannose, glucose, and galactose,[41] and the elaboration of these components seems likely to be involved in their accumulation on surfaces.

V. BACTERIAL ATTACHMENT TO DISSIMILAR ORGANISMS

Electron microscopic observations of sections of human dental plaques often reveal morphologically different bacterial cells directly attached to each other.[7,14] This suggests that the surface components of dissimilar organisms can interact with one another. In vitro experiments have shown that specific pairs of oral bacteria will coaggregate when mixed together. Such coaggregation reactions occur between a wide range of bacteria, and their incidence suggests that they probably also occur in vivo. One of the best known examples of bacterial coaggregation involves the so-called "corn-cob" formations which have been observed by scanning electron microscopic examinations of gingival plaques.[42] These formations consist of densely packed cocci (which comprise the "kernels") attached to the surface of a filamentous organism which forms the "cob". Corn cobs have been isolated by micromanipulation, and the organisms involved are quite specific.[43,44] *Bacterionema matruchotii* cells are the filaments, while cells of a specific serotype of *S. sanguis* are the cocci.

Coaggregation reactions also commonly occur between human strains of *A. viscosus* or *A. naeslundii* and cells of *S. sanguis* or *S. mitis*, but they are infrequent with other *Actinomyces* species.[26] Two groups of *Actinomyces* and four groups of streptococci have been delineated by their coaggregation patterns based upon inhibition by β-galactosides such as lactose, and sensitivity to proteolytic enzymes and to heat.[26,45] *Actinomyces* cells have been shown to contain two types of fimbriae, one of which appears to be a lectin which binds to β-galactoside residues on oral streptococci. Many of the coaggregation reactions between *Actinomyces* and streptococci are not affected by the growth medium or by the physiological age of the organisms. This suggests that these reactions may occur in the oral cavity at a significant frequency.[26] On the other hand, the lactose-reversible coaggregations between *Actinomyces* and oral streptococci are strongly inhibited by salivary components;[46] this decreases the likelihood that this type of coaggregation reaction occurs to a large extent in vivo.

Coaggregation reactions also develop between Gram-positive and Gram-negative organisms. Thus, strains of *Veillonella*, which are unable to accumulate on solid surfaces in vitro by themselves, can adhere to previously grown plaques of *Actinomyces* species.[47] Certain strains of *Veillonella* also coaggregate with strains of *S. mutans*. MacBride and Van der Hoeven[48] have recently shown that a strain of *Veillonella* colonized the teeth of gnotobiotic rats monoinfected with a strain of *S. mutans* with which it coaggregated in 1000-fold higher numbers than in rats monoinfected with *S. mutans* strains with which it did not interact. Coaggregation between these organisms was unaffected by saliva, and this probably accounts in part for the marked colonization effect observed.

Cells of *Bacteroides gingivalis, Fusobacterium nucleatum, Capnocytophaga ochraceus, Eikenella corrodens,* and other Gram-negative species also adhere to the surfaces of various *Actinomyces* and streptococcal strains, forming coaggregates.[31] These Gram-negative organisms are primarily found in gingival and subgingival plaques, and they are often associated with different forms of periodontal disease. Their ability to attach to Gram-positive species seems to account for the observation than in periodontal pockets, the Gram-positive organisms which are adherent to the cemental or root surfaces of teeth are often covered by a layer of Gram-negative organisms.[49]

By enabling bacteria to adhere to each other, interbacterial interactions are thought to promote microbial accumulations on the surfaces of teeth. They may also be important for the initial attachment of some organisms in mucosal environments.[31,48] This would be especially likely in densely populated microbial communities such as those on regions of the teeth, the tongue dorsum, and in certain areas of the intestinal canal.

VI. INFLUENCE OF SECRETIONS ON BACTERIAL ATTACHMENT

In the mouth, unattached bacteria are suspended in saliva and the organisms are therefore in contact with salivary macromolecules which are similar or identical to the salivary components comprising the acquired pellicle. Also, some salivary components, especially mucins, are structurally related to the glycoproteins and glycolipids present on mucosal cell surfaces which are associated with bacterial receptors. Thus, free salivary macromolecules can compete for bacterial adhesins and competitively inhibit microbial attachment to oral surfaces.[27] Components of secretions may also bind to bacterial surface components and sterically inhibit the interaction of adhesins with tissue receptors. Antibodies, including those of the IgA isotype, can interact with bacterial surface antigens and inhibit attachment to epithelial cells.[35] This appears to be one way in which secretory antibodies contribute to the disposal of bacteria on mucosal surfaces. In fact, any component which agglutinates bacterial cells suspended in secretions can influence their ability to attach to oral surfaces. For example, Liljemark and co-workers[50] demonstrated that the formation of small aggregates consisting of two or three streptococcal cells increased the numbers of bacteria which associated with experimental pellicles, whereas the formation of large aggregates resulted in fewer cells becoming associated with the surface. Lysozyme, lactoperoxidase, and other components of secretions probably also can affect bacterial adherence since they can bind to bacterial cell surface components.[7,8] The effectiveness of oral secretions in surpressing the association of bacteria with surfaces of the mouth is illustrated by the sparse numbers of bacteria which are found attached to buccal or palatal epithelial cells despite their continuous exposure to saliva which contains in excess of 10^8 bacteria per milliliter. In vitro exposure of these epithelial cells to buffer suspensions of salivary bacteria, such as *S. salivarius*, results in hundreds of bacteria becoming attached to each cell within a few minutes.

That these interactions of salivary components with bacteria can actually affect colonization is strongly suggested from studies of *B. gingivalis*.[31] Cells of this species contain several types of fimbriae, and when the organisms are suspended in buffer, they can attach in appreciable numbers to experimental pellicles, to epithelial cells, and to erythrocytes. However, when *B. gingivalis* cells are suspended in saliva, few, if any, organisms associate with erythrocytes or with epithelial cells, and their numbers which attach to experimental pellicles are markedly reduced. The adherence-inhibiting activity of saliva for this species is mainly associated with the blood group-reactive mucin fraction, though other salivary fractions also possess some activity. As mentioned above, cells of *B. gingivalis* can also attach to *Actinomyces* and certain other Gram-positive bacteria forming coaggregates, but saliva does not inhibit these interactions. When streptomycin-resistant strains of *B. gingivalis* were introduced into the mouths of human volunteers, the bacilli could only be recovered in significant numbers from pre-existing bacterial plaques which contain *Actinomyces* and other Gram-positive bacteria, and not from the surfaces of cleaned teeth or oral mucosa.[31] These observations suggest that the retention of *B. gingivalis* cells introduced into the mouth is greatly fostered by the prior presence of *Actinomyces* and other Gram-positive organisms which promote its adherence and colonization on the teeth. Another way of phrasing this is that poor oral hygiene appears to predispose to infection by this potentially pathogenic organism. This, in turn, may lead to the development of periodontal disease. Though salivary constituents can reduce bacterial adherence and affect colonization by molecular mimicry of the receptors in the pellicle or on epithelial cells, they are not able to mimic some of the surface components of bacteria, and these components may therefore serve as a unique class of receptors. These observations also point out that it may not be possible to correctly interpret in vitro studies of bacterial adherence unless the effects of secretions are taken into account.

VII. AVAILABILITY OF RECEPTORS ON ORAL TISSUES

Oral and intestinal bacteria are known to synthesize an array of glycoside hydrolases and other saccharases and proteases which can alter the structure of glycoproteins and glycolipids on cell surfaces and in secretions. Since saccharides appear to comprise part of the receptors involved in bacterial attachment, enzymes which alter the sugar moieties of glycoproteins or glycolipids could remove terminal sugars from oligosaccharide chains and expose underlying receptors. Alternatively, enzymes could destroy these receptors and thereby reduce bacterial adherence and colonization. An example of the former possibility is available. *A. viscosus* and *A. naeslundii* cells do not attach to erythrocytes unless terminal sialic acid residues are first removed by treatment with neuraminidase.[25] This exposes penultimate galactoside residues on the erythrocytes which can interact with the galactoside-binding fimbriae of the organisms. In regards to the second possibility, we have recently noted that heat-liable substances presumed to be enzymes are present in freshly collected and in deliberately incubated whole human saliva which can modify receptors in experimental pellicles and reduce the attachment of strains of *S. mutans* and *S. sanguis*.[51] Treatment of experimental pellicles with neuraminidase also impaired the attachment of some strains of *S. sanguis*. This supports earlier studies which demonstrated that strains of these organisms can bind to sialic acid residues in salivary glycoproteins.[52] In contrast, the attachment of strains of *S. mutans* was reduced to pellicles which had been treated with galactose-oxidase. This is consistent with the suggestion that galactose residues may comprise part of the pellicle receptors for this organism.[22]

The above observations suggest that receptors present on newly exposed epithelial cells or teeth may only be available for bacterial attachment for finite periods of time. This would be especially so in locations which are exposed to dense microbial populations such as on the surfaces of teeth, or in the large intestine.

In addition to becoming modified or destroyed by enzymatic action, receptors on mucosal surfaces may also become masked by dietary components. This is suggested by the recent observation that many foods contain lectin-like components which can bind to experimental pellicles, to epithelial cells, and to bacteria.[28,54,54] Pretreatment of experimental pellicles with aqueous extracts of bananas, coconuts, carrots, alfalfa, and sunflower seeds markedly reduced the numbers of *S. mutans* cells which subsequently adsorbed, and therefore these interactions have the potential to affect bacterial colonization.[28] Treatment of pellicles with food extracts had little effect upon the attachment of strains of *A. naeslundii* or *A. viscosus*, indicating that the effects can be specific. Apparently, lectin-like food components bind to the pellicle and mask the receptors involved in attachment of *S. mutans*, but not those for *Actinomyces*. In contrast to these observations, it also appears that some lectin-interactions with host tissues can promote bacterial attachment. For example, we have found that pretreatment of buccal epithelial cells with wheat germ agglutinin promotes attachment of *S. sanguis* cells. In this case, the lectin can bind to both the epithelial cells and the streptococci, and it appears to link them together.

Saliva has been shown to reduce the binding of several purified lectins to experimental pellicles and to buccal epithelial cells, but it does not completely prevent it.[54] In fact, using an enzyme-linked immunoassay, we have detected wheat germ agglutinin and peanut agglutinin on epithelial cells derived from salivary sediment, buccal mucosa, and from the tongue dorsum of humans several hours after they ate 10 to 25 g of raw wheat germ or peanuts.[55] In view of the ability of lectins to influence bacterial attachment to experimental pellicles or to buccal epithelial cells, their persistence on host tissues after oral ingestion increases the likelihood that food lectins may affect host-parasite interactions in the mouth and perhaps elsewhere in the gastrointestinal canal. Such reactions would represent a novel way in which the diet may influence the resistance or susceptibility of a host to bacterial infection.

REFERENCES

1. **Gibbons, R. J. and van Houte, J.,** Selective bacterial adherence to oral epithelial surfaces and its role as an ecological determinant, *Infect. Immun.*, 3, 567, 1971.
2. **van Houte, J., Gibbons, R. J., and Banghart, S. B.,** Adherence as an ecological determinant of the presence of *Streptococcus salivarius* and *Streptococcus sanguis* on the tooth surface, *Arch. Oral Biol.*, 15, 1024, 1970.
3. **van Houte, J., Gibbons, R. J., and Pulkkinen, A. J.,** Aderence as an ecological determinant for streptococci in the human mouth, *Arch. Oral Biol.*, 16, 1131, 1971.
4. **Gibbons, R. J.,** Bacterial adherence in infection and immunity, in *New Approaches for Inducing Immunity to Pyogenic Bacteria*, DHEW Publ. No. (NIH) 74-553, National Institutes of Health, Bethesda, Md., 1973, 115.
5. **Gibbons, R. J. and van Houte, J.,** Bacterial adherence in oral microbial ecology, *Annu. Rev. Microbiol.*, 29, 19, 1975.
6. **Gibbons, R. J.,** Adherence of bacteria to host tissue, *Microbiology*, 395, 1977.
7. **Gibbons, R. J. and van Houte, J.,** Bacterial adherence and the formation of dental plaques, in *Bacterial Adherence*, Receptors and Recognition, Series B, Vol. 6, Beachey, E. H., Ed., Chapman and Hall, London, 1980, 62.
8. **Gibbons, R. J.,** Adhesion of bacteria to the surfaces of the mouth, in *Microbial Adhesion to Surfaces*, Berkeley, R. C. W., Lynch, J. M., Melling, J., Rutter, P. R., and Vincent, B., Eds., Soc. Chem. Indust., Ellis Harwood Ltd., 1980, 351.
9. **Socransky, S. S.,** Microbiology of periodontal disease — present status and future considerations, *J. Periodontol.*, 48, 497, 1977.
10. **Gibbons, R. J. and van Houte, J.,** Dental caries, *Annu. Rev. Med.*, 26, 121, 1975.
11. **Bowen, W. H. and Guggenheim, B.,** Therapeutics of caries prevention — concepts and prospects, *Acta Odontol. Scand.*, 36, 185, 1978.
12. **Ericson, T.,** Adsorption to hydroxyapatite of proteins and conjugated proteins from human saliva, *Caries Res.*, 1, 52, 1967.
13. **Hay, D. I.,** The adsorption of salivary proteins by hydroxyapatite and enamel, *Arch. Oral Biol.*, 12, 937, 1967.
14. **Gibbons, R. J. and van Houte, J.,** On the formation of dental plaques, *J. Periodontol.*, 44, 347, 1973.
15. **Hillman, J. D., van Houte, J., and Gibbons, R. J.,** Sorption of bacteria to human enamel powder, *Arch. Oral Biol.*, 15, 899, 1970.

16. **Clark, W. F. and Gibbons, R. J.,** Influence of salivary components and extracellular polysaccharide synthesis from sucrose on the attachment of *S. mutans* 6715 to hydroxyapatite surfaces, *Infect. Immun.*, 18, 514, 1977.
17. **Clark, W. B., Bammann, L. L., and Gibbons, R. J.,** Comparative estimates of bacterial affinities and adsorption sites on hydroxyapatite surfaces, *Infect. Immun.*, 19, 846, 1978.
18. **Gibbons, R. J. and Kapsimalis, B.,** Estimates of the overall rate of growth of the intestinal microflora of hamsters, guinea pigs and mice, *J. Bacteriol.*, 98, 510, 1967.
19. **Gibbons, R. J., van Houte, J., and Liljemark, W. F.,** Some parameters affecting the adherence of *S. salivarius* to oral epithelial surfaces, *J. Dent. Res.*, 51, 424, 1972.
20. **Ellen, R. P., Walker, D. L., and Chan, K. H.,** Association of long surface appendages with adherence-related functions of the Gram-positive species, *Actinomyces naeslundii, J. Bacteriol.*, 134, 1171, 1978.
21. **Wheeler, T. T. and Clark, W. B.,** Fibril-mediated adherence of *Actinomyces viscosus* to saliva-treated hydroxyapatite, *Infect. Immun.*, 28, 577, 1980.
22. **Ellen, R. P. and Gibbons, R. J.,** M-protein associated adherence of *Streptococcus pyogenes* to epithelial surfaces; a prerequisite for virulence, *Infect. Immun.*, 5, 826, 1972.
23. **Gibbons, R. J. and Qureshi, J. V.,** Inhibition of adsorption of *Streptococcus mutans* strains to saliva-treated hydroxyapatite by galactose and certain amines, *Infect. Immun.*, 26, 1214, 1979.
24. **Kondo, W., Sato, M., and Azawa, H.,** Haemagglutinating activity of *Leptotrichia buccalis* cells and their adherence to saliva-coated enamel powder, *Arch. Oral Biol.*, 21, 363, 1976.
25. **Ellen, R. P., Fillery, E. D., Chan., K. H., and Grove, D. A.,** Sialidase-enhanced lectin-like mechanism for *Actinomyces viscosus* and *Actinomyces naeslundii* hemagglutination, *Infect. Immun.*, 27, 335, 1980.
26. **Cisar, J. O., Kolenbrander, P. E., and McIntire, F.,** Specificity of coaggregation reactions between human oral streptococci and strains of *Actinomyces viscosus* and *Actinomyces naeslundii, Infect. Immun.*, 24, 742, 1979.
27. **Williams, R. C. and Gibbons, R. J.,** Inhibition of streptococcal attachment to receptors on human buccal epithelial cells by antigenically similar salivary glycoproteins, *Infect. Immun.*, 11, 711, 1975.
28. **Gibbons, R. J. and Dankers, J.,** Lectin-like components of foods which react with constituents of serum, saliva and *Streptococcus mutans, Appl. Environ. Microbiol.*, 41, 880, 1981.
29. **Applebaum, B., Golub, E., Holt, S. C., and Rosan, N.,** *In vitro* studies of dental plaque formation. Adsorption of oral streptococci to hydroxyapatite, *Infect. Immun*, 25, 717, 1979.
30. **Liljemark, W. F. and Schauer, S. V.,** Competitive binding among oral streptococci to hydroxyapatite, *J. Dent. Res.*, 56, 157, 1977.
31. **Slots, J. and Gibbons, R. J.,** Attachment of *Bacteroides melaninogenicus* subsp. *asaccharolyticus* to oral surfaces and its possible role in colonization of the mouth and of periodontal pockets, *Infect. Immun.*, 19, 254, 1978.
32. **Qureshi, J. V. and Gibbons, R. J.,** Differences in the adsorptive behavior of human strains of *Actinomyces viscosus* and *Actinomyces naeslundii* to saliva-treated hydroxyapatite surfaces, *Infect. Immun.*, 31, 261, 1981.
33. **Gibbons, R. J. and Qureshi, J.,** Selective binding of blood group reactive salivary mucins by *Streptococcus mutans* and other oral organisms, *Infect. Immun.*, 22, 665, 1979.
34. **Ericson, T. and Magnusson, J.,** Affinity for hydroxyapatite of salivary substances inducing aggregation of oral streptococci, *Caries Res.*, 10, 8, 1976.
35. **Williams, R. C. and Gibbons, R. J.,** Inhibition of bacterial adherence by secretory immunoglobulin A: a mechanism of antigen disposal, *Science*, 177, 697, 1971.
36. **Hamada, S. and Slade, H. D.,** Biology, immunology and cariogenicity of *Streptococcus mutans, Microbiol. Revs.*, 44, 331, 1980.
37. **van Houte, J., Burgess, R. C., and Onose, H.,** Oral implantation of human strains of *Streptococcus mutans* in rats fed sucrose or glucose diets, *Arch. Oral Biol.*, 21, 561, 1976.
38. **Clark, W. F., Bammann, L. L., and Gibbons, R. J.,** Ability of *Streptococcus mutans* and a glucosyltransferase defective mutant to colonize rodents and attach to hydroxyapatite surfaces, *Infect. Immun.*, 21, 681, 1978.
39. **McCabe, R. M., Keyes, R. H., and Howell, A.,** An *in vitro* method for assessing the plaque forming ability of oral bacteria, *Arch. Oral Biol.*, 12, 1653, 1967.
40. **Rosan, B. and Hammond, B. F.,** Extracellular polysaccharides of *Actinomyces viscosus, Infect. Immun.*, 10, 304, 1974.
41. **Wicken, R. J., Broady, K. W., Evans, J. D., and Knox, K. W.,** New cellular and extracellular amphipathic antigen from *Actinomyces viscosus* NY1, *Infect. Immun.*, 22, 615, 1978.
42. **Jones, S. J.,** A special relationship between spherical and filamentous micro-organisms in mature human dental plaque, *Arch. Oral Biol.*, 17, 613, 1972.
43. **Mouton, C., Reynolds, H., and Genco, R. J.,** Combined micro-manipulation, culture and immunofluorescent techniques for isolation of the coccal organisms comprising the "corn cob" configuration of human dental plaque, *J. Biol. Buccale*, 5, 321, 1977.

44. **Lancy, P., Appelbaum, B., Holt, S. C., and Rosan, B.,** A quantitative assay for corn-cob formation, *J. Dent. Res.*, 58, (special issue), Abstr. #59, 107, 1979.
45. **McIntire, F. C., Vatter, A. E., Baros, J., and Arnold, J.,** Mechanism of coaggregation between *Actinomyces viscosus* T14 and *Streptococcus sanguis* 34, *Infect. Immun.*, 21, 978, 1978.
46. **Komiyana, K. and Gibbons, R. J.,** Inhibition of lactose-reversible coaggregation between oral streptococci and *A. viscosus* by saliva, *Caries Res.*, in press.
47. **Bladen, H., Hageage, G., Pollock, F., and Harr, R.,** Plaque formation *in vitro* on wires by Gram-negative oral micro-organisms *(Veillonella), Arch. Oral Biol.*, 15, 127, 1970.
48. **McBride, B. C. and van der Hoeven, J. S.,** Role of interbacterial adherence in colonization of the oral cavities of gnotobiotic rats infected with *Streptococcus mutans* and *Veillonella alcalesens, Infect. Immun.*, 33, 467, 1981.
49. **Listgarten, M. A.,** Structure of the microbial flora associated with periodontal health and disease in man. A light and electron microscopic study, *J. Periodontol.*, 47, 1, 1976.
50. **Liljemark, W. F., Bloomquist, C. S., and Germaine, G. R.,** Effect of bacterial aggregation on the adherence of oral streptococci to hydroxyapatite, *Infect. Immun.*, 31, 935, 1981.
51. **Gibbons, R. J. and Etherden, I.,** Enzymatic modification of bacterial receptors on saliva-treated hydroxyapatite surfaces, *Infect. Immun.*, 36, 52, 1982.
52. **McBride, B. C. and Gisslow, M. T.,** Role of sialic acid in saliva-induced aggregation of *Streptococcus sanguis, Infect. Immun.*, 18, 35, 1977.
53. **Nachbar, M. S. and Oppenheim, J. D.,** Lectins in the United States diet: a survey of lectins in commonly consumed foods and a review of the literature, *Am. J. Clin. Nutr.*, 33, 2338, 1980.
54. **Bratthall, D.,** Daucus carrota (carrot) — a selective bacteriosorbent, *Adv. Exp. Biol. Med.*, 107, 327, 1978.
55. **Dankers, I. and Gibbons, R. J.,** Association of food lectins with human oral epithelial cells, *in vitro, Arch. Oral Biol.*, 28, 561, 1983.

Chapter 3

THE K88 ADHERENCE SYSTEM IN SWINE

R. Sellwood

TABLE OF CONTENTS

I. INTRODUCTION

The pig, like many other species, is at extreme risk from infection immediately after birth. In contrast with the situation in humans, the piglet at birth is almost entirely devoid of any maternal immunoglobulins because there is no placental transfer of macromolecules.[1]

Important enteric pathogens in pigs are strains of *Escherichia coli* which possess two important properties: (1) the ability to colonize the intestinal tract[2] and (2) the means of stimulating fluid and electrolyte secretion.[3,4] The ability to colonize the small intestine of the piglet by allochthanous bacteria requires mechanisms which usually confer an advantage on the organism over the normal intestinal flora. This chapter is concerned with an important mechanism which enables enteropathogenic *E. coli* to colonize the intestinal tract of pigs, i.e., the adhesin K88.

II. THE K88 ANTIGEN AND INTESTINAL COLONIZATION

The K88 antigen is a surface antigen of many porcine enterotoxigenic strains of *E. coli*.[5] It appears under electron microscope as a filamentous structure radiating from the surface of the bacterium.[6] The filaments are approximately 4 to 6 nm in diameter and vary considerably in length. The cell-free antigen can readily be prepared[7] and has been shown to be a polymeric protein composed of equal subunits of approximately 23,500 Daltons[8] with considerable hydrophobic properties.[9] It is coded for by a large 70 kilobasepair plasmid[10] which can easily be transferred in conjugation experiments.[11]

It became apparent that the presence of K88 enabled the organisms to colonize the small intestine,[12] which they did by attaching to the mucosal surface[13-16] and the adhesin was subsequently demonstrated to be the K88 antigen.[17] The attachment mediated by this antigen prevented removal of the organisms by the normal passage of intestinal contents. Studies on the infection of piglets with K88-positive *E. coli* have revealed the extent of colonization that occurs. Usually, when infection experiments are carried out in conventional piglets with 10^7 to 10^9 organisms given orally the lower small intestine is colonized first and upon rapid proliferation of the organisms the infection spreads up into the proximal part of the intestine. Table 1 shows the results of infecting piglets with a nonenteropathogenic *E. coli* strain 08:K40:H9 possessing various combinations of K88 and Ent plasmids.[12] The intestinal tract of the pig infected with the $K88^-$ Ent^+ strain was not colonized to a degree such that diarrhea was caused. However, the other two pigs given the $K88^+$ Ent^+ and $K88^+$ Ent^- developed diarrhea and successful colonization of the intestine occurred. Therefore it appeared that the presence of K88 was important for the successful colonization of the intestinal tract whether it was in combination with enterotoxin production or not.

The K88 antigen can exist in three serologically distinct types K88ab, K88ac, and K88ad.[18] They all appear to have similar adhesive properties attaching to the brush borders of pig intestinal epithelial cells.

A property of the K88 antigen which is common to many *E. coli* is the ability to cause hemagglutination of erythrocytes.[19] The K88 antigen specifically causes hemagglutination (H.A.) of guinea pig erythrocytes[20] and a comparison with the H.A. properties of type 1 pili are shown in Table 2. It has several important properties. Hemagglutination caused by K88 antigen is resistant to inhibition by D-mannose and its analogs, but is destroyed when the bacteria are heated at 65°C for 30 min. Hemagglutination is also destroyed (or eluted) when the hemagglutination test (normally performed at 4°C) is warmed to 37°C. The hemagglutination has therefore been termed mannose-resistant and eluting (MRE) and differs from hemagglutination which is inhibited by D-mannose and does not elute (MS), a property usually mediated by type 1 (or common) pili. Other properties which differentiate the K88 antigen from type 1 pili are its absence of production when grown at 18°C,[21] (the plasmid

Table 1
CONCENTRATION OF *ESCHERICHIA COLI* POSSESSING DIFFERENT COMBINATIONS OF K88 AND ENT PLASMIDS IN THE ALIMENTARY TRACT OF PIGLETS

Organ	Log_{10} number of infecting organisms per gram content of stated organ, in a piglet given the K88⁺ Ent⁻ form	K88⁺ Ent⁺ form	K88⁻ Ent⁺ form
Stomach	5.7	6.3	4.8
Small intestine at level			
1	8.3	9.3	4.4
3	9.3	9.3	5.6
5	9.8	9.8	5.6
7	9.3	10.2	6.0
Colon	9.6	9.9	6.9

From Smith, H. W. and Linggood, M. A., *J. Med. Microbiol.*, 4, 467, 1971. With permission.

Table 2
HEMAGGLUTINATION PROPERTIES OF K88 ANTIGEN AND TYPE 1 PILI

	K88 antigen	Type 1 pili
Hemagglutinates guinea pig erythrocytes	+	+
H.A. at 4°C	+	+
H.A. at 50°C	−	+
Cultural conditions for optimum production	Buffered agar 18 hr 37°C	Static broth
Presence of 0.5% D-mannose	+	−
Type of H.A.	MRE (mannose resistant and eluting)	MS (mannose sensitive)

is turned off) and its production on agar plate cultures. Type 1 pili, on the other hand, prefer aerobic culture in static broth. It is also interesting that these properties of the K88 antigen also apply to the attachment to the brush borders of the pig intestinal epithelium with the exception that adhesion unlike hemagglutination occurs not only at 4°C but also at all temperatures up to 56°C.[22]

III. HOST SUSCEPTIBILITY

A. Genetics of Resistance

It has been the impression for many years that variation in susceptibility to infection either naturally or experimentally by K88-positive *E. coli* often occurs.[23-25] Results of in vivo infection experiments have demonstrated resistance to infection and possible reasons have been discussed.[26]

Table 3
PHENOTYPES OF PIGS OF WHOLE LITTERS (PHENOTYPES OF BOTH PARENTS KNOWN RETROSPECTIVELY)

Type of mating	No. of matings	Litters		Phenotype +	Phenotype −
− x −	41	S[a]	0	0	0
		NS[b]	41	0	340
+ x −	31	S	24	107	103
		NS	7	73	0

Note: − = Nonadherent phenotype.
+ = Adherent phenotype.

[a] S segregating litter.
[b] NS nonsegregating litter.

From Sellwood, R. and Kearns, M., *Proc. Int. Colloq. Frontiers of Knowledge in the Diarrhoeal Diseases*, Janowitz, H. D. and Sachar, D. B., Eds., Projects in Health, Inc., Upper Montclair, N.J., 1979. With permission.

An in vitro study of the adherence of K88-positive *E. coli* to the epithelial cells of the small intestine revealed that these organisms were unable to attach to the brush border surface of some pigs.[27] Clearly, if attachment was important for the colonization of the intestine and also production of the disease then the inability to adhere in vivo to the brush border surface would render the pig resistant. This hypothesis was subsequently tested and confirmed.[28] Both adhesive and nonadhesive conventional piglets were infected with a K88-positive enterotoxigenic *E. coli* strain (0149:K91, K88:H10). Of the 23 adhesive piglets, 21 died or were killed with severe diarrhea and dehydration brought about by the infection, whereas only 1 of 24 nonadhesive animals died. The infecting strain was only recovered in low numbers ($\sim 10^6/m\ell$) from the posterior end of the intestinal tract of the nonadhesive animals that survived. On the other hand it was recovered in very high numbers ($\sim 10^9/m\ell$) from the contents throughout the intestine of the adhesive pigs that died or were killed with severe diarrhea.

This phenotypic difference between pigs has been shown to be genetic in origin.[29] Using the in vitro adhesion test with brush borders prepared from the intestinal epithelial cells and K88-positive *E. coli* it was established that adherence was a dominant characteristic of the host, being inherited in a simple Mendelian way. The simplest interpretation was that of one locus, two alleles S (receptor positive, disease susceptible) and s (receptor negative, disease resistant) with S dominant over s. The results of testing pigs over an extended period of time supported this initial hypothesis (Table 3)[30] and it has been possible to breed two herds of pigs, one of which is homozygous dominant (susceptible) and the other homozygous recessive (resistant). Studies have shown that K88-positive *E. coli*, either wild type or normal commensals that have received the K88 plasmid, colonize the intestinal tract and multiply to higher numbers in susceptible piglets than in resistant piglets.[31] Indeed it also appears that colonization by K88-positive *E. coli* in the resistant pig is much worse than by the K88-negative recipient strain (Table 4). The reasons for this are not understood. It is also interesting to note that a laboratory K12 strain possessing the K88 plasmid, although having the ability to adhere in vitro to epithelial cell surfaces does not survive in the small intestine

Table 4
CONCENTRATION OF ORGANISMS IN DIFFERENT PARTS OF THE ALIMENTARY TRACT OF A SUSCEPTIBLE AND A RESISTANT PIGLET GIVEN A NONPATHOGENIC 09:K36:H19 *E. COLI* STRAIN POSSESSING K88 AND ENT PLASMIDS

		No. of organisms ($\log_{10}$ per gram)					
			Small intestine (scrapings)				
Pig type	*E. coli* strain	Stomach (contents)	1	3	5	7	Colon (contents)
Susceptible	$K88^+$ Ent^+	6.4	9.0	9.2	9.4	9.4	8.5
	$K88^-$ Ent^+	5.2	4.0	4.2	4.6	5.0	5.3
Resistant	$K88^+$ Ent^+	3.0	<2.0	2.3	3.2	3.0	4.7
	$K88^-$ Ent^+	4.5	3.0	5.2	6.4	5.6	6.8

From Smith, H. W. and Huggins, M. B., *J. Med. Microbiol.*, 11, 471, 1978. With permission.

but is gradually removed from the intestinal tract. This suggests that the possession of K88 alone does not necessarily mean successful colonization; other antigens e.g., "O" and "K" and factors such as Col. V, may also be important particularly in respect of the survival of the organism in the gut. Another inconsistency with the over-simplified approach, that an adhesin plus an enterotoxin are the two important virulence determinants that are required for pathogenicity, is that if the K88 plasmid is transferred to a nonpathogenic *E. coli* strain 09:K36:H19:Ent^- diarrhea[12] and death[22] can be produced in conventional piglets orally infected with this strain. High numbers (~10^9 cfu/mℓ) of the strain were recovered throughout the intestinal tract. The questions raised by these results need to be resolved.

B. The Weak-Adhesive Phenotype

An additional phenotype of adhesive pig has more recently been discovered which has been called a "weak-adhesive (±) phenotype".[32] This phenotype is also an inherited character. Usually only 3 to 4 bacteria/brush border can be counted in adhesion tests whereas the normal-adhesive phenotype has 20 bacteria/brush border attached and the brush borders are agglutinated. It is not a feature of the weak-adhesive phenotype that massive agglutination occurs in the brush border test. The results of testing four litters of four different boars mated to homozygous recessive sows are shown in Table 5. One boar (1) sired only weak-adhesive (±) piglets, one boar sired a mixture of all three phenotypes, and two boars sired either weak-adhesive and negative phenotypes or weak-adhesive and positive phenotypes. The genetics of the inheritance of this character have not yet been elucidated, but it does appear that it is a dominant character. Further litters from boar 1 have been produced and the results have been consistent.

Infection of a litter of weak-adhesive piglets with an enterotoxigenic *E. coli* strain 0149:K91, K88ac:H10 was performed. 10^9 organisms were given orally to piglets at 3 days of age. No diarrhea was recorded in any of the eight piglets infected. The strain was excreted in low numbers for 3 days when the pigs were necropsied. Low numbers of the strain (log cfu/mℓ contents <2.0) were isolated from the intestinal contents of all the pigs. The pig phenotypes were also confirmed at this stage.

It would, therefore, appear that another receptor for the K88 antigen exists on pig intestinal epithelial cells. It is, however, either less efficient at binding K88 antigen or there are fewer receptors per cell. It is also not known whether this receptor enables attachment of the organisms to occur in vivo and fluorescent antibody studies have given no further information.

Table 5
PHENOTYPES OF PROGENY OF HOMOZYGOUS RECESSIVE DAMS AND BOARS POSSESSING THE WEAK-ADHESIVE ALLELE

	Phenotypes		
Boar	+	±	−
1	0	9	0
2	5	2	1
3	4	1	0
4	0	3	4

Note: + = Adhesive phenotype.
− = Nonadhesive phenotype.
± = Weak-adhesive phenotype.

From Sellwood, R., *Genetics and Heterogeneity of Common Gastrointestinal Disorders*, Rotter, J. I., Samloff, I. M., and Rimoin, D. L., Eds., Academic Press, New York, 1980, 535. With permission.

It is, however, important when studying susceptibility and resistance in vitro to recognize this phenotype since it could be mistaken for the normal adhesive phenotype.

C. Passive Protection of the Neonate

The outcome of infection is closely related, not only to the genetic susceptibility of the piglets but also to the maternal protection afforded by the colostrum and milk of the sows; a factor which is also complicated by the same genetic system. Support for this view came when the genotyped herd at Compton suffered a spontaneous outbreak of diarrhea attributed to a pathogenic *E. coli* strain, subsequently serotyped as 09; K87; K88ac.[33]

In a total of 38 litters diarrhea occurred in 11 litters all of which were the progeny of homozygous recessive (resistant) dams by homozygous dominant or heterozygote (susceptible) boars (Table 6). Progeny of homozygous recessive dams by homozygous recessive boars were not affected confirming the earlier experimental results. Concurrent infection by a 987P *E. coli* strain affecting pigs of all genotypes also confirmed in vitro adhesion studies that genetically resistant pigs to K88 *E. coli* infection were not resistant to *E. coli* strains possessing the 987P antigen since all pigs tested have the receptor for 987P.

There was also no diarrhea attributed to K88 *E. coli* in any of the litters of susceptible dams. Although diarrhea is not usually associated with the colonization of the intestinal tract in adult pigs, the excretion of large numbers of organisms can occur particularly in the parturient sow.

The level of colonization in the susceptible sows may have been such that an immune response was produced which was sufficient for the passive protection of the piglets. The resistant sows, on the other hand, may have been colonized by K88-positive *E. coli* to a lesser degree and an immune response was not produced. Their susceptible piglets therefore remained unprotected.

Table 6
INCIDENCE OF DIARRHEA AND MORTALITY IN PROGENY FROM DIFFERENT TYPES OF MATING

Type of mating	Litters	Born	Diarrhea	Deaths Total (%)	Deaths Diarrhea (%)
♀ ♂					
SS x ss	8	64	7[a](11)	4(6)	0(0)
SS x ss	4	45	0 (0)	1(2)	0(0)
ss x ss	15	153	19[a](12)	10(7)	3(2)
ss x SS	4	32	32[b](100)	22(69)	22(69)
ss x Ss	7	68	30[b](44)	25(37)	24(35)

Note: S = Dominant allele conferring susceptibility.
s = Recessive allele conferring resistance.

[a] Diarrhea attributed to the 987P *E. coli* strain.
[b] Diarrhea attributed to the K88-positive *E. coli* strain.

From Sellwood, R., *Vet. Res.*, 105, 228, 1979. With permission.

Protective properties of colostrum and milk from vaccinated and nonvaccinated sows have been studied,[34-36] including the type of antibody response produced by various vaccination schedules. It has been concluded that an anti-adhesive function of the colostrum is an important mechanism in addition to a possible bactericidal or bacteristatic action; but the results have been equivocal. Studies have been made of the possible protective mechanisms of the colostrum of both the susceptible and resistant dams collected during the period of this outbreak. The major findings of these studies showed that the colostrum of susceptible dams had significantly higher levels of (1) immunoglobulins IgG and IgM but not IgA,[37] (2) anti-adhesive ability, and (3) opsonic properties of the colostrum in the in vitro killing of the infecting *E. coli* strain by polymorphonuclear leukocytes.[38]

IV. FURTHER CONSIDERATIONS

The role of the K88 antigen in the establishment of K88 *E. coli* infection in the intestinal tract of the pig is now well documented. However, there are some interesting points which still require elucidation. The presence of K88 alone does not ensure successful colonization, although enterotoxin production is not important. An organism must not only be able to attach in vivo but must also survive in an environment which may for several reasons be hostile to it. Antagonistic interactions between strains may regulate the numbers of certain organisms, pH or redox potential may restrict growth of the infecting organism, and indigenous bacteria may elicit a substance or compete for a particular nutrient both of which may affect the growth of the enteropathogenic strain. It is also important to understand why a K88-positive *E. coli* strain, provided it can adhere and colonize the gut causes diarrhea. Are there factors other than enterotoxin that will initiate a cellular response deranged function similar to that brought about by enterotoxin?

ADDENDUM

It now appears that *E. coli* strains possessing either K88ab, K88ac, or K88ad can have different specifications regarding attachment of the bacteria to intestinal brush borders of pigs. Bijlsma et al.[39] described several phenotypes of pig which were either susceptible to adherence by (1) all three K88 variants, (2) K88ab and K88ac, (3) K88ab and K88ad, (4) K88ad, and (5) none of the variants.

REFERENCES

1. **Sterzl, J., Rejnik, J., and Travnicek, J.,** Impermeability of pig placenta for antibodies, *Folia Microbiol.*, 11, 7, 1966.
2. **Smith, H. W. and Jones, J. E. T.,** Observations on the alimentary tract and its bacterial flora in healthy and diseased pigs, *J. Pathol. Bacteriol.*, 86, 387, 1963.
3. **Moon, H. W. and Whipp, S. C.,** Systems for testing the enteropathogeneity of *Escherichia coli*, *Ann. N.Y. Acad. Sci.*, 176, 197, 1971.
4. **Kantor, H. S., Tao, P., and Gorbach, S. L.,** Stimulation of intestinal adenyl cyclase by *Escherichia coli* enterotoxin: comparison of strains from an infant and an adult with diarrhoea, *J. Infect. Dis.*, 129, 1, 1974.
5. **Sojka, W. J.,** *Escherichia coli* in domestic animals and poultry, *Review Series no. 7, Commonwealth Bureau of Animal Health*, Farnham Royal, 1965.
6. **Stirm, S., Ørskov, F., Ørskov, I., and Birch-Andersen, A.,** Episome-carried surface antigen K88 of *Escherichia coli*. III. Morphology, *J. Bacteriol.*, 93, 740, 1967.
7. **Stirm, S., Ørskov, F., Ørskov, I., and Mansa, B.,** Episome-carried surface antigen K88 of *Escherichia coli*. II. Isolation and chemical analysis, *J. Bacteriol.*, 93, 731, 1967.
8. **Mooi, F. R. and De Graaf, F. K.,** Isolation and characterization of K88 antigens, *FEMS Microbiol. Lett.*, 5, 17, 1979.
9. **Wadström, T., Smyth, C. J., Faris, A., Jonsson, P., and Freer, J. F.,** Hydrophobic adsorptive and haemagglutinating properties of enterotoxigenic *Escherichia coli* with different colonization factors: K88 antigen, K99 antigen, colonization factor antigens and adherence factor, in *Proc. 2nd Int. Symp. Neonatal Diarrhoea*, Acres, S. D. Ed., VIDO, Saskatoon, Canada 1978, 29.
10. **Kehoe, M., Sellwood, R., Shipley, P., and Dougan, G.,** Genetic analysis of K88-mediated adhesion of enterotoxigenic *Escherichia coli*, *Nature (London)*, 291, 122, 1981.
11. **Ørskov, I. and Ørskov, F.,** Episome-carried surface antigen K88 of *Escherichia coli*. I. Transmission of the determinant of the K88 antigen and the influence on the transfer of chromosomal markers, *J. Bacteriol.*, 91, 69, 1966.
12. **Smith, H. W. and Lingood, M. A.,** Observations on the pathogenic properties of the K88, Hly and Ent plasmids of *Escherichia coli* with particular reference to porcine diarrhoea, *J. Med. Microbiol.*, 4, 467, 1971.
13. **Arbuckle, J. B. R.,** The location of *Escherichia coli* in the pig intestine, *J. Med. Microbiol.*, 3, 333, 1970.
14. **Drees, D. T. and Waxler, G. L.,** Enteric colibacillosis in gnotobiotic swine: a fluorescent microscopic study, *Am. J. Vet. Res.*, 31, 1147, 1970.
15. **Drees, D. T. and Waxler, G. L.,** Enteric colibacillosis in gnotobiotic swine: an electron microscopic study, *Am. J. Vet. Res.*, 31, 1159, 1970.
16. **Bertschinger, H. U., Moon, H. W., and Whipp, S. C.,** Association of *Escherichia coli* with the small intestinal epithelium. I. Comparison of enteropathogenic and nonenteropathogenic porcine strains in pigs, *Infect. Immun.*, 5, 595, 1972.
17. **Jones, G. W. and Rutter, J. M.,** Role of the K88 antigen in the pathogenesis of neonatal diarrhoea caused by *Escherichia coli* in piglets, *Infect. Immun.*, 6, 918, 1972.
18. **Guinée, P. A. M. and Jansen, W. H.,** Behaviour of *Escherichia coli* K antigens K88ab, K88ac and K88ad in immunoelectrophoresis, double diffusion and haemagglutination, *Infect. Immun.*, 23, 700, 1979.
19. **Duguid, J. P., Smith, I. W., Dempster, G., and Edmunds, P. N.,** Non-flagellar filamentous appendages ('fimbriae') and haemagglutinating activity in *Bacterium coli*, *J. Pathol. Bacteriol.*, 70, 335, 1955.
20. **Duguid, J. P. and Old, D. C.,** Adhesive properties of Enterobacteriaceae, in *Bacterial Adherence*, Beachey, E. H. Ed., Chapman and Hall, London, 1980, chap. 7.
21. **Ørskov, I., Ørskov, F., Sojka, W. J., and Leach, J. M.,** Simultaneous occurrence of *Escherichia coli* B and L antigens in strains from diseased swine, *Acta Pathol. Microbiol. Scand.*, 53, 404, 1961.
22. **Sellwood, R.,** unpublished data, 1981.
23. **Wijeratne, W. V. S., Crossman, P. J., and Gould, C. M.,** Evidence of sire effect on piglet mortality, *Br. Vet. J.*, 126, 94, 1970.

24. **Sweeney, E. J.,** *Escherichia coli* in enteric disease of swine: observations on herd resistance, *Ir. Vet. J.*, 22, 42, 1968.
25. **Smith, H. W. and Halls, S.,** The production of oedema disease and diarrhoea in weaned pigs by the oral administration of *Escherichia coli:* factors that influence the course of the experimental disease, *J. Med. Microbiol.*, 1, 45, 1968.
26. **Rutter, J. M. and Anderson, J. C.,** Experimental neonatal diarrhoea caused by an enteropathogenic strain of *Escherichia coli* in piglets: a study of the disease and the effect of vaccinating the dam, *J. Med. Microbiol.*, 5, 197, 1972.
27. **Sellwood, R., Gibbons, R. A., Jones, G. W., and Rutter, J. M.,** Adhesion of enteropathogenic *Escherichia coli* to pig intestinal brush borders: the existence of two pig phenotypes, *J. Med. Microbiol.*, 8, 405, 1975.
28. **Rutter, J. M., Burrows, M. R., Sellwood, R., and Gibbons, R. A.,** A genetic basis for resistance to enteric disease caused by *E. coli*, *Nature (London)*, 257, 135, 1975.
29. **Gibbons, R. A., Sellwood, R., Burrows, M. R., and Hunter, P. A.,** Inheritance of resistance to neonatal *E. coli* diarrhoea in the pig: examination of the genetic system, *Theor. Appl. Genet.*, 51, 65, 1977.
30. **Sellwood, R. and Kearns, M. J.,** Inherited resistance to *Escherichia coli* diarrhoea in pigs: the genetics and nature of the intestinal receptor, in *Proc. Int. Colloq. Frontiers of Knowledge in the Diarrhoeal Diseases*, Janowitz, H. D. and Sachar, D. B., Eds., Projects in Health, Inc., Upper Montclair, New Jersey, 1979.
31. **Smith, H. W. and Huggins, M. B.,** The influence of plasmid-determined and other characteristics of enteropathogenic *Escherichia coli* on their ability to proliferate in the alimentary tracts of piglets, calves and lambs, *J. Med. Microbiol.*, 11, 471, 1978.
32. **Sellwood, R.,** Genetic susceptibility to intestinal infection — animal models, in *Genetics and Heterogeneity of Common Gastrointestinal Disorders*, Rotter, J. I., Samloff, I. M., and Rimoin, D. L., Eds., Academic Press, New York, 1980, 535.
33. **Sellwood, R.,** *Escherichia coli* diarrhoea in pigs with or without the K88 receptor, *Vet. Rec.*, 105, 228, 1979.
34. **Jones, G. W. and Rutter, J. M.,** Contribution of the K88 antigen of *Escherichia coli* to enteropathogenicity; protection against disease by neutralizing the adhesive properties of K88 antigen, *Am. J. Clin. Nutr.*, 27, 1441, 1974.
35. **Nagy, L. K., Bhogal, B. S., and Mackenzie, T.,** Duration of anti-adhesive and bactericidal activities of milk from vaccinated sows on *Escherichia coli* 0149 in the digestive tract of piglets during the nursing period, *Res. Vet. Sci.*, 27, 289, 1979.
36. **Rutter, J. M., Jones, G. W., Brown, G. T. H., Burrows, M. R., and Luther, P. D.,** Antibacterial activity in colostrum and milk associated with protection of piglets against enteric disease caused by K88-positive *Escherichia coli*, *Infect. Immun.*, 13, 667, 1976.
37. **Sellwood, R. and Metzger, J. J.,** Resistance to *E. coli* diarrhoea in pigs, in *Proc. Int. Pig Vet. Soc. Congress*, Nielsen, N. C., Hogh, P., and Bille, N., Eds., Reproset, Copenhagen, Denmark, 1980, 160.
38. **Sellwood, R.,** *Escherichia coli* associated porcine neonatal diarrhoea: antibacterial activities of colostrum from genetically susceptible and resistant sows, *Infect. Immun.*, 35, 396, 1982.
39. **Bijlsma, I. G. W., De Nijs, A., Van Der Meer, C., and Frik, J. F.,** Different pig phenotypes affect adherence of *Escherichia coli* to jejunal brush borders by K88ab, K88ac, or K88ad antigen, *Infect. Immun.*, 37, 891, 1982.

Chapter 4

THE K99 ADHERENCE SYSTEM IN CATTLE

Harley W. Moon and Paul L. Runnels

TABLE OF CONTENTS

I. INTRODUCTION

Both K88 and K99 are plasmid-mediated pilus antigens which contribute to the virulence of enterotoxigenic *Escherichia coli* (ETEC). Both antigens contribute to virulence by facilitating adhesion and colonization in the small intestine.[1] Some (most?) $K88^+$ or $K99^+$ ETEC require these adhesins in order to colonize intensively enough to cause diarrhea. However, after removal of the K99 plasmid, some strains can still colonize intensively enough to cause diarrhea in some circumstances.[2] Both K88 and K99 are nonmannose sensitive hemagglutinins, both are produced at 37°C but not at 18°C. They were initially recognized serologically and designated as K antigens because they both have the surface location and heat-lability characteristics of K antigens in the Kauffman-White schema.[3,4] However, in contrast to the polysaccharide capsules or envelopes of classical K antigens, K88 and K99 are pili (minute-filamentous-protein appendages).[3,5,6]

The K99 adherence system differs from the K88 adherence system in some respects. These differences will be the basis for the remainder of this chapter.

II. DIFFERENCES BETWEEN $K88^+$ AND $K99^+$ ETEC

A. Bacterial

1. Enterotoxins

Most $K99^+$ ETEC produce heat stable enterotoxin of the type demonstrable in infant mice (STa). Most $K88^+$ ETEC produce heat labile enterotoxin (LT) and STa or heat stable enterotoxin of the type demonstrable in pigs but not infant mice (STb).[7]

2. Capsules

Most $K99^+$ ETEC produce distinct polysaccharide capsular K antigens of the A variety.[4] These capsules enhance the colonizing abilities of $K99^+$ ETEC. The acapsular form of some strains can still colonize intensively enough to cause diarrhea,[2,8,9] while acapsular forms from some other $K99^+$ ETEC strains cannot.[2,10] Most $K88^+$ ETEC lack distinct capsules,[11] and if they produce polysaccharide K antigens, they produce those of the L or B varieties. It is not known if polysaccharide K antigens of the L or B varieties affect colonization by $K88^+$ ETEC.

3. Molecular Charge

Purified K99 has an alkaline isoelectric point (cationic) and purified K88 has an acidic isoelectric point.[6,12]

4. Antigenic Variants

No antigenic subtypes of K99 have been reported. Many $K99^+$ ETEC also produce an antigen and adhesin designated ''anionic K99.''[13] Some ETEC which are enteropathogenic for neonatal pigs produce ''anionic K99'' but do not produce K99.[26] The role of ''anionic K99'' in the disease process is not known. ''Anionic K99'' is apparently not antigenically related to K99 and is probably a distinct separate entity, rather than a variant of K99.[14] In contrast, K88 antigens contain an antigenically constant component (designated K88a) and a variable component. Three antigenic subtypes (b, c, and d) of the variable component have been reported.[15]

The antigen previously referred to as ''anionic K99'' has been shown to be a separate entity, now designated F41.[27]

B. Host Range

1. Species

Pilus-mediated adhesion to intestinal epithelium apparently accounts for much of the host species specificity characteristic of ETEC. Naturally occurring diarrheal disease caused by K99$^+$ ETEC is common in calves, lambs, and pigs. Infant mice are susceptible to experimental challenge if they are maintained at temperatures which permit expression of K99.[16] In contrast, diarrheal disease due to infection with K88$^+$ ETEC is confined to certain genotypes of swine.[2,17]

2. Tissue

Intensive colonization of the small intestine ($\geq 10^8$/g) by most K99$^+$ ETEC tends to be confined to the distal one-half of the organ. In contrast, K88$^+$ strains tend to colonize intensively throughout the small intestine.[2] Jejunal epithelium is susceptible to adhesion by both K88$^+$ and K99$^+$ ETEC in ligated intestinal loops and in vitro.[8] Thus, the tendency for K99$^+$ strains to be restricted to the ileum in intact animals may be due to the comparatively sluggish motility of this region,[18] rather than to a lack of receptors in the jejunum. If this is correct, it implies that the adhesive interactions between host and bacterial cells are stronger for K88$^+$ strains, than for K99$^+$ strains.

3. Age

Pigs are susceptible to ETEC-induced diarrhea during the immediate neonatal period (before they are 1 week old) and again after weaning when they are several weeks old. K88$^+$ ETEC are commonly associated with the disease in both age groups, but K99$^+$ ETEC appear to be confined to the immediate neonatal period. Innate age-dependent resistance to some types of STa accounts in part for the resistance of older pigs to K99$^+$ ETEC.[19] However, it takes several weeks for pig small intestine to become fully resistant to K99$^+$ ETEC enterotoxin-induced secretion 19 and some resistance to disease caused by oral infection develops rapidly during the first few hours after birth. For example, pigs challenged with K99$^+$ ETEC when they were 24 to 36 hr old were more resistant than those challenged when they were 1 to 7 hr old (Table 1). The 1- to 7-hr-old pigs were highly susceptible to challenge both before and after suckling. All pigs in the experiment were from the same herd, were challenged with the same dose and strain of ETEC, and were observed for 5 days after challenge. There were fewer deaths, as well as a lower incidence and duration of diarrhea, and greater weight gains among survivors in the $\geq$ 24-hr age group than in the $\leq$ 7-hr age group.

Age-dependent resistance to diarrhea caused by K99$^+$ ETEC also develops rapidly in calves and lambs. Naturally occurring clinical disease in these two species is confined to the first week after birth. Calves and lambs are susceptible to intensive colonization of small intestine as the result of experimental challenge by K99$^+$ ETEC at birth, but become resistant to intensive colonization (and disease) by the time they are 2 days old (Table 2).[20]

III. MECHANISMS OF AGE-DEPENDENT RESISTANCE TO COLONIZATION

A. Gastric pH, Microflora, and Intestinal Motility

The drop in gastric pH during the first day after birth in pigs establishes a partial barrier against colonization of the intestine by *E. coli*.[21] Endogenous secretion of hydrochloric acid is usually delayed until several weeks of age in pigs.[22] Therefore, the neonatal drop in gastric pH is probably due to organic acids produced by lactobacilli and streptococci which colonize the stomach at this time.[21] Increased intestinal motility stimulated by colostrum and possibly by the developing microflora is another mechanism speculated to contribute to the development of resistance in the immediate neonatal period.

Factors other than colostrum and microflora must also contribute to the development of resistance of calves and lambs to intestinal colonization by K99$^+$ ETEC. This seemed apparent when colostrum-deprived, germ-free calves were challenged with the K99$^+$ ETEC

Table 1
PIGS GIVEN 10^{10} *E. COLI* STRAIN 431 ($K99^+$ ENT^+) ORALLY

Pigs				Response (%)			
		Number of			Watery diarrhea		
Age (hr)[a]	Suckled	Litters	Pigs	Deaths	Day 1	Day 5	Weight change
1 — 7	After[b]	7	56	46	98	23	3
1 — 7	Before[c]	5	38	74	100	10	2
24 — 36	Before	5	46	4	78	0	43

[a] Age when given *E. coli*.
[b] Not allowed to suckle until after *E. coli* were given.
[c] Allowed to suckle at will from birth.

Table 2[a]
CONVENTIONAL CALVES GIVEN 10^{10} *E. COLI* ORALLY

E. coli		Calves		
Type	Strains	Age-days[b]	Watery diarrhea	Ileal *E. coli*[c]
$K99^-$ Ent^-	31	<1	0/31[d]	10^{5-8}
$K99^+$ Ent^+	3[e]	<1	10/10	10^{8-11}
$K99^+$ Ent^+	3[e]	2—4	0/18	10^{3-6}

[a] Modified from Smith, H. W. and Halls, S., *J. Pathol. Bacteriol.*, 93, 499, 1967.
[b] Age when given *E. coli*.
[c] One day after calves were given *E. coli*.
[d] Number positive per number tested.
[e] One of the three *E. coli* strains tested was strain B44.

strain B44 (Table 3). Calves were observed clinically for 48 hr and then the intensity of ileal colonization determined at necropsy. One of two calves challenged at 2 days of age was susceptible, developed profuse, acute, fatal diarrhea and died. The number of ileal *E. coli* in this calf was not determined because of postmortem change. All calves in the 5- to 8-day-old group and the other calf in the 2-day-old group were resistant to intensive colonization and diarrhea. Strain B44 is enteropathogenic in conventional calves challenged before they are 1 day old (Table 2).[20] Age-dependent resistance in colostrum-deprived, germ-free calves occurs with $K99^+$ ETEC other than strain B44,[23] and colostrum-deprived, germ-free lambs also became resistant to challenge with $K99^+$ ETEC by 5 days of age.[24] In summary, changes in gastric pH, microflora, and intestinal motility apparently do not entirely account for age-dependent resistance to colonization by $K99^+$ ETEC.

B. Receptor Availability

We tested the hypothesis that epithelial cells of the neonate lose K99 receptors during the neonatal period.[25] This was done by an in vitro microscopic comparison of adhesion of $K99^+$ ETEC, $K88^+$ ETEC, and nonenterotoxigenic *E. coli* to epithelial cells from the small

Table 3
RESPONSE OF GERM-FREE CALVES GIVEN 10^9 STRAIN B44 ($K99^+$ Ent^+)*E. COLI* ORALLY

Age (days)[a]	Watery diarrhea	Died	Ileal *E. coli*[b]
2	1/1	1/1	?
2	1/1	0/1	10^7
5 — 8	3/5	0/5	10^7

[a] Age when given *E. coli*.
[b] Two days after calves were given *E. coli*.

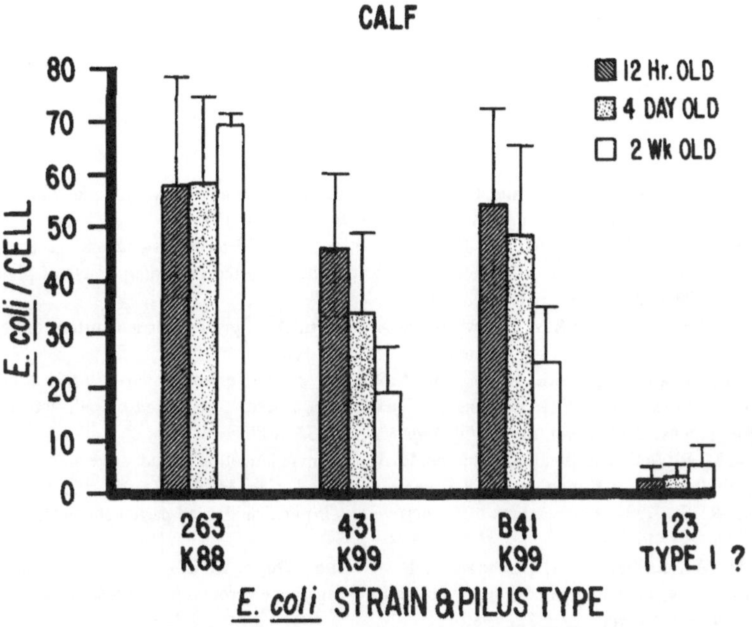

FIGURE 1. Number of *E. coli* adherent to intestinal epithelial cells isolated from calves of different ages. Results with epithelial cells isolated from 1 calf (12 hr old), 2 calves (4 days old), and 1 calf (2 weeks old) are presented. Bars represent the mean ± 1 standard deviation of the number of adherent bacteria per epithelial cell. Bacteria were used at 5 × 10^9/mℓ and epithelial cells at 2 × 10^6/mℓ. (From Runnels, P. L., et al., *Infect. Immun.*, 28, 298, 1980. With permission.)

intestines of calves, pigs, and mice of different ages. The results were similar in all three species in that cells from older animals were more resistant to adhesion by $K99^+$ ETEC than were cells from neonates. The results in Figure 1 for calves are shown as an example. The $K88^+$ strain 263 adhered in comparable high numbers to cells from calves in all three age groups, and the nonenterotoxigenic strain 123 consistently adhered in low numbers. In contrast, the $K99^+$ strains 431 and B41 both adhered to cells from the 12-hr-old calf in significantly greater numbers than to cells from the 2-week-old calf. However, cells from 4-day-old calves were still susceptible to adhesion of $K99^+$ strains in greater numbers than the nonenterotoxigenic strain 123. The results were consistent with the hypothesis. However, the changes during the first few days after birth shown in vitro by this technique, seem intuitively, to be of insufficient magnitude to explain the marked resistance that develops by 2 days of age in calves in vivo.

IV. CONCLUSIONS

There are several host range differences in adhesion or colonization by $K88^+$ and $K99^+$ ETEC. $K88^+$ strains tend to be restricted to specific genotypes in one species, to intensively colonize the entire small intestine, and to affect individuals up to several weeks old. K99 strains tend to affect all genotypes in several species, to be restricted to the ileum, and to be restricted to individuals less than 1 week old. The mechanisms of the innate age-dependent resistance characteristic of $K99^+$ ETEC infections in calves, lambs, and pigs are not well defined. Existing data do not appear to be adequate to support any of the mechanisms proposed. Decreased availability of K99 receptors with age is the hypothesis most attractive to the authors. However, the resistance will probably ultimately be shown to be multifactorial.

REFERENCES

1. **Moon, H. W., Isaacson, R. E., and Pohlenz, J.,** Mechanisms of association of enteropathogenic *Escherichia coli* with intestinal epithelium, *Am. J. Clin. Nutr.*, 32, 119, 1979.
2. **Smith, H. W. and Huggins, M. B.,** The influence of plasmid-determined and other characteristics of enteropathogenic *Escherichia coli* on their ability to proliferate in the alimentary tracts of piglets, calves and lambs, *J. Med. Microbiol.*, 11, 471, 1978.
3. **Strim, S., Ørskov, F., Ørskov, I., and Birch-Andersen, A.,** Episome-carried surface antigen K88 of *Escherichia coli*. III. Morphology, *J. Bacteriol.*, 93, 740, 1967.
4. **Ørskov, I., Ørskov, F., Smith, H. W., and Sojka, W. J.,** The establishment of K99, a thermolabile, transmissible *Escherichia coli* K antigen, previously called "Kco", possessed by calf and lamb enteropathogenic strains, *Acta Pathol. Microbiol. Scand. B*, 83, 31, 1975.
5. **Burrows, M. R., Sellwood, R., and Gibbons, R. A.,** Haemagglutinating and adhesive properties associated with the K99 antigen of bovine strains of *Escherichia coli*, *J. Gen. Microbiol.*, 96, 269, 1976.
6. **Isaacson, R. E.,** K99 surface antigen of *Escherichia coli:* purification and partial characterization, *Infect. Immun.*, 15, 272, 1977.
7. **Moon, H. W., Kohler, E. M., Schneider, R. A., and Whipp, S. C.,** Prevalence of pilus antigens, enterotoxin types, and enteropathogenicity among K88-negative enterotoxigenic *Escherichia coli* from neonatal pigs, *Infect. Immun.*, 27, 222, 1980.
8. **Nagy, B., Moon, H. W., and Isaacson, R. E.,** Colonization of porcine small intestine by *Escherichia coli:* ileal colonization and adhesion by pig enteropathogens that lack K88 antigen and by some acapsular mutants, *Infect. Immun.*, 13, 1214, 1976.
9. **Nagy, B.,** Vaccination of cows with a K99 extract to protect newborn calves against experimental enterotoxic colibacillosis, *Infect. Immun.*, *27, 21, 1980.*
10. **Sojka, W. J., Wray, C., and Morris, J. A.,** Passive protection of lambs against experimental enteric colibacillosis by colostral transfer of antibodies from K99-vaccinated ewes, *J. Med. Microbiol.*, 11, 493, 1978.
11. **Ørskov, F. and Ørskov, I.,** Immunoelectrophoretic patterns of extracts from *Escherichia coli* 0 antigen test strains 01 to 0157 examinations in homologous OK sera, *Acta Pathol. Microbiol. Scand. B*, 80, 905, 1972.
12. **Strim, S., Ørskov, F., Ørskov, I., and Mansa, B.,** Episome-carried surface antigen K88 of *Escherichia coli*. II. Isolation and chemical analysis, *J. Bacteriol.*, 93, 731, 1967.
13. **Morris, J. A., Thorns, C. J., and Sojka, W. J.,** Evidence for two adhesive antigens on the K99 reference strain *Escherichia coli* B41, *J. Gen. Microbiol.*, 118, 107, 1980.
14. **Isaacson, R. E.,** K99 surface antigen of *Escherichia coli*: antigenic characterization, *Infect. Immun.*, 22, 555, 1978.
15. **Guinée, P. A. M. and Jansen, W. W.,** Behavior of *Escherichia coli* K antigens K88ab, K88ac, and K88ad in immunoelectrophoresis, double diffusion, and hemagglutination, *Infect. Immun.*, 23, 700, 1979.

16. **Moon, H. W., Fung, P. Y., Isaacson, R. E., and Booth, G. D.,** Effects of age, ambient temperature, and heat-stable *Escherichia coli* enterotoxin on intestinal transit in infant mice, *Infect. Immun.*, 25, 127, 1979.
17. **Sellwood, R., Gibbons, R. A., Jones, G. W., and Rutter, J. M.,** Adhesion of enteropathogenic *Escherichia coli* to pig intestinal brush borders: the existence of two pig phenotypes, *J. Med. Microbiol.*, 8, 405, 1975.
18. **Summers, R. W., Kent, T. H., and Osborne, J. W.,** Effects of drugs, ileal obstruction, and irradiation on rat gastrointestinal propulsion, *Gastroenterology*, 59, 731, 1970.
19. **Whipp, S. C., Moon, H. W., and Argenzio, R. A.,** Comparison of entero-toxic activities of heat-stable enterotoxins from class 1 and class 2 *Escherichia coli* of swine origin, *Infect. Immun.*, 31, 245, 1981.
20. **Smith, H. W. and Halls, S.,** Observations by the ligated intestinal segment and oral inoculation methods on *Escherichia coli* infections in pigs, calves, lambs and rabbits, *J. Pathol. Bacteriol.*, 93, 499, 1967.
21. **Smith, H. W. and Jones, J. E. T.,** Observations on the alimentary tract and its bacterial flora in healthy and diseased pigs, *J. Pathol. Bacteriol.*, 86, 387, 1963.
22. **Kutas, F. and Szabó, J.,** A study of gastric acid production in pigs in relation to early weaning, *Acta Vet. Acad. Sci. Hung.*, 24, 133, 1974.
23. **Tzipori, S. R., Makin, T. J., Smith, M. L., and Krautil, F. L.,** Clinical manifestations of diarrhea in calves infected with rotavirus and enterotoxigenic *Escherichia coli*, *J. Clin. Microbiol.*, 13, 1011, 1981.
24. **Tzipori, S., Sherwood, D., Angus, K. W., Campbell, I., and Gordon, M.,** Diarrhea in lambs: experimental infections with enterotoxigenic *Escherichia coli*, rotavirus, and *Cryptosporidium sp.*, *Infect. Immun.*, 33, 401, 1981.
25. **Runnels, P. L., Moon, H. W., and Schneider, R. A.,** Development of resistance with host age to adhesion of K99⁺ *Escherichia coli* to isolated intestinal epithelial cells, *Infect. Immun.*, 28, 298, 1980.
26. **Morris, J. A.,** personal communication.
27. **Morris, J. A., Thorns, C. J., Wells, G. A. H., Schott, A. C., and Sojka, W. J.,** The production of F41 fimbriae by piglet strains of enterotoxigenic *Escherichia coli* that lack K88, K99, and 987P fimbriae, *J. Gen. Microbiol.*, 129, 2753, 1983.

Chapter 5

THE RABBIT MODEL OF *ESCHERICHIA COLI* (STRAIN RDEC-1) DIARRHEA

J. Robert Cantey

TABLE OF CONTENTS

I. ETIOLOGIC AGENT

Escherichia coli strain RDEC-1 was originally isolated in this laboratory in 1976 from a postweanling rabbit with diarrhea. Sporadic diarrhea had been noted prior to that time among rabbits purchased from a local rabbitry. In subsequent experiments, rabbits given the bacterium by the orogastric route developed diarrhea.[1] The pathogenesis of this novel disease has been the subject of much study since that initial observation.

The RDEC-1 strain is an O serogroup 15 *E. coli*[1] which lacks flagella. It does not have a K type of capsule[20] but definitely has a negatively charged surface polysaccharide.[2] It is also piliated under defined cultural conditions. Type 1 pili can be demonstrated by serial passage of bacteria growing in the surface scum of a trypticase soy broth (Difco, Inc., Detroit, MI) culture.[3] Serial passage is important as RDEC-1 strain pili are ordinarily suppressed in trypticase soy broth culture.[2] Pili are also expressed when the bacterium is grown in penassay (Difco, Inc.) broth.[4] The latter pili are probably a type of colonization factor antigen as they mediate adherence of the bacterium to partially purified brush borders of rabbit ileal mucosal epithelial cells.[4,5]

The RDEC-1 bacterium contains four plasmids.[21] The plasmid sizes are approximately 85, 60, 37, and 2.5 megadaltons. Dr. Samuel Formal at Walter Reed Army Institute of Research, Washington, D.C., was able to transfer the 85 megadalton plasmid to an avirulent shigella. The shigella expressed RDEC-1 pili after receiving the plasmid. The shigella was able to adhere to partially purified brush borders[4] so that the function of at least one of the plasmids is known. None of the plasmids has any marker as far as is known, making further investigation of their function quite difficult.

II. DISEASE MANIFESTATIONS

The diarrhea occurs 6 to 7 days (range 1 to 12 days) postinoculation with 2×10^6 bacteria, which is the standard dose used in this laboratory.[1] The diarrhea persists for a mean of 4 days and has a tendency to recur. During the diarrhea, the rabbits stop drinking water and eating. Their fur becomes dirty and ruffled, and in the perirectal area, plastered with adherent wet stool. The diarrheal stools are watery, occasionally mucoid, and do not contain blood. A Giemsa stain of the stool may reveal small numbers of polymorphonuclear leukocytes. Blood cultures taken at the time of diarrhea are negative. The mortality rate was not determined in the initial description of the disease but in 83 rabbits recently inoculated with the RDEC-1 strain the mortality was 27%. The rabbits appear to die of dehydration but the cause of death has not been systematically determined.

The attack rate in the first reported series was 77%[1] and in a second report was 78%.[6] However, in the more recent group of 83 rabbits mentioned above it was 61%. At times the bacterium has appeared to manifest a lower attack rate. Strains have been systematically examined for decreased virulence and we unsuccessfully tried to produce such strains by serial passage, prolonged maintenance on agar slants at room temperature, etc. Recently groups of rabbits were given four different RDEC-1 isolates which, in preliminary studies appeared to have differing attack rates for diarrhea. The group sizes were 40, 15, 21, and 15. The diarrhea attack rates, which were not statistically significantly different by Chi Square testing, were 70, 53, 71, and 61%, respectively. Differences in attack rate among small groups of rabbits are probably a function of the subject population, rather than any change in virulence of the strain.

III. CHARACTERISTICS OF GUT COLONIZATION

The RDEC-1 bacterium is species specific and does not colonize or cause diarrhea in guinea pigs or rats.[7] Also, it does not adhere to guinea pig, rat, or human ileal brush borders.[7]

The diarrheal attack rate does not appear to vary significantly with an orogastric dose varying from 10^2 to 10^{10} organisms. Rabbits have not been inoculated with fewer than 100 bacteria. A dose of 2×10^6 has been chosen for most of the studies.

The number of bacteria in the gut lumen of rabbits postinoculation with the RDEC-1 bacterium has been examined. The ileum, cecum, and colon, together with their contents, were tied off, excised, homogenized, and the number of colony forming units (CFU) per gram of homogenate ascertained.[6] The greatest number of CFU were found in the cecum. The median number of CFU in the ileum, cecum, and colon at the time of diarrhea was 2×10^6, 3×10^7, and 9×10^6, respectively. The values are somewhat lower than those obtained by Cheney et al., perhaps due to differences in technique.[7] The numbers of CFU were lower prior to and following diarrhea. At the same time the rabbits were sacrificed for gut cultures, a rectal swab culture was taken. The rectal swabs were streaked onto MacConkey agar plates (Difco, Inc., Detroit, MI) and the numbers of lactose positive colonies semi-quantitated according to the following schema: 0 = no growth, 1 = widely spaced colonies, 2 = closely spaced colonies, and 3 = confluent colonies. Groups of 5 to 10 colonies per plate were confirmed as the RDEC-1 bacterium by slide agglutination using anti-RDEC-1 OK antigen sera. The rectal swab culture grades were correlated with the numbers of CFU per gram of gut and gut contents ($P < 0.005$ by Kruskal-Wallis analysis). The rectal swab culture grade was then used to follow the colonization of the rabbit gut from the beginning to the end of colonization. The results were surprising. The rabbits did not become heavily colonized until 4 or 5 days postinoculation. They remained heavily colonized through the time of diarrhea. Colonization began to decrease in intensity about 15 days postinoculation and disappeared entirely by 24 to 35 days postinoculation. The time of disappearance is compatible with a protective immune response, but the 4 to 5 day interval required for colonization to reach maximum intensity has only recently been explained (see Section V).

IV. IMMUNOLOGY

The protection that follows RDEC-1 colonization is striking. Fifteen rabbits that had previously been colonized with the RDEC-1 strain were reinoculated at 6 to 23 day intervals after RDEC-1 colonization had ceased. Only 1 of the 15 was colonized, and for just 1 day. Thirteen control rabbits were inoculated and all became colonized, 12 developed diarrhea and 8 died.[8] Thus, the protection provided by previous colonization was complete.

The serum antibody response to the O antigen of the RDEC-1 bacterium following RDEC-1 diarrhea was quantitated using a microtiter agglutination technique.[1] Thirty-two of thirty-six rabbits with diarrhea had titers of antibody in undiluted sera or sera diluted as much as 1:64. The geometric mean ($\pm$ SD) of the reciprocal titers of serum drawn 2 to 28 days postinoculation was 13 ± 15.1. Rabbits immunized in this lab with live bacteria i.v., in contrast, develop titers as high as 1:2048. The absent to low titers that follow colonization and diarrhea are probably due to the fact that the bacterium does not invade the gut mucosa (see Section V).

The local immune response in the gut lumen has not been closely examined but in preliminary experiments and using a microtiter agglutination technique, antibody to O antigen was detected in the gut lumen.[22] Secretory IgA is the only immunoglobulin present in significant quantity in the rabbit gut lumen,[9] so that the conclusion that the antibody detected was sIgA antibody seems justified.

The gut is part of a common mucosal immune system which includes the mammary gland. Thus, it was not surprising to find large amounts of high titer sIgA antibody postdelivery in the milk of doe rabbits that had been immunized during gestation by the orogastric route with live RDEC-1 bacteria.

Immune sIgA in the rabbit milk was purified and given by the orogastric route to rabbits prior to and for several doses following an orogastric inoculum of 2×10^6 bacteria. The immune sIgA prevented colonization and diarrhea due to the RDEC-1 bacterium, confirmation of the role of sIgA in the protection against infection with this *E. coli* strain.[9]

V. PATHOGENESIS

RDEC-1 Enterotoxin — The RDEC-1 strain was initially assumed to produce one of the usual enterotoxins, but efforts to demonstrate either a labile or stable enterotoxin failed.[1] Live bacteria did not cause fluid secretion in the 18-hr ileal loop. Culture supernates, even when concentrated 200-fold, also failed to cause fluid secretion. The adrenal and Chinese hamster ovary cell assays for labile toxin were negative. There was no evidence of the presence of stable toxin in the 6-hr rabbit ileal loop or the suckling mouse assay. Preliminary investigation at Walter Reed Army Institute of Research suggested that a shiga-like toxin might be present.[10] However, in subsequent experiments in our laboratory, we were unable to reproduce this initial observation.

Several strains of enteropathic *E. coli*, which cause diarrhea in human volunteers, also do not give a positive reaction in any of the classic tests for *E. coli* enterotoxins.[11] However, labile or stable toxin-like materials prepared from them have been shown to cause secretion in the rat jejunal perfusion assay.[12] Encouraged by these results we tested similar materials prepared from the RDEC-1 strain in this assay. Technical difficulties when using the rat and the desirability of testing a rabbit specific strain in the rabbit, led us to develop a rabbit jejunal perfusion assay. The assay has been found to be both sensitive and reproducible, and, in preliminary work, secretion was produced with a RDEC-1 whole cell lysate of molecular weight greater than 30,000 megadaltons in a concentration of protein as low as 10 ng/mℓ.[13] Further investigations into the nature of this material are in progress.

VI. HISTOPATHOLOGY

Fluorescent, light, and electron microscopic histopathological studies have provided the bulk of current knowledge of the pathogenesis of RDEC-1 diarrhea. RDEC-1 can be found, by fluorescent and light microscopic study of intestinal tissues, adhering to the lymphoid follicle epithelium of ileal Peyer's patches as early as 4 to 8 hr postinoculation with 2×10^6 tryptic soy broth grown bacteria (Figure 1).[14] The bacteria have begun to spread over the dome of the lymphoid follicle and form microcolonies 24 hr postinoculation. The microcolonies enlarge and by 48 hr (Figure 2) appear to be shedding bacteria into the gut lumen. About this time bacteria can be found in increasing numbers in the gut lumen, and by 3 days postinoculation small patches of adherent bacteria can be found in ileum, cecum, and colon. Six to ten days postinoculation large numbers of bacteria can be found adhering to the absorptive epithelial cells in the above-mentioned gut segments and diarrhea occurs. The cecum is the most heavily involved tissue. The RDEC-1 rarely adheres to jejunal tissues although cultures of jejunal contents are always positive.[1]

A mild to moderate subepithelial inflammatory response can be seen adjacent to patches of adherent bacteria.[1] Polymorphonuclear leukocytes can also be found in the gut lumen at this time.

Electron microscopic observations have provided additional insights into the pathogenesis of RDEC-1 diarrhea. Our discovery that the RDEC-1 bacterium adheres to the Peyer's patch is recent, thus the electron microscopic examination of that phenomenon is at an early stage. However, the interaction of the RDEC-1 bacterium with the absorptive epithelium has previously been the subject of intense scrutiny. Electron microscopic examination of absorptive epithelium of rabbits with RDEC-1 diarrhea revealed bacteria adhering to mucosal epithelial cells that had lost their microvillus border.[15] This is well shown in Figure 3, a

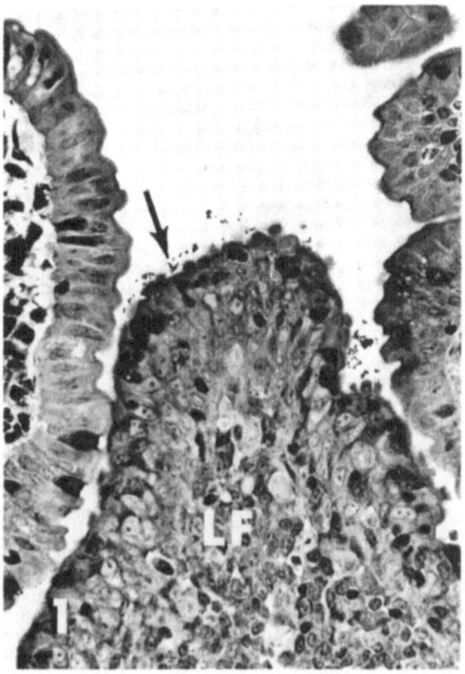

FIGURE 1. Section of formalin-fixed, methacrylate-embedded ileal Peyer's patch from a rabbit 8 hr after inoculation with 2×10^6 strain RDEC-1 *Escherichia coli* (toluidine blue stain). Small numbers of bacteria (arrow) are attached to the coarse brush borders of single epithelial cells of the tip of the lymphoid follicle (LF). No bacteria are attached to the smooth brush border of adjacent villar epithelium.

high voltage electron micrograph. The distance between the cell wall of the bacterium and the plasma membrane of absorptive epithelial cell was 11 nm.

The elucidation of further details of the bacterial-mucosal interaction required the use of the high voltage electron microscope, which produces an accelerating voltage 10 times greater than the usual transmission electron microscope. Its power allows penetration of sections greater than 0.5 μm thick. The result is that the thickness of a single section may contain several layers of microvilli and, if the bacterium is laying flat within the section, a complete bacterium. The examination of thick sections is less subject to errors of interpretation and sampling than is true for thin sections. Stereo views of such sections can be dramatic and they facilitate three-dimensional conceptualization. Most of these high voltage electron microscopic studies were done at the N.I.H. supported facility at the University of Colorado, Boulder, Colorado.

The high voltage electron microscopic studies resulted in the concept of early and late stage adherence of the RDEC-1 bacterium to absorptive epithelial cells.[2] The recognition of these stages was based on an interpretation of spatial relationships between bacteria and mucosal epithelial cells. It would have been preferable to have developed these concepts by examination of tissues taken at timed intervals postinoculation, but the time and effort required to perform such studies using the high voltage electron microscope would have been prohibitive.

The sequence of events in bacterial-mucosal interaction can be described in the following manner. The bacterium approaches and closely associates with intact, normal appearing

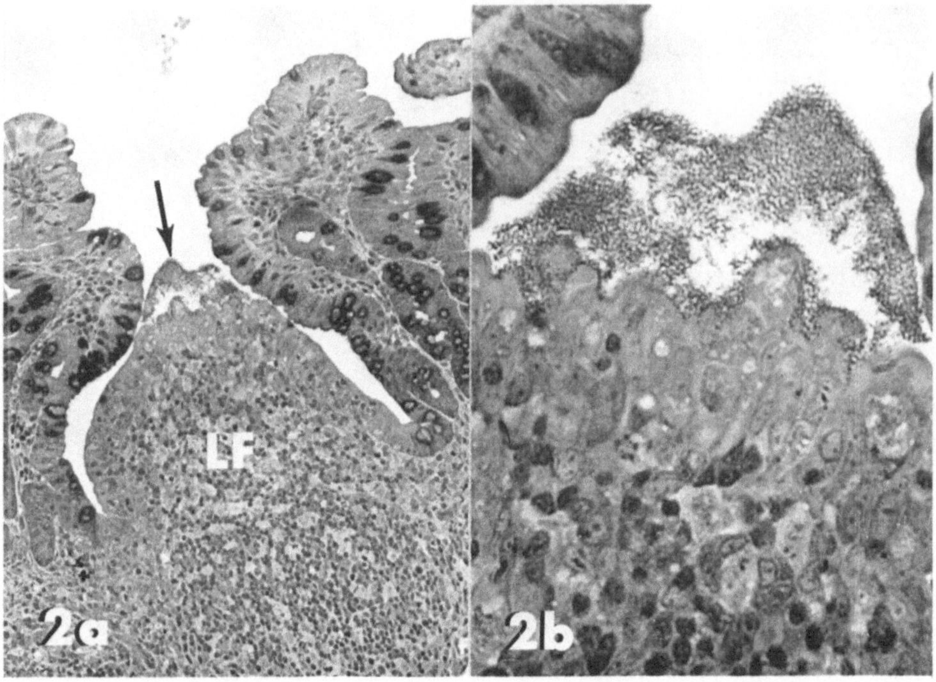

FIGURE 2. Section of formalin-fixed, methacrylate-embedded ileal Peyer's patch from a rabbit 48 hr after inoculation with 2×10^6 strain RDEC-1 *Escherichia coli* (toluidine blue stain). (a). The lymphoid follicle (LF) has a microcolony (arrow) of bacteria on its tip. No bacteria are adherent to the adjacent villar epithelium, even though the rabbit had been inoculated 2 days earlier. (b) Enclosed view of the tip of the lymphoid follicle. The microcolony is composed of many bacteria which extend into the gut lumen and cover a number of epithelial cells. Part of the microcolony seems to have broken away, probably as an artifact of preparation. Heavily colonized follicles such as this one are probably the source of bacteria attached to absorptive gut epithelium beginning 3 days after inoculation with RDEC-1 bacteria.

microvilli (Figure 4). The microvilli appear to be pushed aside and there is a clear space between the surface of the bacterium and the microvilli. As the bacterium approaches even more closely towards the cell surface the microvilli become notched and membrane-bounded vesicles or round bodies appear in the vicinity of the bacterial-mucosal interaction, both among the microvilli and in the gut lumen. Next, the bacteria can be seen resting on extrusions of epithelial cell plasma membrane which, in the stereopairs can be seen to cup the bacteria, and which we have termed "pedestals".[2] The pedestals can be found among intact microvilli (Figure 4) and on cells which have lost the majority of their microvilli (Figure 3). In the final stage, the bacteria are adherent to the surfaces of cells that have lost all or most of their microvilli (Figure 3). Although the bacteria do not invade the cell, degenerative changes are seen in cells with many adherent bacteria.[15] Cisternae of rough and smooth endoplasmic reticulum as well as Golgi apparatus are dilated. Vesicles and vacuoles of different sizes can be seen. Mitochondria may be swollen and have deranged cistae with opacified matrix. The cells have increased numbers of phagosomes which contain altered host cytoplasmic components. It is likely that such cells are sloughed from the villi more rapidly than would otherwise be the case. Increased sloughing would account for the blunting of ileal villi seen in the light micrographs.[15] In fact, cells lacking microvilli and covered with adherent bacteria are often seen free in the lumen.

The possibility that surface polysaccharides or pili may play a role in the adherence process in RDEC-1 adherence will be discussed in subsequent articles in this book. The article on the role of surface polysaccharide contains many micrographs pertinent to this discussion.

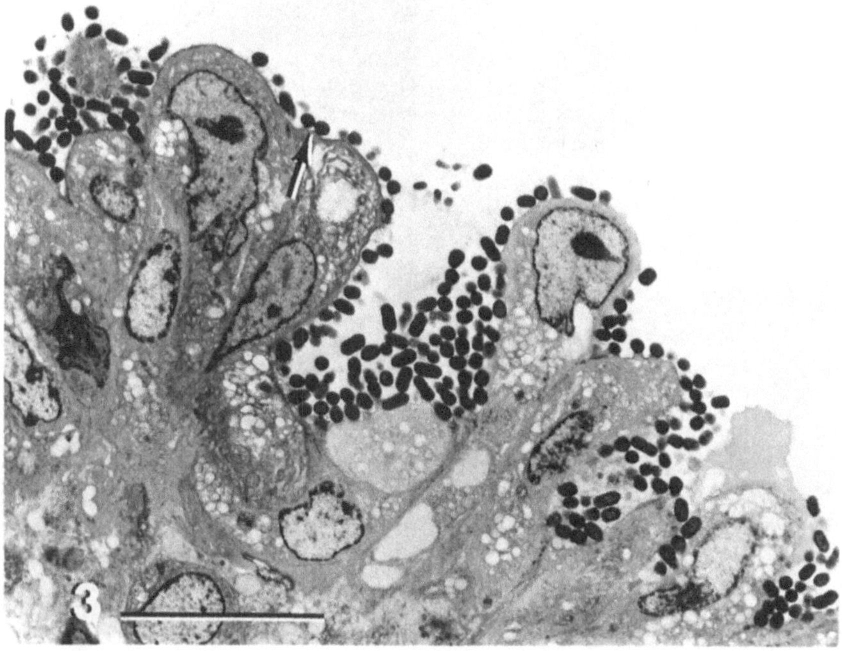

FIGURE 3.. High voltage electron micrograph of 0.5 μm section of cecal mucosa of a rabbit with strain RDEC-1 *Escherichia coli* diarrhea. Large numbers of bacteria are closely adherent to the absorptive epithelial cells, most of which have lost their microvillus border. Bacteria adherent to epithelial cells appear to anchor bacteria in the adjacent lumen. Close inspection reveals that some of the bacteria are resting on adherence pedestals (arrow). The epithelial cells are beginning to degenerate as evidenced by vesiculation, dilatation of the endoplasmic reticulum, etc. Bar = 20 μm.

VII. CONCLUSIONS

Current knowledge of the pathogenesis of RDEC-1 diarrhea can be summarized in the following manner. Immediately following inoculation, the RDEC-1 bacteria adhere to the lymphoid follicle epithelium of the Peyer's patch. The single adherent bacteria increase in numbers until patches of bacteria are formed which, over a period of 2 or 3 days, enlarge into microcolonies. The microcolonies shed bacteria into the lumen which presumably go on to adhere to absorptive epithelial cells in the ileum, cecum, and colon. Why the RDEC-1 bacterium is unable to adhere to absorptive epithelium immediately postinoculation is unknown. The bacteria adhere first to intact microvilli and somehow cause them to disappear from the surface of the absorptive epithelial cell. Once the bacteria are closely adherent to epithelial cells lacking microvilli diarrhea occurs, usually 6 to 10 days postinoculation. The diarrhea may be due to a novel enterotoxin which can be detected only in the perfused jejunal loop. The necessity for the bacteria to be closely adherent to the cells for diarrhea to occur suggests that the enterotoxin is cell bound, at least, until the moment of close adherence. Rabbits that survive the diarhea and no longer carry the organism in their stools are not easily reinfected, probably due to a mucosal immune response.

The pathogenesis of the diarrhea caused by the RDEC-1 strain of *E. coli* is unique among diarrheal diseases and among infectious diseases. The adherence to lymphoid follicle epithelium, the ability to cause absorptive epithelial cell microvillar loss, the close attachment to the surface of absorptive epithelial cells lacking microvilli, the pedestal formation by the epithelial cells and the enterotoxin that can be detected only by the jejunal perfusion technique are all novel and likely areas of future investigation.

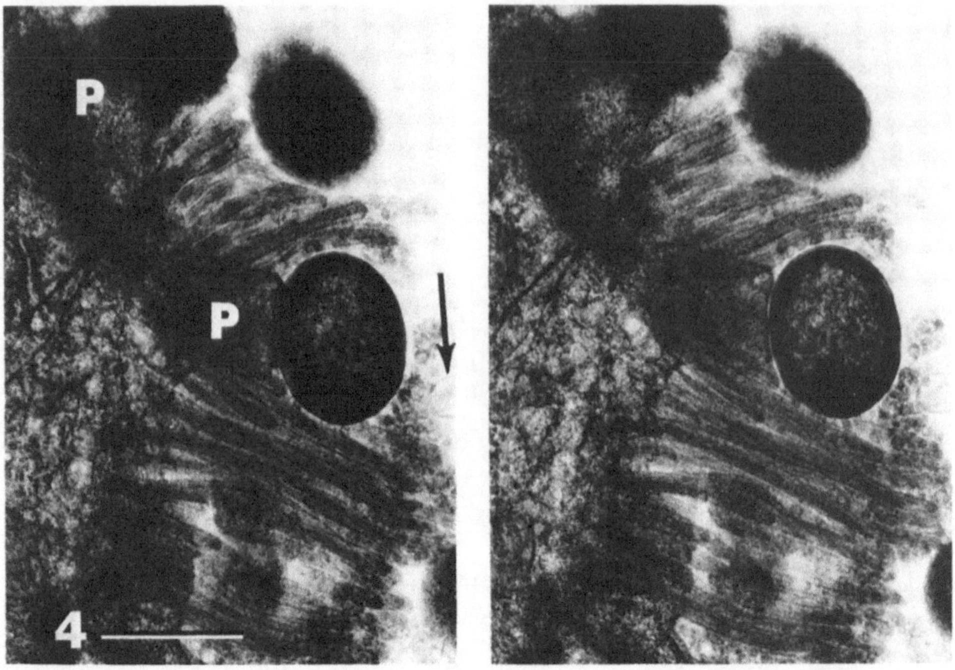

FIGURE 4. Stereopair high voltage electron micrographs of 0.5 μm section of cecal tissue of a rabbit with strain RDEC-1 *Escherichia coli* diarrhea. The tissue was stained en bloc with uranyl acetate and double stained with uranyl acetate and lead citrate on the section. The figures may be viewed in depth with a magnifying stereoviewer or by crossing the eyes until the two figures appear as three. The center figure will be seen in depth. Several stages of adherence are evident in this micrograph. One bacterium is cupped by intact microvilli. An adjacent bacterium is resting on a pedestal (P) and is separated from the plasmalemma of the epithelial cell by a narrow 11 nanometer space. Portions of other bacteria at various stages of adherence can be seen through the depth of the section. A bleb has formed at the tip of one microvillus (arrow) and many round bodies (presumably vesicles formed from damaged microvilli) are present in the lumen adjacent to the microvilli. Bar = 1 μm.

The importance of the disease is strengthened by three published reports of four human cases[16-18] of a histopathologically similar diarrheal disease. Another 12 human cases have been reported in abstract[19] and Dr. Giannella has reported to this workshop[23] and elsewhere[24] that the number of cases has increased to 30. The O serogroups of the causative *E. coli* include 0111, 0119, 0125, and 0128. The cases reported so far have been in infants and young children. The difficulty in identifying human cases has been that the only way to make the diagnosis is by jejunal or rectal biopsy. Until there is some epidemiologically applicable technique to identify cases of RDEC-1-like diarrhea in humans its true incidence in human populations will remain unknown.

REFERENCES

1. **Cantey, J. R. and Blake, R. K.,** Diarrhea due to *Escherichia coli* in the rabbit: a novel mechanism, *J. Infect. Dis.*, 135, 454, 1977.
2. **Cantey, J. R., Lushbaugh, W. B., and Inman, L. R.,** Attachment of bacteria to intestinal epithelial cells in diarrhea caused by *Escherichia coli* strain RDEC-1 in the rabbit: stages and role of capsule, *J. Infect. Dis.*, 143, 219, 1981.
3. **O'Hanley, P. D. and Cantey, J. R.,** Surface structures of *Escherichia coli* that produce diarrhea by a variety of enteropathic mechanisms, *Infect. Immun.*, 21, 874, 1978.

4. **Cheney, C. P., Schad, P. A., Boedeker, E. C., and Formal, S. B.,** The role of pili in adherence of enteropathogenic *Escherichia coli* to rabbit brush borders. Abstracts of the Annual Meeting, American Society for Microbiology, Washington, D.C., 1979, 19.
5. **Cheney, C. P., Boedeker, E. C., and Formal, S. B.,** Quantitation of the adherence of an enteropathogenic *Escherichia coli* to isolated rabbit intestinal brush borders, *Infect. Immun.*, 26, 736, 1979.
6. **Cantey, J. R. and Hosterman, S. D.,** Characterization of colonization of the rabbit gastrointestinal tract by *Escherichia coli* RDEC-1, *Infect. Immun.*, 26, 1099, 1979.
7. **Cheney, C. P., Schad, P. A., Formal, S. B., and Boedeker, E. C.,** Species specificity of *in vitro Escherichia coli* adherence to host intestinal cell membranes and its correlation with *in vivo* colonization and infectivity, *Infect. Immun.*, 28, 1019, 1980.
8. **Cantey, J. R.,** Prevention of bacterial infections of mucosal surfaces by immune secretory IgA, in *Secretory Immunity and Infection, Advances in Experimental Medicine and Biology*, Vol. 107, McGhee, J. R., Mestecky, J., and Babb, J. F., Eds., Plenum Press, New York, 1978, 461.
9. **O'Hanley, P. D. and Cantey, J. R.,** Immune response of the ileum to invasive *Escherichia coli* diarrheal disease in rabbits, *Infect. Immun.*, 31, 316, 1981.
10. **O'Brien, A. D., Thompson, M. R., Cantey, J. R., and Formal, S. B.,** Production of a *Shigella dysenteriae*-like toxin by pathogenic *Escherichia coli*. Abstracts of the Annual Meeting, American Society for Microbiology, New Orleans, LA, 1977, 32.
11. **Levine, M. D., Bergquist, E. J., Nalin, D. R., Waterman, D. H., Hornick, R. B., Young, C. R., Sotman, S., and Rowe, B.,** *Escherichia coli* strains that cause diarrhea but do not produce heat-labile or heat-stable enterotoxins and are non-invasive, *Lancet*, 1, 1119, 1978.
12. **Klipstein, F. A., Rowe, B., Engert, R. F., Short, H. B., and Gross, R. J.,** Enterotoxigenicity of enteropathogenic serotypes of *Escherichia coli* isolated from infants with epidemic diarrhea, *Infect. Immun.*, 21, 171, 1978.
13. **Mullaney, D. T. and Cantey, J. R.,** Detection of enterotoxin-like activity in the whole-cell lysate of an enteroadherent *Escherichia coli* (Strain RDEC-1), *Clin. Res.*, 30, 374A, 1982.
14. **Cantey, J. R. and Inman, L. R.,** Diarrhea due to *Escherichia coli* strain RDEC-1 in the rabbit: the Peyer's patch as the initial site of attachment and colonization, *J. Infect. Dis.*, 143, 440, 1981.
15. **Takeuchi, A., Inman, L. R., O'Hanley, P. D., Cantey, J. R., and Lushbaugh, W. B.,** Scanning and transmission electron microscopic study of *Escherichia coli* 015 (RDEC-1) enteric infection in rabbits, *Infect. Immun.*, 19, 686, 1978.
16. **Ulshen, M. H. and Rollo, J. L.,** Pathogenesis of *Escherichia coli* gastroenteritis in man — another mechanism, *N. Engl. J. Med.*, 302, 99, 1980.
17. **Phillips, A. D.,** Small intestinal mucosa in childhood in health and disease, *Scand. J. Gastroenterol.*, 16, 65, 1981.
18. **Clausen, C. R. and Christie, D. L.,** Chronic diarrhea in infants caused by adherent enteropathogenic *Escherichia coli*, *J. Pediatr.*, 100, 358, 1982.
19. **Rothbaum, R. J., Partin, J. C., McAdams, A. J., Shah, D. B., and Giannella, R. A.,** Enterocyte adherent, *E. coli* 0119:B14: a novel mechanism of infant diarrhea, *Gastroenterology*, 80, 1265, 1981.
20. **Ørskov, F., Ørskov, I., and Cantey, J. R.,** unpublished observations.
21. **John, J. F. and Cantey, J. R.,** unpublished observations.
22. **Cantey, J. R. and Inman, L. R.,** unpublished observations.
23. **Giannella, R. A.,** Observations on infantile diarrhea caused by enteropathogenic *E. coli*, in *Attachment of Organisms to the Gut Mucosa*, Vol. I, Boedeker, E. C., Ed., CRC Press, Boca Raton, Fla., 1984.
24. **Rothbaum, R. J., McAdams, A. J., Giannelli, R. A., and Partin, J. C.,** A clinicopathologic study of enterocyte adherent *E. coli*: a cause of protracted diarrhea in infants, *Gastroenterology*, 83, 441, 1982.

Chapter 6

ADHERENCE IN INFANTILE DIARRHEAS

Martin H. Ulshen

TABLE OF CONTENTS

I. INTRODUCTION

Escherichia coli is known to cause diarrhea in man through two well-described pathogenic mechanisms: production of enterotoxin (both heat labile (LT) and heat stable (ST)) and invasion of the intestinal mucosa.[1,2] Studies will be presented in this chapter suggesting another mechanism by which *E. coli* can cause gastroenteritis in man. The pathogenesis of this diarrheal disease involves apparent adherence of organisms to the intestinal mucosa and damage to luminal microvilli in the absence of either conventional toxin production or invasion of the intestinal wall. The electron micrographs of bowel mucosa from patients with this illness have been strikingly similar to electron micrographs of intestine from rabbits with RDEC-1 associated diarrhea.[3] Therefore, this diarrheal illness appears in a number of respects to be analogous to the prevously described RDEC-1 diarrhea in rabbits.[4]

II. PREVIOUS STUDIES

Epidemiologic studies beginning in the 1940s have associated strains of *E. coli* with epidemics of nursery or infantile enteritis.[5] These serotypes became known as enteropathogenic *E. coli* or EPEC. The etiologic role of these strains in epidemics of infantile diarrhea was confirmed when ingestion of organisms by volunteers produced diarrhea.[6] With the recent development of conventional methods to identify mechanisms of *E. coli* diarrhea (i.e., Y1 adrenal cell or Chinese hamster ovary cell assay for LT, suckling mouse assay for ST, as well as the Serény test for invasiveness), EPEC strains stored during epidemics have been reevaluated.[7,8] However, these organisms generally have not demonstrated production of conventional enterotoxins or the ability to invade mucosa. This observation has suggested that (1) the organisms have lost their virulence in storage or (2) that the diarrhea has been caused by some other previously unidentified mechanism(s). More recently, it has been shown that EPEC strains which have been stored for as long as 9 years can cause diarrhea when ingested by volunteers, despite the absence of pathogenicity by conventional tests.[9] Therefore, the pathogenesis of this *E. coli* diarrhea has not been explained by established mechanisms. It is possible that toxin may be produced although not apparent on routine testing. Klipstein, using a rat jejunal perfusion model, demonstrated that some strains of EPEC can produce toxin analogous to LT and ST which is not identified by conventional methods.[10]

III. CASE DESCRIPTION

Little information is available concerning the role of bacterial adherence in the pathogenesis of infantile diarrhea. It has been shown that a number of strains of *E. coli* which produce LT also have the ability to adhere to human fetal intestine in vitro.[11] In 1980, Dr. John Rollo, a pathologist, and Dr. Martin H. Ulshen, a pediatric gastroenterologist, reported studies of an infant with chronic diarrhea which appeared to be secondary to an adherent *E. coli*.[12] The unique feature of this illness was the remarkable similarity to RDEC-1 enterocolitis in rabbits. This infant was a 7 week old who had had several weeks of severe diarrhea. No conventional pathogens were found in the infant's stool by culture or evaluation for ova and parasites. No white blood cells or gross blood were seen in the stool, although occult blood was present. At the onset of the illness four family members had had diarrhea which resolved within 24 hr. Because of the persistence of the diarrhea, further evaluation was completed. A duodenal aspirate was cultured and grew pure *E. coli* (greater than 10^6 organisms/mℓ). This organism did not appear to be pathogenic by the conventional tests (i.e., the Y1 adrenal and suckling mouse assay as well as the Serény test). However, it was serotyped as 0125ac:H21, a strain previously identified with outbreaks of infantile diarrhea.[12]

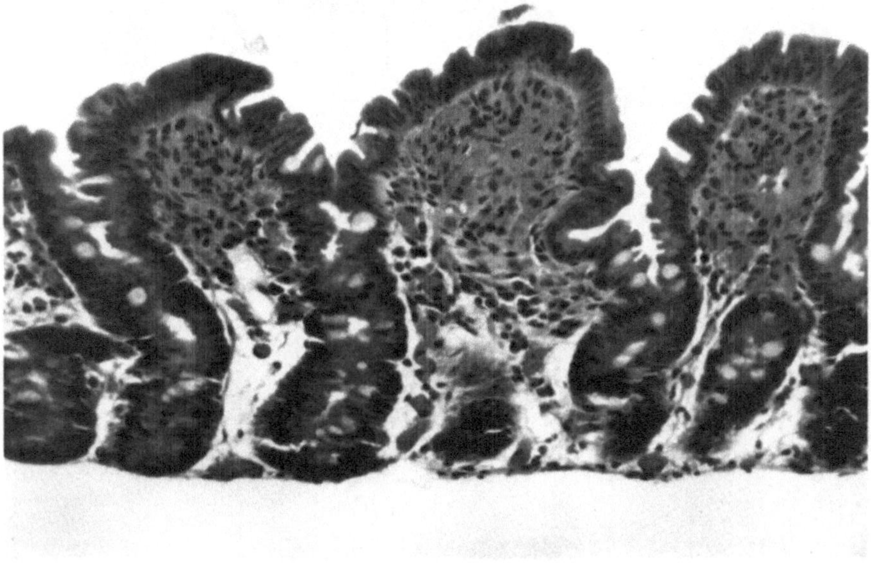

FIGURE 1. A biopsy from the distal duodenum demonstrates blunted villi and crypt hypertrophy. The lamina propria is broadened by an infiltrate rich in histiocytes, present predominantly in the upper villi (H & E stain magnified × 200).

A small bowel biopsy demonstrated short, blunted villi and deep crypts (Figure 1). The villus to crypt ratio was about 1:1 compared to a normal ratio of 4:1. At higher magnification, the surface epithelium was well preserved except for some loss of the normal basilar position of the nuclei. Cellularity of the lamina propria was increased with an increase in the proportion of tissue histiocytes. On electron microscopy, regions of normal small bowel mucosa were interrupted by regions in which organisms which were consistent with Gram-negative bacilli appeared to be closely adherent to the luminal surface (Figure 2). An increase in intracytoplasmic vacuoles suggested cellular damage. Although an occasional surface epithelial cell appeared to contain organisms, these may, in fact, have been adjacent to these surface cells. Serial sectioning was required to demonstrate rare organisms in the lamina propria; in contrast, in the animal model for invasive *E. coli* organisms are found predominantly at this level.[13] In regions where Gram-negative bacilli appeared to adhere to the luminal plasma membrane, the glycocalyx had disappeared and microvilli were either displaced or damaged. The disaccharidase activity of the intestinal mucosa was uniformly low, consistent with brush border damage. On rectal biopsy, the lamina propria was expanded with inflammatory cells, despite the absence of gross abnormalities of the mucosa on sigmoidoscopy. As in the small bowel biopsy, this included a proportionate increase in histiocytes. In some areas surface epithelial cells had lost their columnar appearance and occasional crypt abscesses were present (Figure 3). In addition, rare giant cells were noted in the lamina propria which occasionally contained calcium deposits.

Further evaluation for toxin production was kindly carried out by Drs. Sam Formal and Christopher Cheney (Walter Reed Army Institute for Research, Washington, D.C.). The rabbit ileal loop model demonstrated no secretion using culture supernatants (at 8 and 16 hr after inoculation), Van Heyningen extract of the culture, and broth culture (approximately 2×10^9 organisms). In addition, there was no cytotoxicity with HeLa cells using either

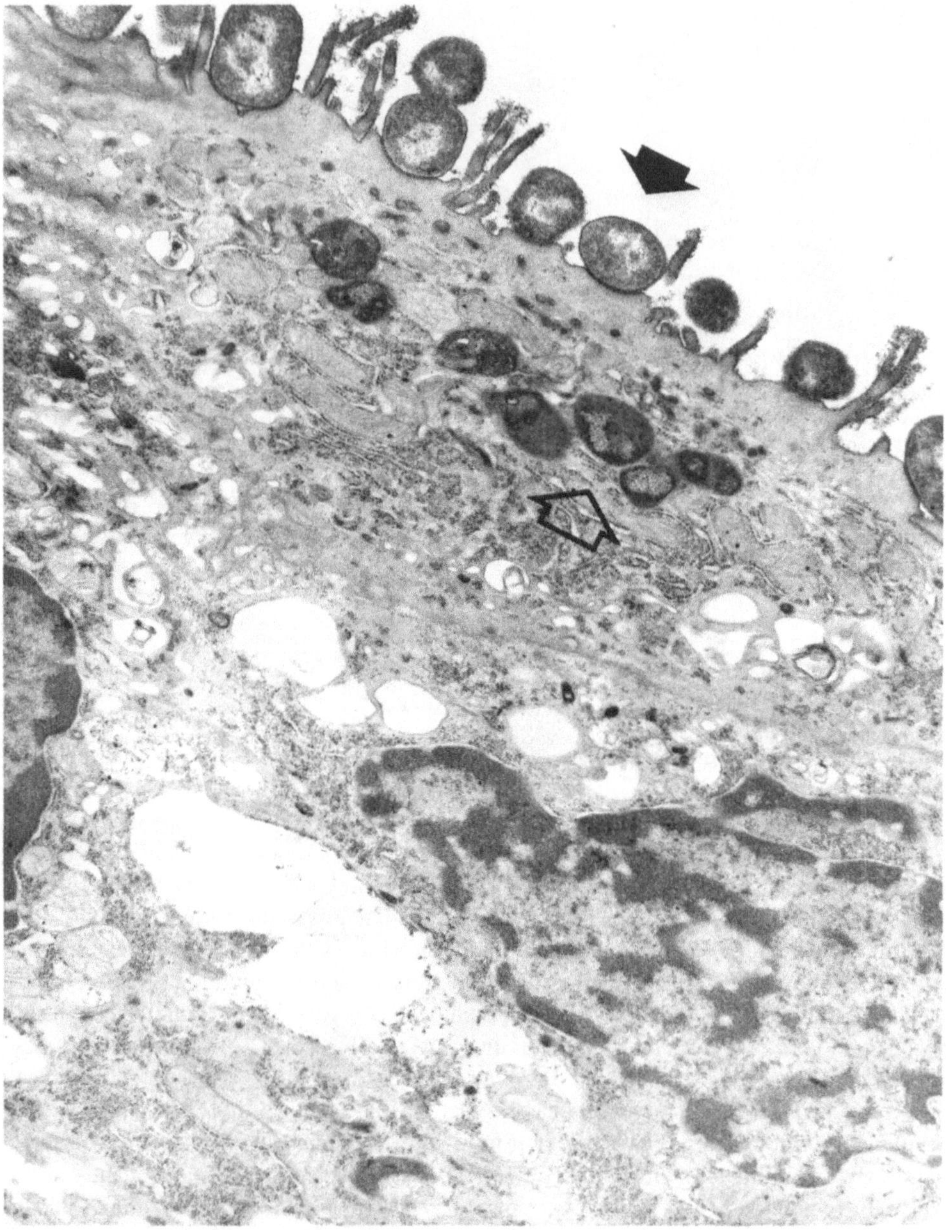

A

FIGURE 2. (A) On electron microscopy, Gram-negative bacteria appear to be adherent to the luminal surface of the duodenal epithelium (closed arrow) and microvilli are either damaged or absent. Bacteria appear to be seen within occasional enterocytes (open arrow). An increase in intracytoplasmic vacuoles is noted, as well (magnification × 16,500). (B) At higher magnification the luminal surface of the enterocyte is seen to closely cup a portion of a bacterium (magnification × 48,000).

crude supernatant or Van Heyningen extract. Although the organisms were piliated, unlike the RDEC-1 strain they did not adhere to hydrophobic columns.[12]

Rothbaum et al. recently presented preliminary and complete descriptions of an epidemic of infantile diarrhea associated with *E. coli* 0119: B14 which was quite similar to the illness in the previously reported patient.[14,14a] Twelve infants (age 3 weeks to 7 months) with severe

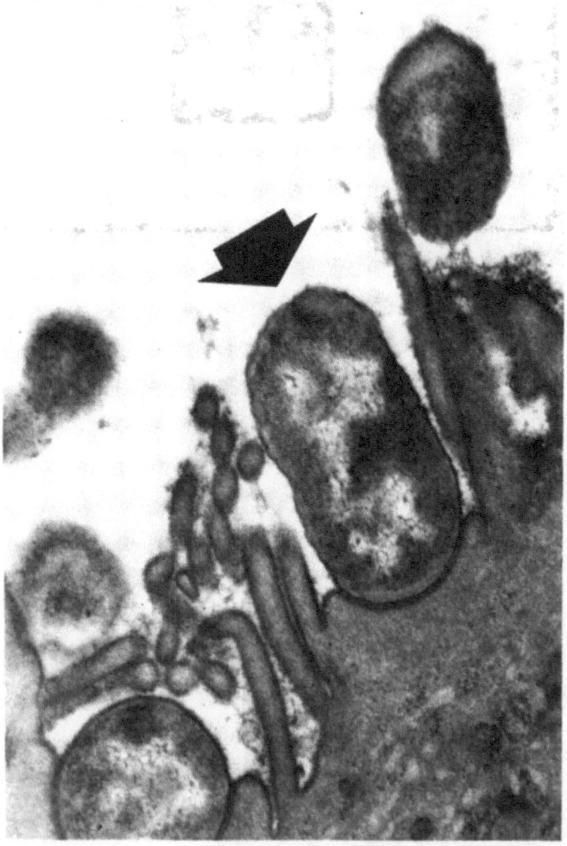

FIGURE 2B

diarrhea were described. Although no conventional enteropathogens were isolated from these patients, *E. coli* 0119: B14 was present in both stool and duodenal aspirate. This is a strain which has been identified previously with epidemic infantile diarrhea.[8] There was no evidence of toxin production by conventional or rabbit ileal loop studies. Small bowel biopsies demonstrated varying degrees of villous atrophy. In addition, electron micrographs of both the small bowel and colon were similar to the previous case as well as the RDEC-1 model.[3,12]

IV. SUMMARY

In summary, a form of infantile enterocolitis has been described which has been associated with strains of *E. coli* previously characterized as enteropathogenic. The pathogenesis of this diarrheal illness appears to be related to adherence of organisms to the bowel luminal surface and damage to microvilli. There are a number of similarities between this illness and RDEC-1 diarrhea in rabbits: (1) The organism does not appear to be toxigenic or invasive. (2) Villous atrophy is seen on light microscopy of the small bowel. (3) On electron microscopy of both small bowel and rectal biopsies, there is patchy involvement of the luminal surface including loss of glycocalyx, loss or damage of microvilli, and adherence of organisms to the luminal plasma membrane. (4) In contrast to invasive *E. coli*, few organisms have been seen in the lamina propria. However, dissimilarities between the human illness and RDEC-1 diarrhea in rabbits have also been noted: (1) In the rabbit, involvement is most severe in

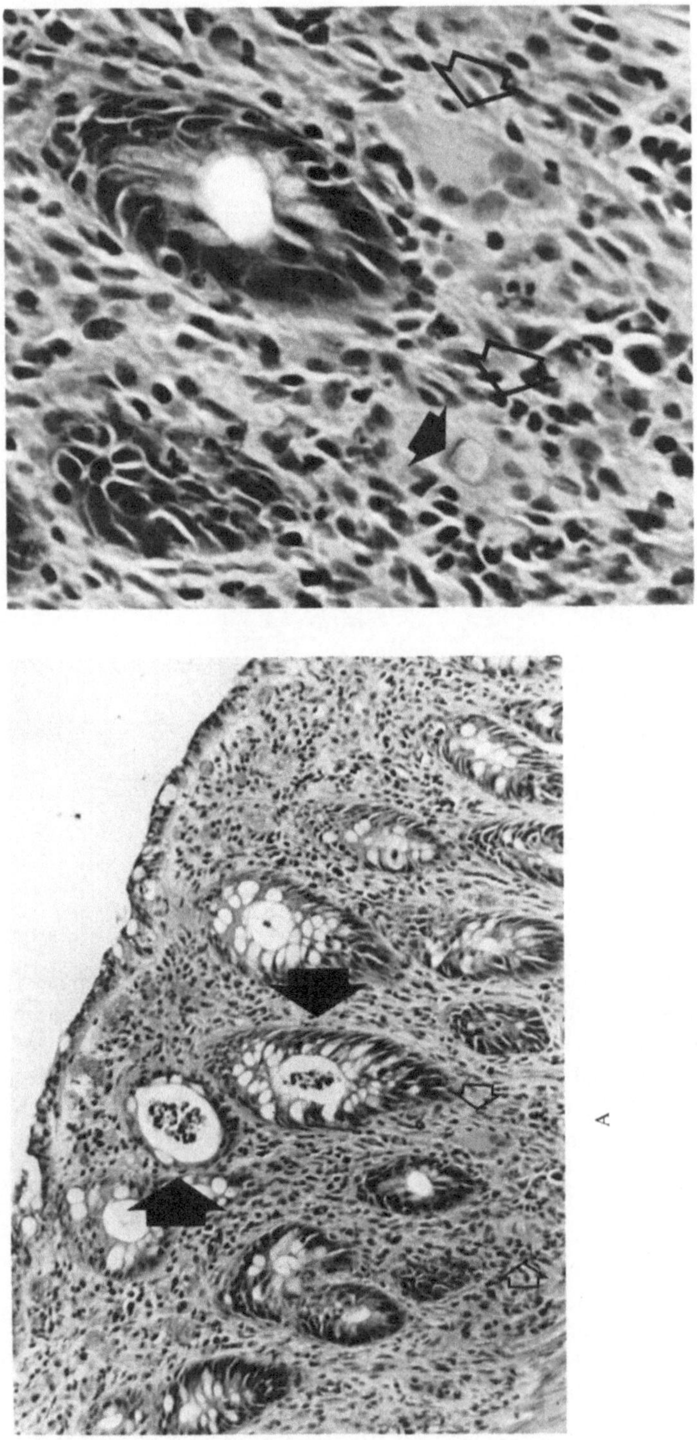

A

B

FIGURE 3. (A) Crypt abscesses (closed arrows) and damaged surface epithelium are noted on rectal biopsy. Two giant cells are present (open arrows, H & E stain × 100). (B) At higher magnification one giant cell is seen to contain a calcium deposit (closed arrow, H & E stain magnified × 400).

the distal small bowel and colon, while the jejunum, as well as the colon, have been clearly involved in the infants.[4] (2) Intestinal tissue from these rabbits has demonstrated a polymorphonuclear inflammatory response whereas in the infants this has been predominately mononuclear. (3) Finally, the RDEC-1 organism binds to hydrophobic columns while the organism isolated from the infant did not.

It is remarkable that these studies in the animal model have suggested a human disease prior to its recognition in man. Because this disease mechanism has only recently been described in man, the frequency of this illness and its overall role in infantile *E. coli* diarrhea remains to be determined. This entity may, in fact, be more common than currently appreciated; however, simpler methods of diagnosis will be necessary in order to establish its prevalence.

REFERENCES

1. **DuPont, H. L., Formal, S. B., Hornick, R. B., Synder, M. J., Libonati, J. P., Sheahan, D. G., LaBrec, E. H., and Kalas, J. P.**, Pathogenesis of *Escherichia coli* Diarrhea, *N. Engl. J. Med.*, 285, 1, 1971.
2. **Guerrant, R. L. Moore, R. A., Kirschenfeld, P. M., and Sande, M. A.**, Role of toxigenic and invasive bacteria in acute diarrhea of childhood, *N. Engl. J. Med.*, 293, 567, 1975.
3. **Takeuchi, A., Inman, L. R., O'Hanley, P. D., Cantey, J. R., and Lushbaugh, W. B.**, Scanning and transmission electron microscopic study of *Escherichia coli* 015 (RDEC-1) enteric infection in rabbits, *Infect. Immun.*, 19, 686, 1978.
4. **Cantey, J. R. and Blake, R. K.**, Diarrhea due to *Escherichia coli* in the rabbit: a novel mechanism, *J. Infect. Dis.*, 135, 454, 1977.
5. **Gangarosa, E. J. and Merson, M. H.**, Epidemiologic assessment of the relevance of the so-called enteropathogenic serogroups of *Escherichia coli* in diarrhea, *N. Eng. J. Med.*, 296, 1210, 1977.
6. **Ferguson, W. W. and June, R. C.**, Experiments on feeding adult volunteers with *Escherichia coli* 111, B_4, a coliform organism associated with infant diarrhea, *Am. J. Hyg.*, 55, 155, 1952.
7. **Gross, R. J., Scotland, S. M., and Rowe, B.**, Enterotoxin testing of *Escherichia coli* causing epidemic infantile enteritis in the U.K., *Lancet*, 1, 629, 1976.
8. **Sack, R. B.**, Human diarrheal disease caused by enterotoxigenic *Escherichia coli*, *Annu. Rev. Microbiol.*, 29, 333, 1975.
9. **Levine, M. M., Bergquist, E. J., Nalin, D. R., Waterman, D. H., Hornick, R. B., Young, C. R., Sotman, S., and Rowe, B.**, *Escherichia coli* strains that cause diarrhea but do not produce heat-labile or heat-stable enterotoxins and are non-invasive, *Lancet*, 1, 1119, 1978.
10. **Klipstein, F. A., Rowe, B., Engert, R. F., Short, H. B., and Gross, R. J.**, Enterotoxigenicity of enteropathogenic serotypes of *Escherichia coli* isolated from infants with epidemic diarrhea, *Infect. Immun.*, 21, 171, 1978.
11. **McNeish, A. S., Turner, P., Fleming, J., and Evans, N.**, Mucosal adherence of human enteropathogenic *Escherichia coli*, *Lancet*, 2, 946, 1975.
12. **Ulshen, M. H. and Rollo, J. L.**, Pathogenesis of *Escherichia coli* gastroenteritis in man — another mechanism, *N. Engl. J. Med.*, 302, 99, 1980.
13. **Cantey, J. R., O'Hanley, P. D., and Blake, R. K.**, A rabbit model of diarrhea due to invasive *Escherichia coli*, *J. Infect. Dis.*, 136, 640, 1977.
14. **Rothbaum, R. J., Partin, J. C., McAdams, A. J., Shah, D. B., and Gianella, R. A.**, *Gastroenterology*, 80, 1265, 1981.

14a. **Rothbaum, R. J., McAdams, A. J., Giannella, R., and Partin, J. C.**, A clinicopathologic study enterocyte-adherent *Escherichia coli*: a cause of protracted diarrhea in infants, *Gastroenterology*, 83, 441, 1982.

Chapter 7

OBSERVATIONS ON INFANTILE DIARRHEA CAUSED BY ENTEROPATHOGENIC *ESCHERICHIA COLI*

Ralph A. Giannella

TABLE OF CONTENTS

I. INTRODUCTION

With respect to the question of human infantile diarrhea, my colleagues at the University of Cincinnati and Children's Hospital Medical Center[1] and I have now identified and studied 25 infants with diarrhea caused by enteropathogenic *E. coli* serotype 0119, and I think these studies allow us to answer some of the questions concerning infantile diarrhea that have been frequently raised.

II. CHARACTERISTICS OF 0119 DIARRHEA

First of all, as to the characteristics of the organism, the organisms we have identified and isolated are piliated, but the only pili that we can identify thus far are type I pili. We cannot yet provide an answer as to whether these organisms are encapsulated.

Second, as to the severity of the diarrhea caused by enteropathogenic *E. coli,* we tend to agree with Dr. Myron Levine that we usually identify only one end of the disease spectrum, namely those patients with severe disease and diarrhea. However when we became aware of the problem of *E. coli* 0119-caused diarrhea, we instituted routine surveillance procedures for this organism in all infants with diarrhea below the age of 6 months entering the hospital and several interesting observations emerged. Those infants with the severe diarrhea, i.e., those with severe diarrhea which would continue for weeks, had *E. coli* 0119 in large numbers in both the small intestine and colon. In contrast, in those infants with mild diarrhea, we could only demonstrate the organisms in the colon and did not find these organisms in the small bowel. So these observations suggest that the severity of the diarrheal process may be related to the extent of the bowel colonized by the pathogenic strain and possibly to the particular area of the bowel colonized. Colonization of the small bowel may be required for a severe, watery diarrheal syndrome.

Third, I would like to demonstrate that the morphologic features of the *E. coli* 0119 adhesive process are very similar to those of the rabbit *E. coli* diarrhea described by Dr. Cantey.[2] Figure 1 shows a rectal biopsy from one such infant. It shows that the surface epithelium is disordered and cuboidal and that the lamina propria is quite edematous. Note that there are clusters of organisms adherent to a number of the surface epithelial cells. Figure 2 shows an electron microscopic view of the same section. One can see a bacterial microcolony adherent to the surface and the ''cupping'' caused by the individual organisms. These morphological features are very similar to those shown by Dr. Cantey in his rabbit model. It is a very similar adhesive process with obliteration of the underlying microvilli by the bacteria.

III. SUMMARY

As a result of these and other studies, there is no doubt in our minds that these organisms have the ability to colonize not only the small intestine, but also the large intestine. Furthermore, enteropathogenic *E. coli* probably have the ability to colonize the entire small bowel, i.e., the ileum as well as the jejunum. We have seen one infant who had a colectomy and an ileostomy because of neonatal necrotozing enterocolitis. Several weeks later he developed severe diarrhea with *E. coli* 0119 and we had the opportunity of doing an ileal biopsy through the ileostomy. This biopsy demonstrated that these organisms had also colonized the ileum. So I presume that these organisms can colonize virtually the entire intestinal tract.

Finally, as to whether these organisms elaborate an enterotoxin, I am afraid that we cannot give a positive testimonial. We have looked in various systems and we have been unable to demonstrate an enterotoxin thus far. Perhaps there is a host specificity and enterotoxin activity will only be detectable in human intestine.

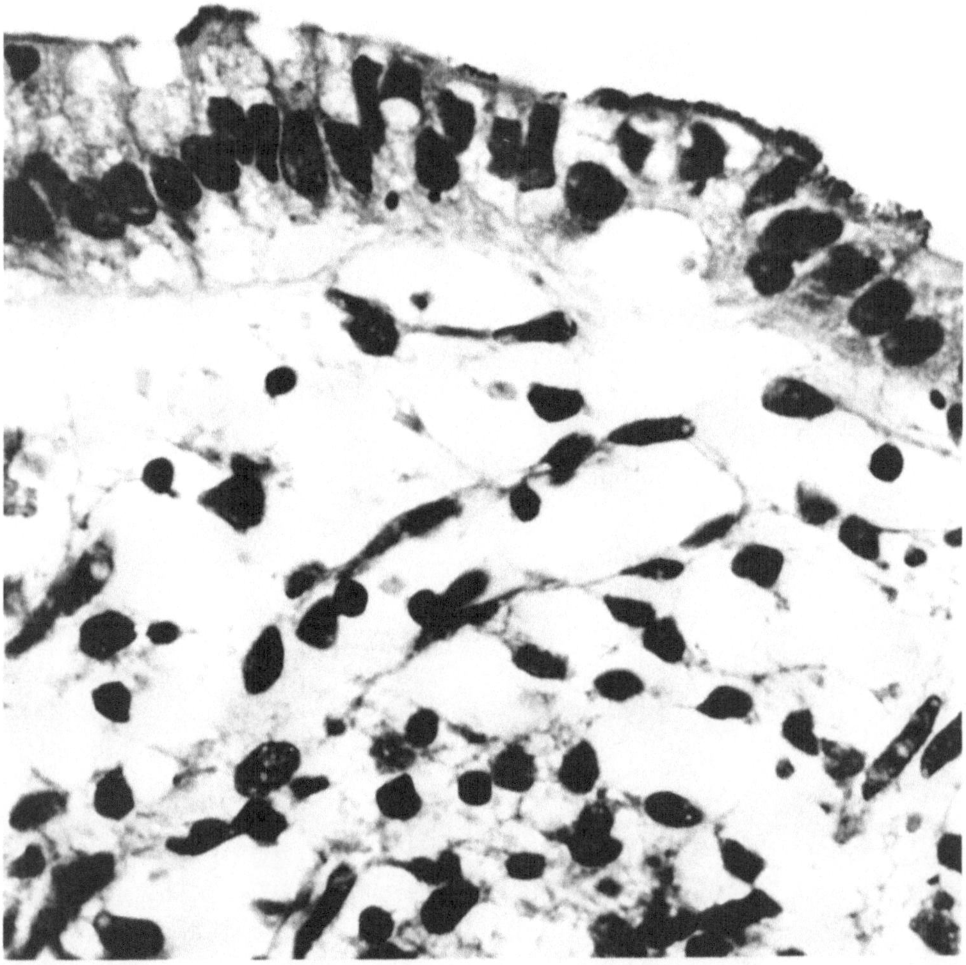

FIGURE 1. Light microscopic section of a rectal biopsy specimen from an infant with *E. coli* O119-diarrhea. Note basophilic caps of adherent bacteria on several of the surface epithelial cells. Affected cells are shortened and often vacuolated. The lamina propria is edematous and contains a mononuclear cell infiltrate. (H & E stain magnified × 1000).

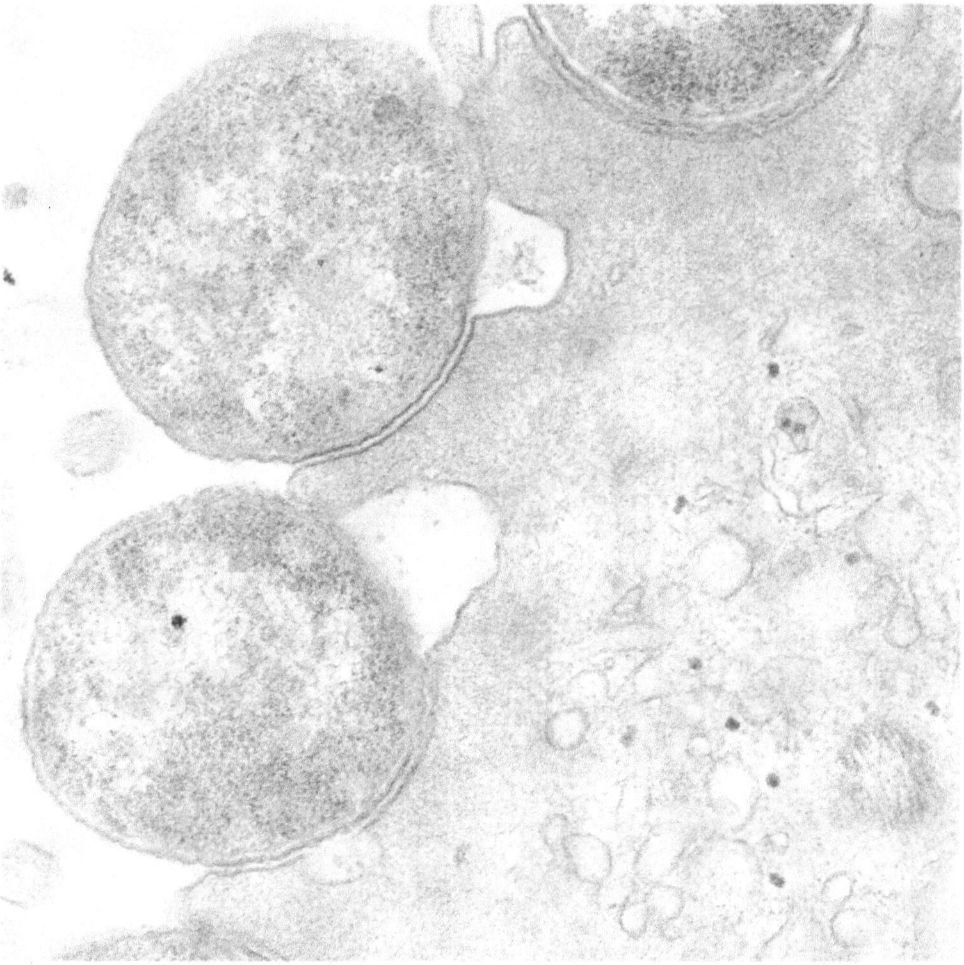

FIGURE 2. An electron micrograph of a rectal biopsy specimen from an infant with *E. coli* 0119-diarrhea. Note that at areas of bacterial adherence, microvilli and their core of actin filaments have disappeared. The cellular plasma membrane "cups" the adherent organisms. Intracellular organelles are disorganized (magnification × 51,000).

REFERENCES

1. **Rothbaum, R. J., McAdams, A. J., Giannella, R., and Partin, J. C.,** A clinicopathologic study of enterocyte adherent *Escherichia coli:* a cause of protracted diarrhea in infants, *Gastroenterology,* 83, 441, 1982.
2. **Cantey, J. R.,** The rabbit model of *Escherichia coli* (strain RDEC-1) diarrhea, in *Attachment of Microorganisms to the Gut Mucosa,* Vol. I, Boedeker, E. C., Ed., CRC Press, Boca Raton, Fla., 1984, chap. 5.

Pili

Chapter 8

ENTEROTOXIGENIC *ESCHERICHIA COLI* PATHOGENIC FOR MAN: BIOLOGICAL AND IMMUNOLOGICAL ASPECTS OF FIMBRIAL COLONIZATION FACTOR ANTIGENS

Dolores G. Evans, Doyle J. Evans, Jr., David A. Sack, and Steven Clegg

TABLE OF CONTENTS

I. INTRODUCTION

Enterotoxigenic *Escherichia coli* (ETEC) are defined as those *E. coli* which produce a heat-stable toxin (ST), a heat-labile toxin (LT), or both ST and LT. LT is very similar to the heat-labile enterotoxin produced by *Vibrio cholerae*. The similarities in subunit structure, mechanism of action (enzymatic activation of intestinal adenylate cyclase) and antigenicity are remarkable.[1-5] In fact the major difference is that cholera enterotoxin is encoded by chromosomal genes while LT of the ETEC is encoded by genes residing on plasmids.[6,7] ST production by *V. cholerae* has not been detected although in the ETEC the production of ST is determined by genes which are also carried by plasmids and some plasmids encode for both ST and LT production.[6]

In the natural infection, ETEC do not simply multiply in the lumen of the intestine; these bacteria possess specialized fimbriae by which the bacteria attach to the surface of the intestinal epithelial cells. We have called these specialized fimbriae colonization factor antigens, or CFAs, because attachment results in colonization of the mucosal surface by the ETEC and thereby exposes the intestinal tissue to locally produced enterotoxin(s) with resultant diarrhea.[8,9] The importance of CFAs as primary virulence factors of the ETEC is evidenced by results of many different lines of investigation. Comparison of ETEC isolated from human cases of diarrhea and from diarrhea in domestic animals has revealed that both human-associated and animal-associated ETEC produce essentially the same enterotoxins but produce CFAs which are host-specific in attachment activity and which are antigenically unique.[10-13] Human-associated ETEC produce serotype-associated CFAs; for example, CFA/I is commonly found on ETEC belonging to serotypes 025:H42, 025:H^-, 078:H11, 078:H12, 078:H^-, and 063:H^- and CFA/II is commonly found on ETEC of serotypes 08:H9, 08:H^-, 06:H16 and 06:H^-.[7,9,10,14] It is of interest that *E. coli* belonging to the classical enteropathogenic *E. coli* (EPEC) serogroups are for the most part nonenterotoxigenic but CFA/I-positive enterotoxigenic 0114 and 0128 and CFA/II-positive enterotoxigenic 0111 isolates have been described.[10,15] Most ETEC isolated from human cases of diarrhea which produce both enterotoxins also produce either CFA/I or CFA/II. Animal-associated ETEC produce fimbrial CFAs which have been characterized as K88, K99, and 987P.[11,12,16] CFA/I, CFA/II, K88, and K99 are all fimbriae which mediate mannose-resistant hemagglutination (MR-HA) of erythrocytes in a manner (erythrocyte species specificity) which demonstrates that each CFA has different receptor activity.[10,11,13,17-19]

In vivo production of CFAs has been evidenced by confirmation of seroconversion in naturally occurring and artificially induced ETEC diarrhea and by the serological detection of CFAs directly in diarrheal stool specimens.[20-24] In vivo experiments performed with the suckling rabbit have confirmed the biological function of CFA/I and CFA/II; analogous studies with animal-associated ETEC, using appropriate animal species, have been equally successful in the hands of other investigators.[8,-12,24,25] The unique role of CFAs in ETEC virulence indicates that anti-CFA antibody should prevent colonization and thus prevent ETEC diarrheal disease. Preliminary results involving experiments on passive protection against ETEC colonization and diarrhea in the infant rabbit and on active immunity in adult rabbits in response to perorally administered purified CFA/I support this view.[8,10,26]

This report is mainly concerned with work stimulated by the idea that CFA/I and CFA/II may play the most important role as ETEC-specific antigens in terms of immune responses protective against ETEC infection in man. The results reported here concerning the attachment function of CFA/I and CFA/II and antibody responses of man to these antigens does indicate that it may be possible to develop a safe and effective anti-ETEC vaccine using a combined preparation of purified CFA/I and CFA/II.

II. MATERIALS AND METHODS

A. Bacterial Strains

ETEC strains used in this work were isolated from cases of acute diarrhea, primarily from adults and primarily in Bangladesh and from cases of travelers' diarrhea in Mexico.[8,10,27,28] Serotyping, assay for production of LT by the tissue-cultured Y-1 adrenal cell assay and by the passive immune hemolysis assay technique and assay for production of ST by the suckling mouse assay have been described elsewhere.[28-30] HA tests and HA-typing using human, bovine, adult chicken, African green monkey, and guinea pig erythrocytes have been described in detail.[17-19] Mannose-sensitive HA (MS-HA) is defined as HA which is inhibited in the presence of mannose; MR-HA is defined as HA which occurs both in the presence and absence of mannose.[19] CFA agar,[17] which contains Casamino Acids (Difco) and yeast extract (Difco), salts and 2% agar, was used to grow the bacteria for HA and adherence tests, for administration of bacteria to volunteers, and for large-scale production of CFA/I and CFA/II for purification.

B. Tissue Culture Assay for Adhesiveness of *E. coli*

Human fetal intestinal epithelial (FIE) cells (Flow Laboratories, Rockville, Md.) were maintained in Earle's MEM (Flow Labs.) plus 10% fetal bovine serum and 1.5% L-glutamine, plus antibiotic and antimycotic, at 37°C in a 5% CO_2 atmosphere. FIE cells were harvested by trypsinization and adjusted to 1.5×10^6 cells per milliliter in MEM for use in the assay. Bacterial cells were prepared as follows. Five milliliters of Brain Heart Infusion Broth (BHIB) were inoculated with two loops from a peptone stock slant and this broth was incubated at 37°C for 4 hr; 0.2 mℓ of the BHIB culture was used to inoculate a CFA agar plate for confluent growth. The CFA agar plates were incubated for 18 hr at 37°C and the bacterial cells then suspended in phosphate-buffered saline (PBS) at approximately 1×10^{10} cells per milliliter. Incubation mixtures consisted of 0.5 mℓ FIE cells and 0105 mℓ bacterial cells, each prepared as above, plus 0.05 mℓ PBS with or without 10% mannose, as specified. These 0.6 mℓ reaction mixtures were incubated for 30 min at 37°C with shaking at 90 rpm. After incubation, the reaction mixtures, prepared in duplicate, were centrifuged for 7 min at 1000 rpm and the supernatant fluid discarded. The cells were washed twice by centrifugation, as above, in 0.5 mℓ PBS with or without mannose (1.0%) and the resultant pellets resuspended in 0.3 mℓ PBS plus or minus 1.0% mannose. For each reaction mixture, duplicate smears were prepared on microscope slides using 0.15 mℓ/slide. After 15 min air-drying, the smears were fixed with 100% methanol for 10 min and stained by the standard Gram stain technique. The number of bacterial cells attached to the FIE cells was determined by light microscopy. For all tests, a minimum of 100 intact FIE cells were observed and each FIE cell was recorded according to the number of bacteria attached. The percent of FIE cells having 0, 1 to 5, and greater than 5 bacteria attached was calculated as an index of adhesiveness.

C. Determination of Anti-CFA/I IgG Antibody Responses to Subcutaneously Injected Antigen.

A group of five volunteers was parenterally immunized with CFA/I using a primary subcutaneous dose of 100 μg antigen followed by two booster injections, of 50 μg each, 23 and 36 days after the primary dose. Serum samples were collected before the first injection and again 58 and 75 days after the second booster injection. The CFA/I used was antigen purified to homogeneity according to biophysical and serological criteria, as reported elsewhere.[25] Also, safety (toxicity) tests of the antigen preparation were carried out using mice and guinea pigs and sterility was assured. Anti-CFA/I IgG was determined as recently described by the ELISA technique;[21] total IgG levels were determined by the radial immunodiffusion technique (Highland Laboratories). In the ELISA, microtiter plate wells were coated with an optimum concentration of purified CFA/I and after appropriate washing the wells received dilutions of the test serum (prepared in PBS-Tween 20). After 2 hr incubation, goat antihuman IgG conjugated to alkaline phosphatase was added to each well. Following another incubation and appropriate washing of the wells, alkaline phosphatase substrate was added and the reaction product determined after 100 min incubation. Purified anti-CFA/I IgG was used in the assay to obtain a standard curve for calculating the concentration of anti-CFA/I IgG in the samples. These methods and appropriate controls are described in detail elsewhere.[21,24]

D. Oral Challenge and Re-Challenge of Volunteers with ETEC Strain H-10407

Challenge and re-challenge of volunteers with strain H-10407 (078:H11:CFA/I) at a dose of 1×10^8 CFU per volunteer was performed as previously described.[22] Of 18 volunteers undergoing primary challenge, 8 were re-challenged, in an identical fashion, 9 weeks later and four were similarly re-challenged 1 year after the primary challenge. Detection of the test strain in fecal specimens at 3 days and 7 days after ingestion of the bacteria and titration of serum samples, collected after 15 to 30 days, for anti-078, anti-LT, and anti-CFA/I antibodies was done as originally described.[22] Medical criteria for admission to the study, informed consent, and medical surveillance and treatment during and after the study are also described, including overview by appropriate Committees for the Protection of Human Subjects.[22,31]

E. Quantitation of Total SIgA and CFA-Specific SIgA in Milk Specimens by the ELISA Technique

For quantitating total SIgA in milk, microtiter plate wells were first coated with an optimum concentration of monomeric IgA (2.0 μg/mℓ) and reacted with goat antihuman alpha chain antibody. The next reagent added, after appropriate washing, was SIgA which had been purified by affinity chromatography. The final antibody reagent was alkaline-phosphatase-conjugated rabbit antisecretory component (anti-SC), producing a standard curve. Total SIgA in the milk specimens was determined by substituting dilution of the milk in place of the purified SIgA. The specificity of the anti-SIgA ELISA was determined by using urine from a patient with IgA-specific immunodeficiency as a control to determine whether free SC interfered with the assay, using serum from a patient with IgA myeloma to eliminate the possibility that the assay might measure IgA rather than SIgA, and by substituting antihuman gamma and antihuman μ-chain for anti-SC antibody to ensure immunoglobulin class specificity of the assay; i.e., to eliminate the possibility that the assay might detect IgG or IgM. A human milk specimen was also used in a similar control to ensure anti-SIgA specificity.

III. RESULTS

A. Adhesiveness of CFA-Positive Enterotoxigenic *E. coli*

Previous experiments with the infant rabbit model for intestinal colonization by CFA-positive ETEC and studies with volunteers have shown that the ability to colonize the intestine and cause acute diarrhea correlates with possession of a colonization factor antigen, either CFA/I or CFA/II, by the test strains. However, adhesiveness of CFA-positive ETEC can be demonstrated in vitro using tissue-cultured human fetal intestinal epithelial (FIE) cells.[32] This tissue culture method is useful for testing large numbers of strains and for comparing adhesiveness of different types of *E. coli*. The data in Table 1 demonstrate several interesting features of ETEC adhesiveness. Ten CFA/I-positive and 10 CFA/II-positive ETEC strains were assayed. None of the test strains produced common fimbriae, as shown by the failure of these strains to produce MS-HA with guinea pig erythrocytes. Two control (non-ETEC)

Table 1
ADHESIVENESS OF CFA-POSITIVE AND CFA-NEGATIVE DERIVATIVE STRAINS OF ETEC FOR HUMAN FETAL INTESTINAL EPITHELIAL (FIE) CELLS IN RELATION TO SEROTYPE, ENTEROTOXIN PRODUCTION AND HEMAGGLUTINATION PATTERN

		Production of:		HA Pattern[b]					Adhesive index[c]	
Strain[a]	Serotype	ST	LT	H	B	C	M	G	−Man	+ Man
H-9192-19	078:H11:CFA/I	+	+	R	R	R	−	−	100	100
H-16098-1	0128:H12:CFA/I	+	−	R	R	R	−	−	100	98
H-16098-1-P	0128:H12:CFA$^-$	−	−	−	−	−	−	−	0	0
H-9195-16	025:H42:CFA/I	+	+	R	R	R	−	−	90	96
H-9190-6	078:H11:CFA/I	+	+	R	R	R	−	−	100	100
H-9190-6-P	078:H11:CFA$^-$	−	+	−	−	−	−	−	12	4
36035-10	078:H12:CFA/I	+	−	R	R	R	−	−	92	96
H-10401-18	078:H11:CFA/I	+	+	R	R	R	−	−	100	100
H-19123-13	063:H$^-$:CFA/I	+	−	R	R	R	−	−	80	97
H-19123-13-P	063:H$^-$:CFA$^-$	−	−	−	−	−	−	−	0	2
PB-200	078:H12:CFA/I	+	+	R	R	R	−	−	100	100
PB-200-P	078:H12:CFA$^-$	−	+	−	−	−	−	−	10	12
GV-33A	063:H$^-$:CFA/I	+	+	R	R	R	−	−	75	67
GV-33A-P	063:H$^-$:CFA$^-$	−	+	−	−	−	−	−	6	3
H-10407	078:H11:CFA/I	+	+	R	R	R	−	−	41	42
H-10407-P	078:H11:CFA$^-$	−	+	−	−	−	−	−	6	3
H-15862-2A	08:H9:CFA/II	+	+	−	R	R	−	−	98	95
H-19796	08:H9:CFA/II	−	+	−	R	R	−	−	92	91
H-19935	085:H7:CFA/II	+	+	−	R	R	−	−	100	100
H-15697-3	08:H$^-$:CFA/II	+	+	−	R	R	−	−	100	100
PB-122B-1	080:H9:CFA/II	+	−	−	R	R	−	−	87	88
CL-9626	06:H16:CFA/II	+	+	−	R	R	−	−	85	87
H-19942	08:H9:CFA/II	+	+	−	R	R	−	−	98	98
H-18646	08:H?:CFA/II	+	+	−	R	R	−	−	76	81
H-10400-7	08:H$^-$:CFA/II	+	+	−	R	R	−	−	57	47
H-10400-7-P	08:H$^-$:CFA$^-$	−	−	−	−	−	−	−	11	2
GV-50B	06:H$^-$:CFA/II	+	+	−	R	R	−	−	56	53
GV-50B-P	06:H$^-$:CFA$^-$	−	−	−	−	−	−	−	0	14
LY-835	07:H6	−	−	S	−	S	S	S	57	2
LY-186	018:H15	−	−	S	−	S	S	S	97	3

[a] "-P" designates a CFA$^-$ derivative of a CFA$^+$ ETEC strain.

[b] Erythrocytes used: H = human group A; B = bovine; C = chicken; M = African green monkey; G = guinea pig; R designates mannose-resistant HA; S designates mannose-sensitive HA; − designates HA-negative.

[c] Adhesive index = percent of FIE cells with five or more bacteria attached after incubation and washing of FIE cells; −Man denotes test performed in absence of mannose; +Man denotes test performed in the presence of mannose.

strains possessing common fimbriae were used for comparison; note the HA pattern of these strains. Although the HA patterns of the CFA/I- and CFA/II-positive ETEC are different, both types hemagglutinate erythrocytes by a mannose-resistant mechanism. Adherence of the ETEC and of the control strains to FIE cells correlates with the type of HA; i.e., adherence of the CFA-positive ETEC strains to FIE cells is mannose-resistant and adherence of the MS-HA-positive ETEC strains to FIE cells is mannose-sensitive. Furthermore, "P"-strains of the ETEC which have spontaneously lost the ability to produce CFA/I or CFA/II are HA-negative and also negative for adhesiveness to FIE cells.

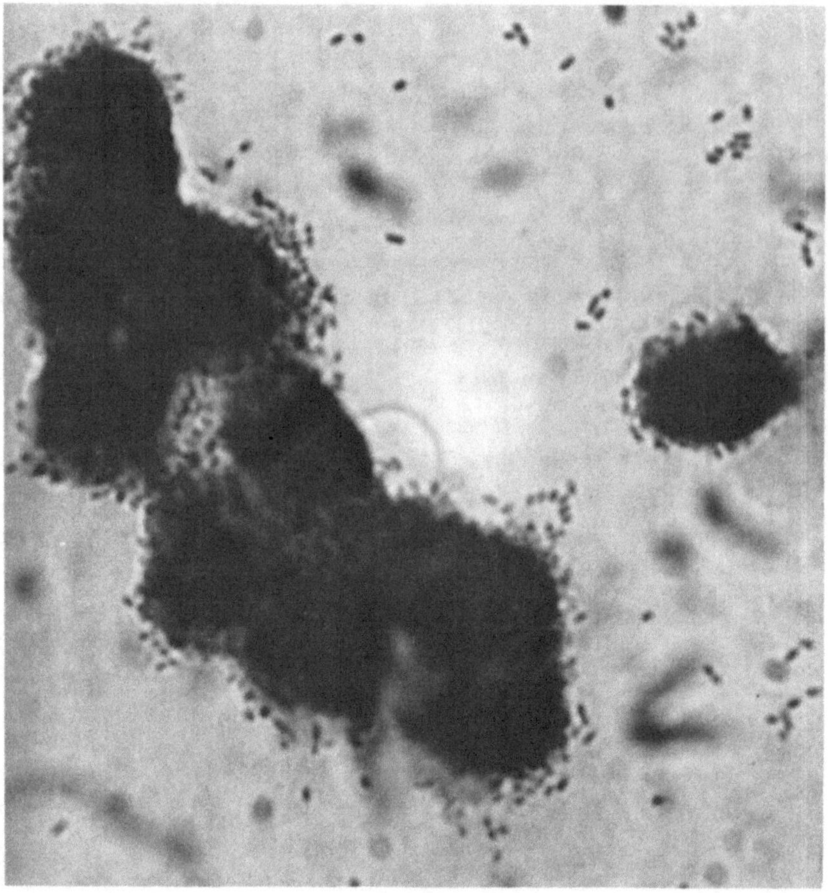

FIGURE 1. Light micrograph of human fetal intestinal epithelial (FIE) cells after incubation with the CFA/I^+ ETEC strain H-9190-6.

The ST and LT assay results shown in Table 1 also demonstrate the genetic interrelatedness which we and others have observed to exist between enterotoxin production and CFA production by ETEC. CFA/I production is controlled by a plasmid which also determines the production of ST while LT production is controlled by another plasmid; therefore ST^+LT^+CFA/I^+ ETEC give rise to LT-only (ST^-CFA/I^-) "P" strains upon loss of the unstable ST-CFA/I plasmid. If the plasmid which controls LT production in such ETEC is also lost, as in the case of strain H-16098-1, the resultant "P" strain produces neither ST nor LT. For most, if not all, CFA/II-positive ETEC a single plasmid controls ST, LT, and CFA/II production and therefore "P" strains of the ETEC do not produce ST, LT, or CFA/II as for example strains H-10400 and GV-50B shown in Table 1. By another mechanism, i.e., deletion of a portion of the ST:LT:CFA/II plasmid, strains lacking only ST, LT, or CFA/II production may arise. The cause of these deletion events is not known. Figure 1 shows a typical result in which large numbers of a CFA/I-positive ETEC strain have adhered to the FIE cells; Figure 2 shows the results obtained with the "P" derivative of the same strain.

B. Immunology of Enterotoxigenic *E. coli* Diarrhea

1. General

ETEC diarrhea is an especially difficult disease from the immunological standpoint since ETEC do not invade the intestinal tissue; therefore the diagnostic value of circulating antibody

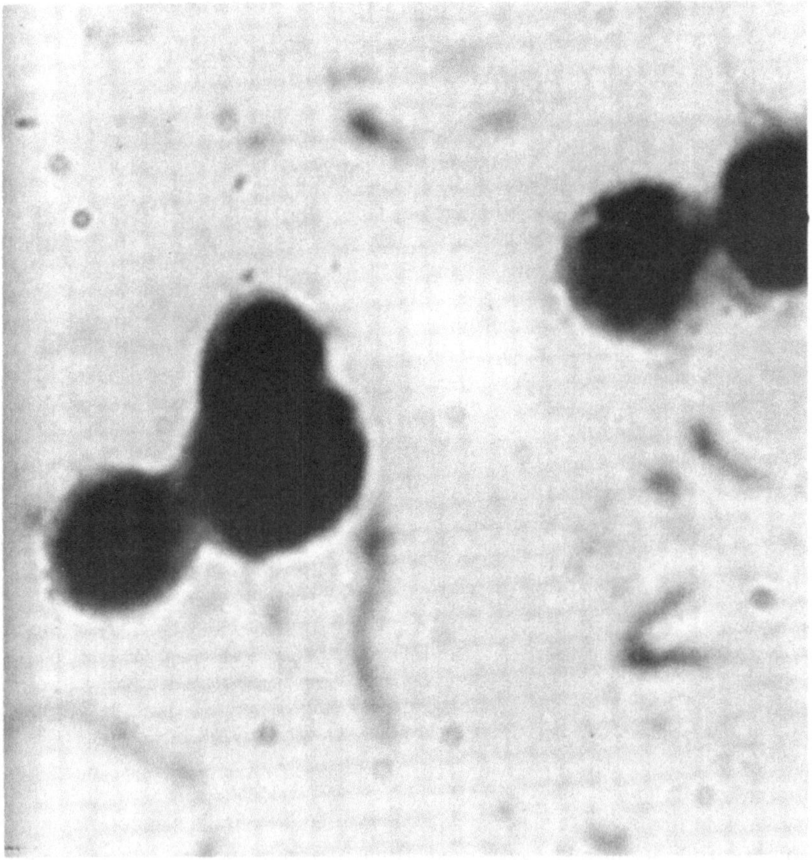

FIGURE 2. Light micrograph of human fetal intestinal epithelial (FIE) cells after incubation with the CFA-negative derivative, or P, strain of H-9190-6-P ETEC.

is to be questioned. It also follows that protective antibody; i.e., antibody functional in the intestinal lumen, should be of the SIgA class of immunoglobulin. In terms of antibody specificity antisomatic, antiflagellar, antitoxin, and antifimbrial could all be important. However in theory anti-CFA antibody would be the most protective in the intestine since this antibody could most effectively block colonization and thus prevent diarrhea. Passive protection with anti-CFA/I antibody has been demonstrated in the infant rabbit model.[8,10] More recently, we have demonstrated in the adult rabbit model that oral immunization with purified CFA/I antigen protects against colonization by CFA/I-positive ETEC and also protects against diarrhea (unpublished results).[26]

2. Antibody Responses in Serum to ETEC Infection and to Parenterally Administered CFA/I.

In prior work we demonstrated antisomatic, anti-LT, and anti-CFA antibody responses in humans, both in cases of naturally occurring ETEC diarrhea and in artificially infected volunteers.[20-23] In travelers' diarrhea we observed a correlation between the type of anti-CFA antibody response (anti-CFA/I vs. anti-CFA/II) and the CFA-type of the ETEC isolated from individual cases.[20]

One major goal is to develop the methodology for preventing ETEC diarrhea by vaccination with ETEC-specific antigens. The CFAs are considered the antigens of choice for reasons already discussed and also because of their ready availability in purified form and the fact that these antigens are inherently nontoxic. Several different approaches have been used in

Table 2
HUMORAL ANTI-CFA/I IgG ANTIBODY RESPONSES TO PARENTERALLY ADMINISTERED CFA/I IN VOLUNTEERS

Volunteer	Serum sample[a]	Total IgG (mg/mℓ)	Anti-CFA/I IgG (mg/mℓ)	Percent CFA/I specific IgG
1	Preimmunized	10.2	0.000	0.00
	1st postimmunized	9.3	1.512	16.25
	2nd postimmunized	8.9	1.008	11.32
2	Preimmunized	9.4	0.000	0.00
	1st postimmunized	10.8	0.014	0.13
	2nd postimmunized	12.2	0.354	2.90
3	Preimmunized	9.8	0.000	0.00
	1st postimmunized	10.4	0.129	1.24
	2nd postimmunized	9.9	0.354	3.58
4	Preimmunized	10.8	0.000	0.00
	1st postimmunized	8.3	0.016	0.19
	2nd postimmunized	7.2	0.068	0.94
5	Preimmunized	12.0	0.000	0.00
	1st postimmunized	12.0	0.042	0.35
	2nd postimmunized	13.8	0.567	4.11

[a] Volunteers were injected with 100 μg purified CFA/I, subcutaneously, and then with 50 μg antigen 23 and 36 days later. The first and second postimmunization serum samples were taken 58 and 75 days after the second booster injection, except for volunteer No. 1 from which serum samples were obtained 88 and 114 days, respectively, after the second booster injection.

preliminary work in this area. Table 2 shows the results of volunteer experiments in which the antigenicity of CFA/I in man was tested using parenteral (subcutaneous) injections of antigen. Antibody responses were determined, as anti-CFA/I IgG, following a primary injection of 100 μg purified antigen and two booster doses of 50 μg each. Anti-CFA/I IgG was quantitated by the ELISA technique which is easily adaptable to determining specific classes of immunoglobulin;[21] we have also employed the ELISA for the detection and quantitation of CFAs directly in stool specimens from cases of ETEC diarrhea.[24] As shown in Table 2, the anti-CFA/I IgG responses to parenterally administered CFA/I varied with the individual but all five of the volunteers did respond.

3. Antibody Responses in Volunteers Orally Exposed to CFA/I-Positive ETEC

Previously, we found that when volunteers ingested CFA/I-positive ETEC at two different doses humoral antibody responses occurred only at the higher dose (1×10^8 CFU) which caused intestinal colonization and illness; the lower dose (1×10^6 CFU) did not cause illness and did not elicit a detectable immune response.[22] Tables 3 and 4 show the results of a subsequent study in which volunteers were re-challenged with the same ETEC strain (078:H11:CFA/I; ST^+LT^+) 9 weeks (8 volunteers) and 1 year (4 volunteers) after a primary challenge with the same dose (1×10^8 CFU) which caused diarrhea and seroconversion in response to the primary challenge. Even after 1 year re-challenge with the same ETEC strain did not cause illness (Table 3) or seroconversion (Table 4). The test strain failed to colonize the volunteers upon re-challenge as indicated by the rapid excretion of the organism (Table 4); this probably accounts for the failure of these volunteers to seroconvert with respect to the ETEC antigens. Although the homologous strain was used for re-challenge in this study the results are compatible with the idea that anti-CFA/I antibody might be responsible for the observed protection from re-challenge. Challenge and re-challenge volunteer studies need to be performed using CFA/I-positive ETEC belonging to different serotypes and/or with

Table 3
DIARRHEA RESPONSE OF VOLUNTEERS AFTER PRIMARY AND SECONDARY CHALLENGES WITH ORALLY ADMINISTERED ETEC STRAIN H-10407 (078:H11:CFA/I; ST^+LT^+) AT A DOSE OF 1 × 10^8 CFU PER INDIVIDUAL

Challenge	Number with liquid stools	Onset (hr) (range)	Duration (hr) (range)	Stool volume (mℓ) (range)
First	14 of 16	58 (41—72)	56 (24—96)	1716 (775—3490)
Second (9 weeks later)	0 of 8	0	0	230 (113—325)
Second (1 year later)	0 of 4	0	0	321 (200—390)

Table 4
CHALLENGE AND RE-CHALLENGE OF VOLUNTEERS WITH ORALLY ADMINISTERED ETEC STRAIN H-10407 (078:H11:CFA/I) AT A DOSE OF 1 × 10^8 CFU PER INDIVIDUAL: ISOLATION OF ETEC FROM STOOL AND SERUM ANTIBODY RESPONSES

	Isolation of strain H-10407 at:		Serum antibody responses to		
Challenge	3 days	7 days	LT	078	CFA/I
First	16 of 16	16 of 16	9/16	15/16	14/16
Second (9 weeks later)	7 of 8	0 of 7	0/8	0/8	0/8
Second (1 year later)	3 of 4	1 of 4	0/4	0/4	1/4

and without the ability to produce LT in order to test for the protective capacity of anti-CFA/I in the intestine.

4. Anti-CFA Secretory IgA in Human Milk as the Result of Natural Exposure to ETEC

Since ETEC diarrhea is a major problem in developing countries, as reflected by the travelers' diarrhea syndrome, and especially in infants and young children in these countries one area of concern is the potential of anti-CFA antibody in breast milk of mothers to protect their infants against ETEC diarrhea. A study was designed to survey anti-CFA SIgA antibody in breast milk of mothers living in an ETEC-endemic area (Dacca, Bangladesh) for comparison with anti-CFA levels in breast milk of mothers living in an area of very low ETEC endemicity (Houston, Texas). First, an ELISA technique was developed for quantitating SIgA specific for CFA/I (or for CFA/II) in terms of micrograms of antibody per milliliter of milk. Figure 3 shows the results of ELISA tests performed to determine the specificity of the assay for SIgA. For the curved labeled SIgA, microtiter plate wells were exposed, in the following order, to purified monomeric Iga, goat antihuman alpha chain antibody, purified SIgA, and finally rabbit anti-SC conjugated to alkaline phosphatase (see Section II). For the curve labeled IgA, serum from an IgA myeloma patient was substituted for purified SIgA. For the curve labeled IgG, antihuman gamma- and antihuman μ-chain an-

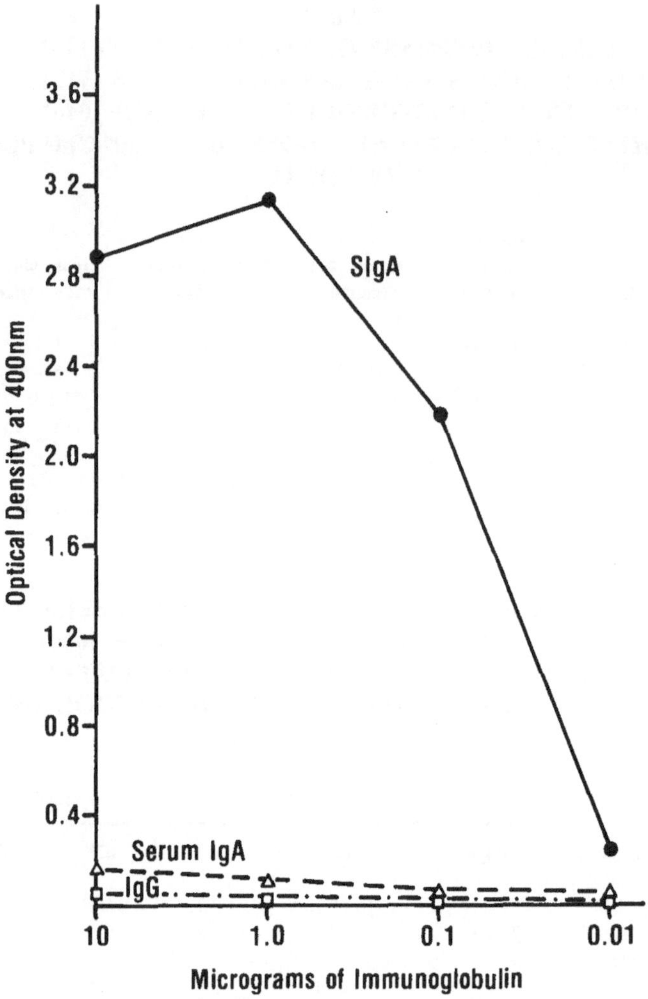

FIGURE 3. Results of double-antibody ELISA test for SIgA. Alkaline phosphatase was conjugated to rabbit antihuman SC. Samples were dilutions of purified human SIgA (●), or serum IgA (see text) (△). Bottom line (□) was a test for possible detection of immunoglobulin G (IgG) (see text for details).

tibody was substituted for anti-SC antibody. Similar results were obtained when appropriate milk dilutions were substituted for purified SIgA; the results shown in Figure 3 show that the assay technique did not detect SC alone, IgA alone, IgG, or IgM. Figure 4 shows the standard curve obtained with purified SIgA; the curve is linear with respect to optical density of the alkaline phosphatase product between 0.039 and 0.623 μg SIgA per milliliter in the test. Similar standard curves were performed in parallel with milk samples assayed for SIgA, anti-CFA/I SIgA, and anti-CFA/II SIgA. Anti-CFA/I and anti-CFA/II were determined by coating the microtiter plate wells with purified CFA/I or purified CFA/II, followed by appropriate dilutions of milk and finally anti-SC-alkaline phosphatase conjugate. Figure 5 shows typical titration curves obtained when milk specimens were assayed for total SIgA and Figure 6 shows typical titration curves when milk samples were assayed for anti-CFA SIgA.

Table 5 shows the data for total SIgA, in micrograms per milliliter in breast milk specimens collected in Dacca vs. those collected in Houston according to the day of collection. It can

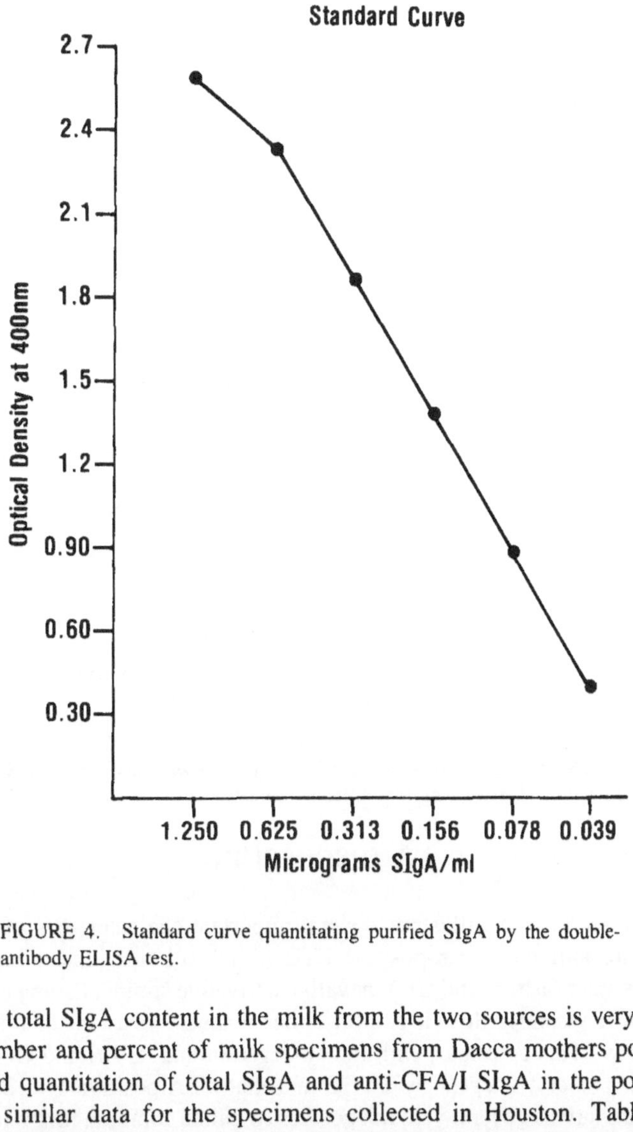

FIGURE 4. Standard curve quantitating purified SIgA by the double-antibody ELISA test.

be seen that the total SIgA content in the milk from the two sources is very similar. Table 6 shows the number and percent of milk specimens from Dacca mothers positive for anti-CFA/I SIgA and quantitation of total SIgA and anti-CFA/I SIgA in the positive samples; Table 7 shows similar data for the specimens collected in Houston. Table 8 shows the comparison of anti-CFA SIgA in breast milk from the two sources. It can be seen that when anti-CFA SIgA was present in a sample there was no significant difference in relation to the source of the sample. However, a major difference can be seen in the percentage of samples containing anti-CFA SIgA, with 74% of the samples from Dacca and 20% of those from Houston containing such antibody.

Two important points can be made about the above results. First, the percent of samples containing anti-CFA SIgA is consistent with the high endemicity of ETEC diarrhea in Bangladesh as compared with the low endemicity in Houston, showing that this antibody might be an important factor for protection of infants in Bangladesh. Second, the finding that 20% of the breast milk samples from Houston are positive for anti-CFA antibody was unexpected because it is higher than would be predicted from data on the occurrence of ETEC diarrhea in Houston. This indicates that ETEC diarrhea might be more common than is indicated by sampling infants with diarrhea for ETEC since such sampling leads to ETEC identification very infrequently.

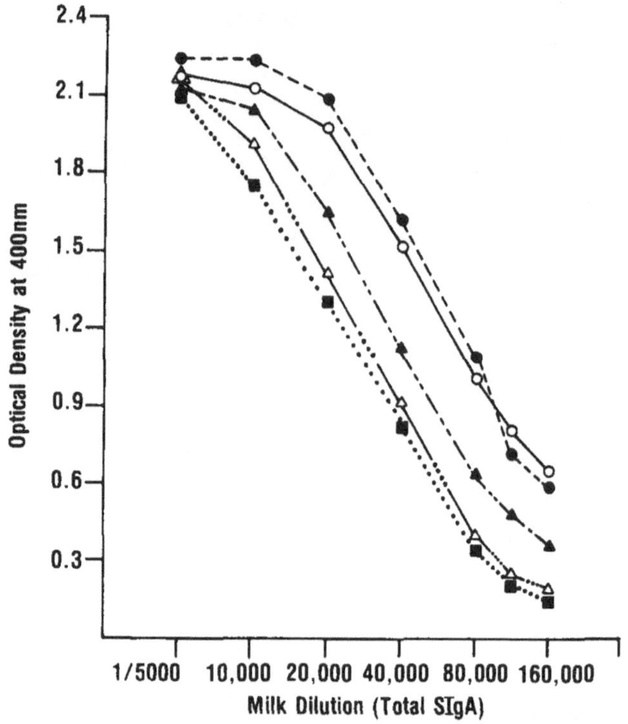

FIGURE 5. ELISA for total SIgA in breast milk specimens; curves show the results obtained with five different samples.

IV. DISCUSSION

The results presented above illustrate several important characteristics of ETEC fimbrial colonization factor antigens. CFA-positive ETEC bind to human intestinal epithelial cells in a mannose-resistant fashion and CFA-negative derivative strains, having lost the ST:CFA/I or ST:LT:CFA/II plasmid, do not exhibit either mannose-resistant or mannose-sensitive attachment activity. This correlates with the observation that CFA/I and CFA/II are responsible for MR-HA of erythrocytes by ETEC and that CFA-positive ETEC and even their CFA-negative derivatives generally do not possess so-called common fimbriae which mediate MS-HA and mannose-sensitive attachment to tissue-cultured cells. When CFA/I and CFA/II are detached from the bacteria and purified these antigens retain their morphological characteristics, antigenicity, and the ability to react with erythrocytes and intestinal cells in a specific fashion. For example, we have demonstrated that purified CFA/I avidly binds to the infant rabbit small intestine and that this bound CFA can block adherence of CFA-positive bacteria and thus prevent colonization and diarrhea.[25] This indicates that protection from colonization and subsequent diarrhea could be achieved efficiently with anti-CFA antibody or with a biochemical reactive with the reactive sites on either the CFA fimbriae or on the intestinal epithelial cells.

It follows from the above that anti-CFA antibody of the SIgA immunoglobulin class should afford protection against ETEC diarrhea. Although it has been difficult, for technical reasons, to demonstrate intestinal anti-CFA SIgA immune responses in man humoral (IgG) anti-CFA antibody responses to ETEC have been demonstrated in volunteers and in natural infection.[20-24] Data presented here show that parenterally administered CFA is safe and antigenic in humans and also that challenge of volunteers with an ETEC strain which is CFA/I-positive affords protection against re-challenge with the homologous strain. Much more work remains

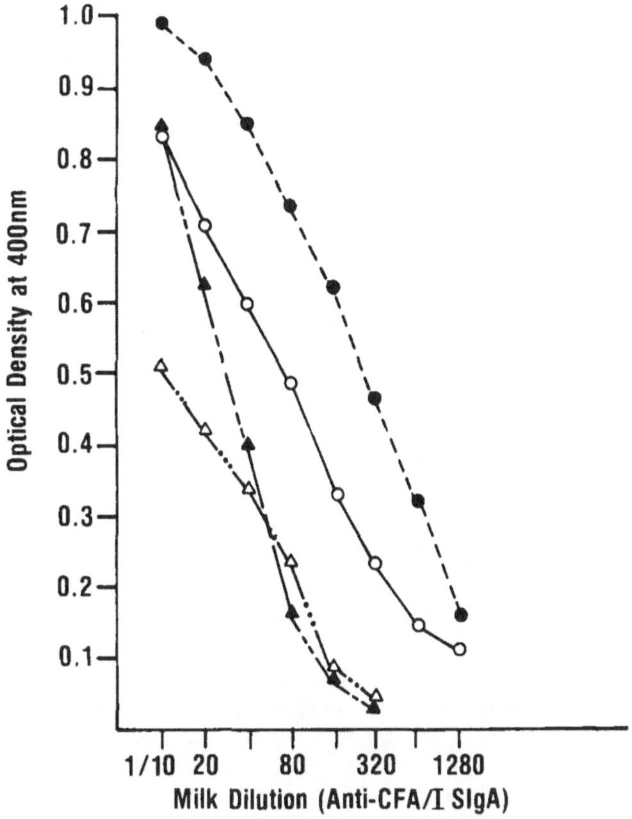

FIGURE 6. ELISA for anti-CFA/I SIgA in brast milk specimens; curves show the results obtained with four different samples. Anti-CFA/I SIgA was calculated in micrograms per milliliter using a standard curve such as that shown in Figure 4.

to be done in order to show conclusively that a protective intestinal anti-CFA SIgA response can be produced by vaccination with purified CFAs but the results so far achieved are encouraging. Human breast milk provides one source of SIgA which is readily obtainable and thus we tested that idea that mothers in a country of high ETEC endemicity should have demonstrably higher levels of anti-CFA SIgA in their milk than those mothers in an area of low endemicity such as Houston. The results presented here showed no appreciable difference in the levels of anti-CFA SIgA in mothers' milk from the two sources but did show a large difference in the percentage of mothers having such antibody in their milk. The majority (74%) of breast milk samples collected in Bangladesh had CFA-specific SIgA as compared to only 20% of those collected in Houston. This confirms the idea that exposure to ETEC in the environment should be reflected in the immune status of mothers and is therefore a factor to be considered in terms of susceptibility of their infants to ETEC diarrhea.

Another approach to the immunology of ETEC diarrhea is the study of intestinal immune response of laboratory animals to exposure to ETEC antigens such as the CFAs. We have used the temporary intestinal ligation technique, with adult rabbits, to study anti-CFA and class-specific immunoglobulin responses to CFA/I given orally in the purified form and the bacteria-associated form. We have shown that in this animal model oral doses of CFA/I elicit significant anti-CFA/I and SIgA antibody responses and that these responses do correlate with protection from challenge with CFA/I-positive bacteria (manuscript in preparation).[26] These studies support the idea that an oral vaccine composed of combined purified CFA/I and CFA/II for protection aainst ETEC diarrhea is a viable goal.

Table 5
TOTAL SIgA IN BREAST MILK SPECIMENS COLLECTED IN DACCA AND IN HOUSTON

Day of collection	Dacca, Bangladesh	Houston, Texas
1	(1)[a] 7,831	(1) 10,481
2	(8) 2,571 ± 1,629[b]	(25) 1,899 ± 806
3	(10) 984 ± 114	(29) 972 ± 302
4	(6) 919 ± 46	(5) 894 ± 33
5	(8) 800 ± 98	
6	(3) 723 ± 62	
7	(2) 610 ± 72	
All specimens	(38) 1,409 ± 409	(60) 1,510 ± 485

[a] Number of specimens.
[b] SIgA in micrograms per mℓ, ± 1.0 standard deviation.

Table 6
BREAST MILK SPECIMENS COLLECTED IN DACCA, BANGLADESH: TOTAL SIgA AND ANTI-CFA/I SIgA IN POSITIVE SAMPLES

Day	Number positive/total	Total SIgA in μg/mℓ	Anti-CFA/I SIgA in μg/mℓ
1	0/1	0	0
2	4/8 (50%)	1818 ± 702	10.12 ± 7.40
3	10/10 (100%)	984 ± 114	5.29 ± 3.25
4	5/6 (83%)	938 ± 28	3.45 ± 1.35
5—7	4/13 (31%)	823 ± 67	2.87 ± 1.09

Table 7
BREAST MILK SPECIMENS COLLECTED IN HOUSTON, TEXAS: TOTAL SIgA AND ANTI-CFA/I SIgA IN POSITIVE SAMPLES

Day	Number positive/total	Total SIgA in μg/mℓ	Anti-CFA/I SIgA in μg/mℓ
1	0/1	0	0
2	8/25 (32%)	1675 ± 651	3.34 ± 2.66
3	4/29 (14%)	1341 ± 272	3.75 ± 1.97
4	1/5 (20%)	900	1.22

Table 8
ANTI-CFA/I SIgA IN BREAST MILK: COMPARISON OF MOTHERS LIVING IN BANGLADESH AND MOTHERS LIVING IN HOUSTON, TEXAS

	Source of breast milk specimens:	
	Dacca	**Houston**
Number of milk specimens	38	60
Mean sampling day	3.76 ± 1.53[a]	2.63 ± 0.66
Total SIgA for all samples (μg/mℓ)	1409.0 ± 409	1510.0 ± 485
Number with detectable anti-CFA (%)	28 (74%)	13 (20%)
Number with anti-CFA/I (%)	23 (61%)	8 (13%)
Number with anti-CFA/II (%)	10[b] (26%)	5 (8%)
Anti-CFA SIgA (μg/mℓ)	5.31 ± 3.18	3.30 ± 2.24
Anti-CFA SIgA (Percent of total SIgA)	0.473 ± 0.296	0.210 ± 0.097

[a] Plus/minus one standard deviation.
[b] Five of these samples also had anti-CFA/I detectable. Total and specific SIgA was measured by the double-antibody ELISA technique.

REFERENCES

1. **Evans, D. G., Evans, D. J., Jr., and Gorbach, S. L.,** Identification of enterotoxigenic *Escherichia coli* and serum antitoxin activity by the vascular permeability factor assay, *Infect. Immun.*, 8, 731, 1973.
2. **Holmgren, J. and Svennerholm, A. M.,** Immunological cross-reactivity between *Escherichia coli* heat-labile enterotoxin and cholera toxin A and B subunits, *Curr. Microbiol.*, 2, 55, 1979.
3. **Holmgren, J., Soderlind, O., and Wadstrom, T.,** Cross-reactivity between heat-labile enterotoxins of *Vibrio cholerae* and *Escherichia coli* in neutralization tests in rabbit ileum and skin, *Acta Pathol. Microbiol. Scand.*, 81, 757, 1973.
4. **Kunkel, S. L. and Robertson, D. C.,** Purification and chemical characterization of the heat-labile enterotoxin produced by enterotoxigenic *Escherichia coli*, *Infect. Immun.*, 25, 586, 1979.
5. **Guerrant, R. L. and Brunton, L. L.,** Characterization of the Chinese hamster ovary cell assay for the enterotoxins of *Vibrio cholerae* and *Escherichia coli* and for antitoxin: differential inhibition by gangliosides, specific antisera, and toxoid, *J. Infect. Dis.*, 135, 720, 1977.
6. **Penaranda, M. E., Mann, M. B., Evans, D. G., and Evans, D. J., Jr.,** Transfer of an ST:LT:CFA/II plasmid into *Escherichia coli* K-12 strain RR1 by cotransformation with PSC301 plasmid DNA, *FEMS Microbiol. Lett.*, 8, 251, 1980.
7. **Smith, H. R., Cravioto, A., Willshaw, G. A., McConnell, M. M., Scotland, S. M., Gross, R. J., and Rowe, B.,** A plasmid coding for the production of colonisation factor antigen I and heat-stable enterotoxin in strains of *Escherichia coli* of serogroup 078, *FEMS Microbiol. Lett.*, 6, 255, 1979.
8. **Evans, D. G., Silver, R. P., Evans, D. J., Jr., Chase, D. G., and Gorbach, S. L.,** Plasmid-controlled colonization factor associated with virulence in *Escherichia coli* enterotoxigenic for humans, *Infect. Immun.*, 12, 656, 1978.
9. **Evans, D. G., Evans, D. J., Jr., Tjoa, W. S., and DuPont, H. L.,** Detection and characterization of colonization factor of enterotoxigenic *Escherichia coli* isolated from adults with diarrhea, *Infect. Immun.*, 19, 727, 1978.
10. **Evans, D. G. and Evans, D. J., Jr.,** New surface-associated heat-labile colonization factor antigen (CFA/II) produced by enterotoxigenic *Escherichia coli* of serogroups 06 and 08, *Infect. Immun.*, 21, 638, 1978.
11. **Jones, G. W.and Rutter, J. M.,** Role of the K88 antigen in the pathogenesis of neonatal diarrhea caused by *Escherichia coli* in piglets, *Infect. Immun.*, 6, 918, 1972.
12. **Isaacson, R. E.,** K99 surface antigen of *Escherichia coli:* antigenic characterization, *Infect. Immun.*, 22, 555, 1978.
13. **Burrows, M. R., Sellwood, R., and Gibbons, R. A.,** Naemagglutinating and adhesive properties associated with the K99 antigen of bovine strains of *Escherichia coli*, *J. Gen. Microbiol.*, 96, 269, 1976.

14. **Back, E., Mollby, R., Kaijser, B., Stintzing, G., Wadstrom, T., and Habte, D.,** Enterotoxigenic *Escherichia coli* and other gram-negative bacteria of infantile diarrhea: surface antigens, hemagglutinins, colonization factor antigen and loss of enterotoxigenicity, *J. Infect. Dis.*, 142, 318, 1980.
15. **Scotland, S. M., Day, N. P., Cravioto, A., Thomas, L. V., and Rowe, B.,** Production of heat-labile or heat-stable enterotoxins by strains of *Escherichia coli* belonging to serogroups 044, 0114, and 0128, *Infect. Immun.*, 31, 500, 1981.
16. **Isaacson, R. E., Dean, E. A., Morgan, R. L., and Moon, H. W.,** Immunization of suckling pigs against enterotoxigenic *Escherichia coli*-induced diarrheal disease by vaccinating dams with purified K99 or 987P pili: antibody production in response to vaccination, *Infect. Immun.*, 29, 824, 1980.
17. **Evans, D. G., Evans, D. J., Jr., and Tjoa, W.,** Hemagglutination of human group A erythrocytes by enterotoxigenic *Escherichia coli* isolated from adults with diarrhea: correlation with colonization factor, *Infect. Immun.*, 18, 330, 1977.
18. **Evans, D. J., Jr., Evans, D. G., and DuPont, H. L.** Hemagglutination patterns of enterotoxigenic and enteropathogenic *Escherichia coli* determined with human, bovine, chicken, and guinea pig erythrocytes in the presence and absence of mannose, *Infect. Immun.*, 23, 336, 1979.
19. **Evans, D. J., Jr., Evans, D. G., Young, L. S., and Pitt, J.,** Hemagglutination typing of *Escherichia coli:* definition of seven hemagglutination types, *J. Clin. Microbiol.*, 12, 235, 1980.
20. **Evans, D. G., Evans, D. J., Jr., and Tjoa, W.,** Relationship between enterotoxigenic *Escherichia coli* virulence factors and travelers' diarrhea, in *Current Chemotherapy and Infectious Disease*, Vol. 2, Nelson, J. D. and Grassi, C., Eds., American Society for Microbiology, Washington, D. C., USA, 1980, 917.
21. **Clegg, S., Evans, D. G., and Evans, D. J., Jr.,** Enzyme-linked immunosorbent assay for quantitating the humoral immune response to the colonization factor antigen of enterotoxigenic *Escherichia coli, Infect. Immun.*, 27, 525, 1980.
22. **Evans, D. G., Satterwhite, T. K., Evans, D. J., Jr., and DuPont, H. L.,** Differences in serological responses and excretion patterns of volunteers challenged with enterotoxigenic *Escherichia coli* with and without the colonization factor antigen, *Infect. Immun.*, 19, 883, 1978.
23. **Deetz, T. R., Evans, D. J., Jr., Evans, D. G., and DuPont, H. L.,** Serologic responses to somatic O and colonization-factor antigens of enterotoxigenic *Escherichia coli* in travelers, *J. Infect. Dis.*, 140, 114, 1979.
24. **Evans, D. G., Evans, D. J., Jr., and Clegg, S.,** Detection of enterotoxigenic *Escherichia coli* colonization factor antigen I in stool specimens by an enzyme-linked immunosorbent assay, *J. Clin. Microbiol.*, 12, 738, 1980.
25. **Evans, D. G., Evans, D. J., Jr., Clegg, S., and Pauley, J. A.,** Purification and characterization of the CFA/I antigen of enterotoxigenic *Escherichia coli, Infect. Immun.*, 25, 738, 1979.
26. **Cabada, F. J., Evans, D. G., And Evans, D. J., Jr.,** Immunoprotection against enterotoxigenic *Escherichia coli* diarrhea in rabbits by peroral administration of purified colonization factor antigen I (CFA/I), *FEMS. Microbiol. Lett.*, 11, 303, 1973.
27. **Evans, D. J., Jr. and Evans, D. G.,** Three characteristics associated with enterotoxigenic *Escherichia coli* isolated from man, *Infect. Immun.*, 8, 322, 1973.
28. **Evans, D. J., Jr., Ruiz-Palacios, R., Evans, D. G., DuPont, H. L., Pickering, L. K., and Olarte, J.,** Humoral immune response to the heat-labile enterotoxin of *Escherichia coli* in naturally acquired diarrhea and antitoxin determination by passive immune hemolysis, *Infect. Immun.*, 16, 781, 1977.
29. **Evans, D. J., Jr. and Evans, D. G.,** A direct serological assay for the heat-labile enterotoxin of *Escherichia coli* using passive immune hemolysis, *Infect. Immun.*, 16, 604, 1977.
30. **Evans, D. J., Jr., Evans, D. G., DuPont, H. L., Ørskov, F., and Ørskov, I.,** Patterns of loss of enterotoxigenicity by *Escherichia coli* isolated from adults with diarrhea: suggestive evidence for an interrelationship with serotype, *Infect. Immun.*, 17, 105, 1977.
31. **Satterwhite, T. K., Evans, D. G., DuPont, H. L., and Evans, D. J. Jr.,** Role of *Escherichia coli* colonisation factor antigen in acute diarrhea, *Lancet*, 2, 181, 1978.
32. **Bergman, M. J., Evans, D. G., Mandell, G. L., Sullivan, J. A., Salit, I. E., and Guerrant, R. L.,** Attachment of *E. coli* to human intestinal epithelial cells: a functional *in vitro* test for intestinal colonization factor, *Trans. Assoc. Am. Phys.*, xci, 80, 1978.

Chapter 9

PILI, HEMAGGLUTINATION, AND ADHERENCE

Grace M. Thorne, Carl F. Deneke, and Sherwood L. Gorbach

TABLE OF CONTENTS

I. INTRODUCTION

The ability of enterotoxigenic *Escherichia coli* (ETEC) to adhere to the small bowel mucosa of both humans and animals is an important factor in the pathogenesis of diarrheal disease. The adherence has been found to be mediated by specialized protein surface antigens (pili) K88,[1] K99,[2] 987P,[3] CFA/I[4] CFA/II.[5]

Because of the apparent host species-specificity, we have studied the attachment of ETEC strains isolated from humans with diarrheal disease to human oral buccal mucosal cells. The rationale for choosing oral buccal cells came from the previous studies of *Vibrio cholerae* by D. Sherwood Gorbach.[6] In India patients with acute cholera were found to harbor vibrios in the mouth for as long as 3 days. Similar studies have documented the ability to culture ETEC from the oropharynx of children with chronic diarrhea.[7]

II. MATERIALS AND METHODS

A. Buccal Cell Assay

In this adherence model system, washed buccal cells are incubated with radiolabeled ETEC strains. The buccal cells are then collected and washed on a 5 μ pore Nucleopore filter. The scanning electron micrograph in Figure 1 displays the filter surface with buccal cells retained under low vacuum pressure (<1 lb/in^2). The buccal cells are then washed free of nonadherent bacteria which are not retained by a 5 μ pore filter. The filters can be put in a liquid scintillation counter in order to determine the number of radiolabeled bacteria retained. In Figure 2 the scanning micrograph shows buccal cells incubated with a nonpiliated, nontoxigenic control strain of *E. coli*. When buccal cells are incubated with a human ETEC strain, bacteria can be seen to adhere singly and in clusters over the buccal cell surface (Figure 3). The scanning electron micrograph in Figure 4 was taken at 20,000 magnification and clearly shows the close proximity between bacterial cells and the surface of the oral buccal mucosal cell.

In our studies using this adherence model, we have found approximately 52 to 60% of ETEC strains examined bound to the buccal cells at levels of 100 to 500 bacteria/buccal cell.[8] Control strains included plasmid-free, nonenterotoxigenic strains, derived in the laboratory from human ETEC isolates, ETEC strains pathogenic for animals capable of producing the K88 or K99 antigens, and the rabbit enteric pathogen RDEC-1.[9] The control strains exhibited no or low level binding (< 100 bacteria/buccal cell). Although the in vitro buccal adherence system cannot duplicate in vivo growth and competitive colonization of the small bowel, the assay does allow for studies of the early events in bacterial attachment. The buccal cell adherence assay also allows for the discrimination of at least some ETEC isolated with the potential to colonize the human small bowel.

B. Pilus Isolation and Antiserum Preparation

ETEC strains isolated from humans with diarrheal disease were found to have numerous rigid, needle-like structures (5 to 10 nm in diameter) on their cell surface when visualized by transmission electron microscopy (Figure 5, A to E).[10] We have isolated and purified these attachment structures, by first shearing them from the bacterial cells using a Sorvall Omnimixer. The bacteria are then mixed with red blood cells in the cold in the presence of mannose. The sugar mannose will block attachment of type 1 pili. The red cells are washed free of other bacterial components and the attached pili are eluted by warming the red cell to 37°C. Finally, the remaining red blood cells are removed by centrifugation. This preparation yields purified pili (Figure 5F). These isolated pili have been used to develop pili-specific antisera. To date, three antigenic groups of pili have been found. These pili-specific sera gave high titers against the type organism and only limited cross-reactivity between

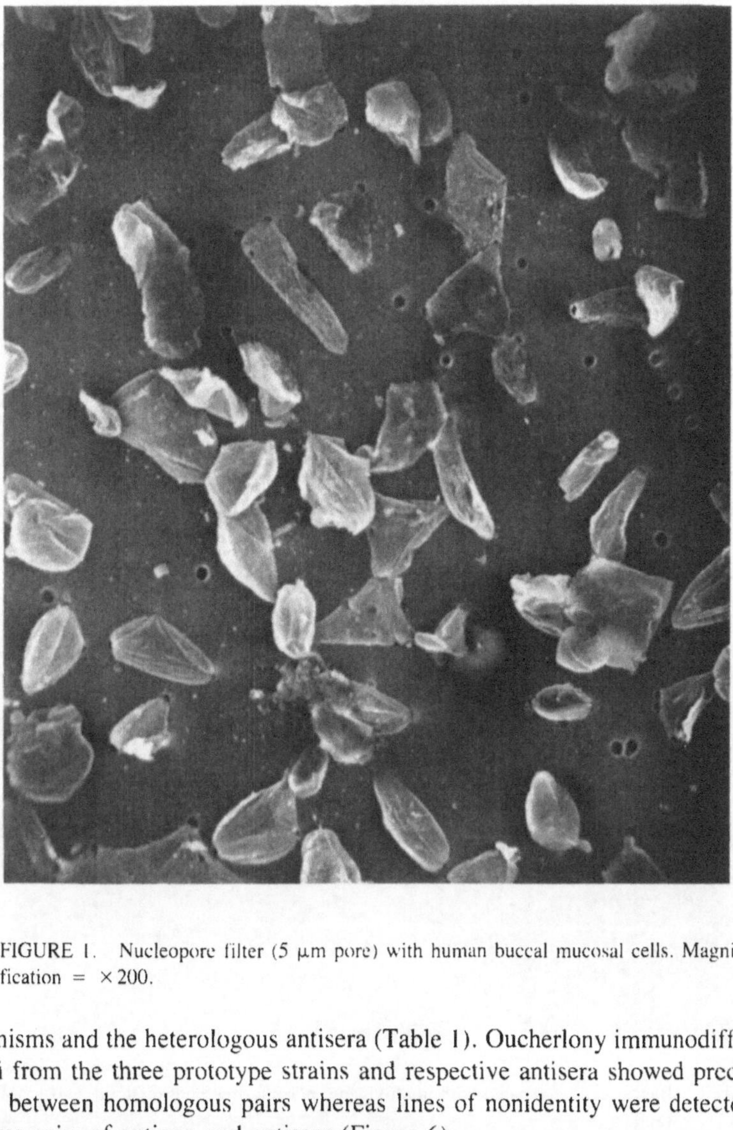

FIGURE 1. Nucleopore filter (5 μm pore) with human buccal mucosal cells. Magnification = ×200.

these organisms and the heterologous antisera (Table 1). Oucherlony immunodiffusion analysis of pili from the three prototype strains and respective antisera showed precipitin lines of identity between homologous pairs whereas lines of nonidentity were detected between heterologous pairs of antigen and antisera (Figure 6).

III. RESULTS

The three pili antisera were used to screen for these antigens on 106 ETEC strains isolated from human diarrheal disease. Sixty (56%) reacted with one or more of the antisera (Table 2). The remaining 46 ETEC strains were nonreactive in these three antisera. One strain, H10407, reacted with a high titer (1:4,096) with antisera against the pili of both strains 334 and D542. Five strains reacted with a titer of 1:1,024 or greater with antisera against the pili of strains 334 and M9800-5. These six strains are listed in Table 2 as 1, 3, and 2, 3 reactive strains, respectively. The antisera were also nonreactive with the 16 strains of nontoxigenic *E. coli* isolated from healthy adults, the rabbit pathogenic strain RDEC-1,[9] and the porcine pathogenic strain P307 (K88+).

The 106 strains in our culture collection were gathered from a variety of geographic sources (Table 3). There did not seem to be any association between pili serogroup and toxin production. The enterotoxin profile of these 106 ETEC are recorded in Table 4. There

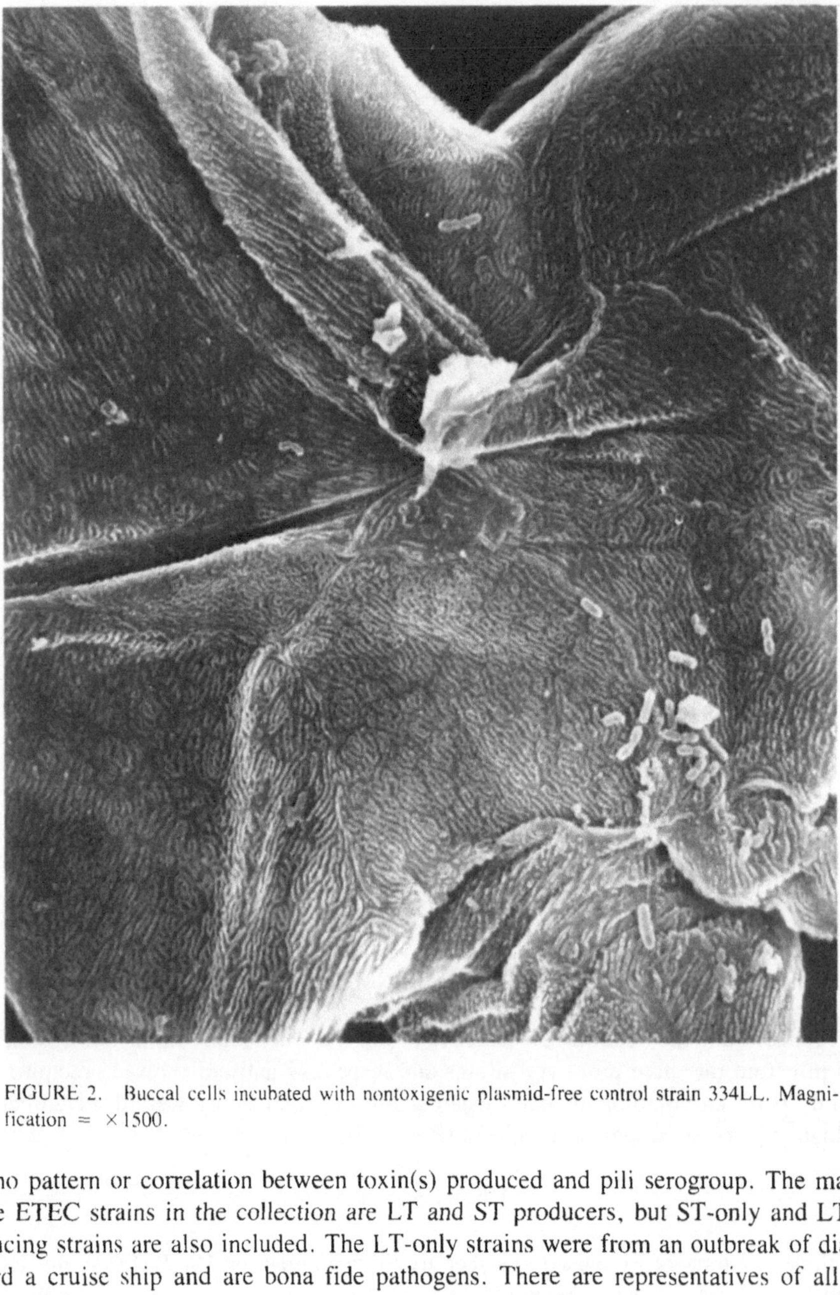

FIGURE 2. Buccal cells incubated with nontoxigenic plasmid-free control strain 334LL. Magnification = ×1500.

was no pattern or correlation between toxin(s) produced and pili serogroup. The majority of the ETEC strains in the collection are LT and ST producers, but ST-only and LT-only producing strains are also included. The LT-only strains were from an outbreak of diarrhea aboard a cruise ship and are bona fide pathogens. There are representatives of all toxin profiles among the three pili serogroups.

The distribution of O antigens within each pili serogroup is given in Table 5. No absolute correlation between pili serogroups and O antigenic groups was observed, although there was some clustering of 015 strains in pili serogroup 1 and 06 strains in serogroup 2. However, no pili group included only a single somatic antigen.

Last, the MR-HA patterns of the strains were tested since this reaction is felt to correlate with the presence of these piliated surface antigens. Strains were tested with human A, B, bovine, and guinea pig red cells and only 38/106 (36%) were reactive with at least one species of red cell (Table 6). Thus, the MR-HA reaction detected fewer strains than did the pili-antisera which, as shown earlier, reacted with 56% of the strains. Of those strains which did react with erythrocytes, there was no pattern of pili serogroups. The few strains which

FIGURE 3. Buccal cells incubated with ETEC strain 334. Magnification = ×1500.

reacted with two pili sera also exhibited varying MR-HA patterns. The majority of strains unreactive in the pili sera were also MR-HA negative. We are currently preparing pili and new pili antisera using strains in the last group which exhibit MR-HA ability, are piliated by EM examination and are adherent to human buccal cells. These strains most likely possess as yet, unclassified pili types. The information in Table 7 summarizes some of the characteristics of adherence pili currently recognized. The three serogroups that we have characterized differ in blood cell reactivity and molecular weights from that reported by the Evan's group.[4-5] The latter may be due to the different methodologies for pili isolation employed which could be explained by monomer-dimer pili structures. Using SDS-PAGE,Wevers et al.[13] have found that the subunits of CFA/I and CFA/II exhibited molecular weights of 12,000 and 13,000, respectively. Serologically, the serogroup 3 strains are similar to CFA/II. Serogroup 3 shows some serologic cross reactivity with CFA/I, but the two antigens are not identical. As stated above, these pili demonstrate different MR-HA patterns from those of CFA/I and CFA/II. Further investigations of the interrelationships of these important surface antigens are in progress for serological identification of attachment pili might be useful in the diagnosis of human diarrheal disease. Ultimately, these antigens may be employed as a vaccine to prevent diarrheal disease caused by ETEC in humans.

Table 1
RECIPROCAL TITERS OF ANTIPILI SERA WITH TYPE STRAINS

	Titer of:			
	Antisera against pili of strain:			
Strain	334	M9800-5	D542	Control serum
334	8192	128	128	256
M9800-5	128	8192	256	256
D542	256	64	4096	64

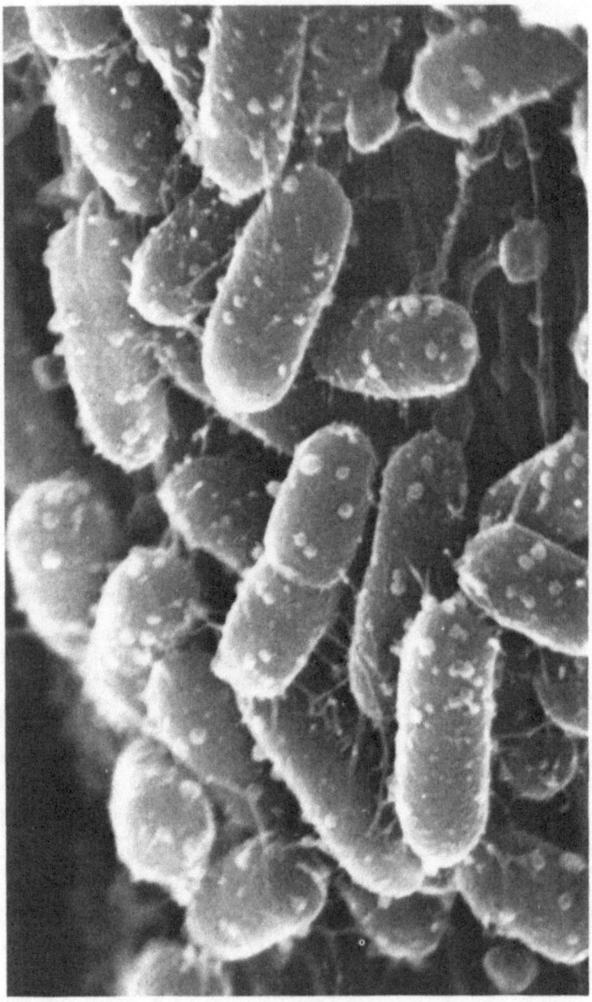

FIGURE 4. Buccal cells incubated with ETEC strain 334. Magnification = ×20,000.

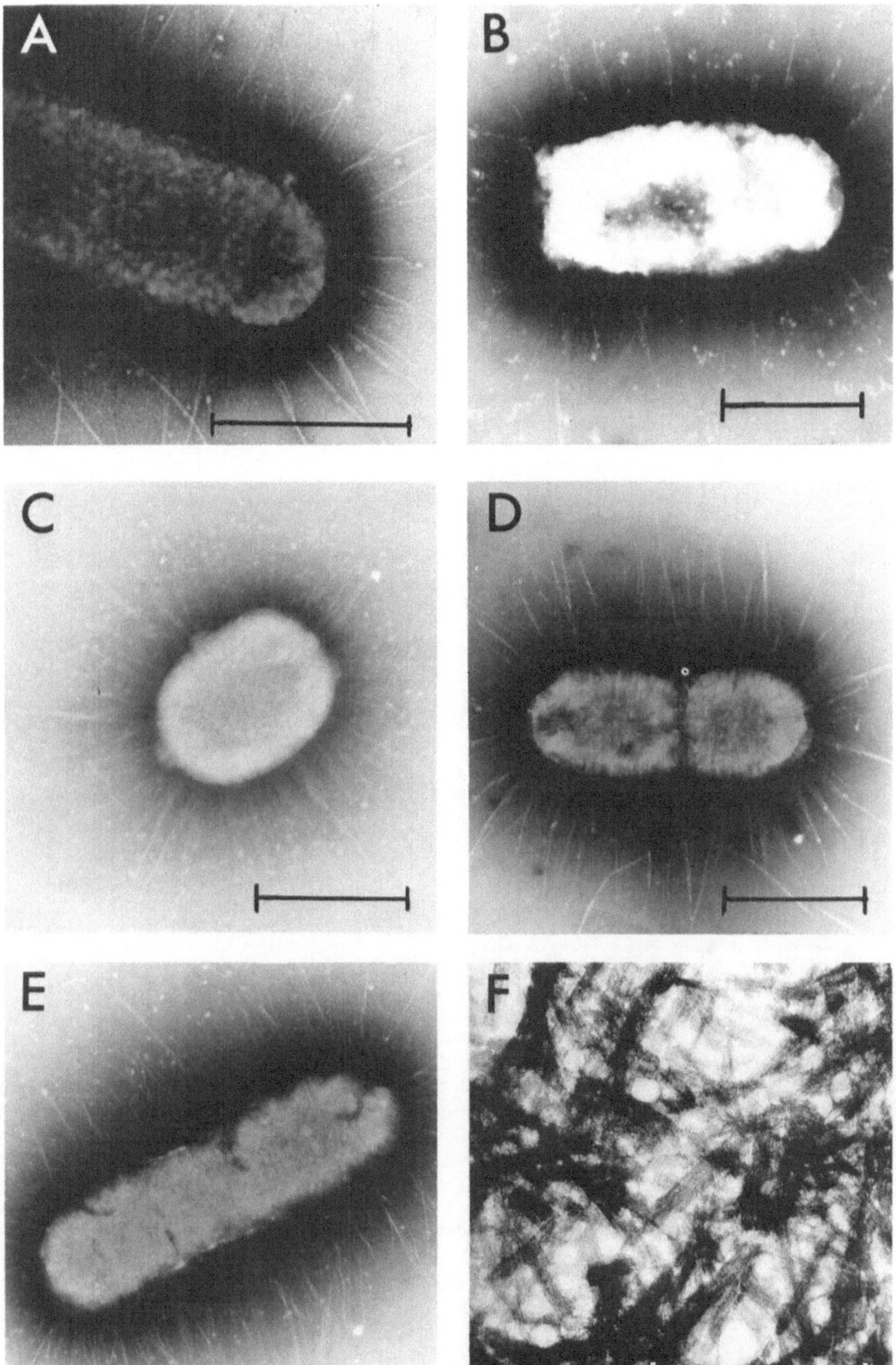

FIGURE 5. Five strains of ETEC negatively stained with 2% phosphotungstic acid (A) 334; (B) 193-4; (C) D563; (D) D481; (E) Tx-1. Bar = 1.0 μm; (F) pili preparation of ETEC strain 193-4 negatively stained with 4% uranyl acetate. Bar = 0.25 μm.

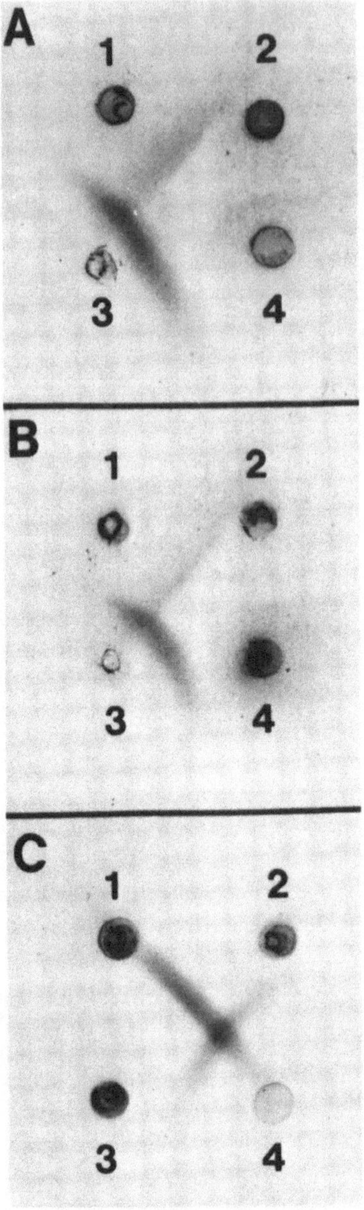

FIGURE 6. Immunodiffusion patterns of pili serogroups 1, 2, and 3. Isolated pili preparations were cross-linked with PDM. (A) Well 1, anti-334 pili; well 2, anti-M9800-5 pili; well 3, M9800-5 pili; well 4, 334 pili. (B) Well 1, anti-D542 pili; well 2, anti-M9800-5 pili; well 3, M9800-5 pili; well 4 D542 pili. (C) Well 1, anti-334 pili; well 2, anti-D542 pili; well 3, D542 pili; well 4, 334 pili.

Table 2
REACTIVITY OF ETEC STRAINS OF HUMAN ORIGIN WITH ANTIPILI SERA[a]

Serogroup	No. of reactive strains (%)
1	12 (11.3)
2	34 (32.1)
3	8 (7.5)
1,3	1 (0.9)
2,3	5 (4.7)
Negative (nonreactive)	46 (43.4)

[a] ETEC strains obtained from throughout the world were titrated with antisera prepared against MR-HA pili of strain 334 (serogroup 1), strain M9800-5 (serogroup 2), and strain D542 (serogroup 3).

Table 3
GEOGRAPHIC SOURCE OF ETEC STRAINS BY PILI SEROTYPE

Pili serotype	Source (no. of isolates)
1	Indian (2), Dacca (2), Pakistan (1), United States (5), unknown (2)
2	U.S. Army (2), foreign (2), United States (6), Kenya (5), Dacca (4), Sweden (2), Ethiopia (2), Honduras (2), Vietnam (1), Mexico (2), unknown (6)
3	Dacca (3), Mexico (2), Honduras (2), Ethiopia (1)
1,3	India (1)
2,3	Texas (2), Dacca (3)
Negative	Pakistan (1), United States (3), Kenya (5), Vietnam (2), Morocco (2), Mexico (6), Ethiopia (4), Honduras (11), cruise ships (2), unknown (10)

Table 4
ENTEROTOXIN PROFILE OF *E.COLI* STRAINS BY PILI SEROGROUP

	No. of strains producing toxin			
Pili serogroup	LT and ST	LT	ST	ND[a]
1	6	3	3	0
2	20	4	8	2
3	4	1	3	0
1,3	1	0	0	0
2,3	3	0	2	0
Negative	11	16	11	8

[a] ND, Toxin profile either not reported or determined by nonstandard methods.

Table 5
SOMATIC O ANTIGENS OF 106 ETEC STRAINS BY PILI SEROGROUP

Pili serogroup	O antigens (no. of strains)
1	015 (8), 078 (1), 0128 (2), ND[a] (1)
2	06 (11), 08 (1), 020 (1), 025 (1), 027 (3), 078 (1), 0148 (2), ND (14)
3	06 (1), 078 (1), 0128 (1), ND (5)
1,3	078 (1)
2,3	078 (2), ND (3)
Negative	06 (1), 015 (1), 025 (4), 027 (2), 078 (5), 0128 (3), 0148 (2), ND (28)

[a] ND, O antigen not determined.

Table 6
MR-HA OF 106 ETEC STRAINS BY PILI SEROTYPE

	MR-HA with erythrocytes[a]			
Pili serotypes	Human A and B	Bovine	Guinea pig	No. of strains
1	+	+	+	4[b]
	+	−	+	2
	−	−	−	6
2	−	+	−	4[b]
	+	−	−	2
	−	−	−	23
	−	−	+	3
	+	+	−	1
3	+	+	−	5[b]
	+	−	−	2
	−	−	−	1
1,3	+	+	−	1
2,3	+	+	−	4
	−	−	+	1
Negative	−	−	−	40
	+	+	−	2
	+	−	+	1
	+	−	−	3

[a] MR-HA was determined at 0°C using 1% mannose and washed erythrocytes.

[b] Includes prototype strain.

Table 7
ADHERENCE ANTIGENS PRODUCED BY ETEC STRAINS

Antigen	Host	Molecular weight	Size width(nm)	MR-HA[a]	Ref.
Serogroup 1	Human	13,100	7—12	HuA ±	
		12,500		HuB ±	11
				GP ±	
				B ±	
Serogroup 2	Human	13,100	7—12	HuA ±	
		12,500		HuB ±	11
				GP ±	
Serogroup 3	Human	13,100	7—12	HuA ±	
		12,500		HuB ±	11
CFA/I	Human	23,800		HuA +	
		(12,000)	8—10	B +	12,13
				C +	
CFA/II	Human	(13,000)	—	B +	5,13
K88	Piglet	23,500	7—13	GP +	14,15
K99	Calf	22,500	7—10	H +	16
	Piglet				
987P	Piglet		7	N.R.	3

[a] Mannose-resistant hemagglutination (MR-HA) of washed human group A (HuA), group B (HuB), guinea pig (GP), bovine (B), horse (H), chicken (C) erythrocytes at 0°C. Non-reactive (N.R.).

REFERENCES

1. **Jones, G. W. and Rutter, J. M.,** The association of K88 with hemagglutinating activity in porcine strains of *Escherichia coli*, *J. Gen. Microbiol.*, 84, 135, 1974.
2. **Ørskov, I., Ørskov, F., Smith, H. W., and Sojka, W. J.,** The establishment of K99, a thermolabile, transmissible *Escherichia coli* K antigen, previously called "K_{co}" possessed by calf and lamb enteropathogenic stains, *Acta Pathol. Microbial Scand. Sect. B.*, 83, 31, 1975.
3. **Isaacson, R. E., Nagy, B., and Moon, H. W.,** Colonization of porcine small intestine by *Escherichia coli* colonization and adhesion factors of pig enteropathogens that lack K88, *J. Infect. Dis.*, 135, 531, 1977.
4. **Evans, D. G., Silver, R. P., Evans, D. J., Jr., Chase, D. G., and Gorbach, S. L.,** Plasmid-controlled colonization factor associated with virulence in *Escherichia coli* enterotoxigenic for humans, *Infect. Immun.*, 12, 656, 1975.
5. **Evans, D. G. and Evans, D. J., Jr.,** New surface-associated heat-labile colonization factor antigen (CFA/II) produced by enterotoxigenic *Escherichia coli* of serogroups 06 and 08, *Infect. Immun.*, 21, 638, 1978.
6. **Gorbach, S. L., Banwell, J. G., Jacobs, B., Chatterjee, B. D., Mitra, R., Brigham, K. L., and Neogy, K. N.,** Intestinal microflora in Asiatic cholera. II. The small bowel, *J. Infect. Dis.*, 121, 38, 1970.
7. **Challacombe, D. N., Richardson, J. M., Rowe, B., and Anderson, C. M.,** Bacterial microflora of the upper gastrointestinal tract in infants with protracted diarrhoea, *Arch. Dis. Child.*, 49, 270, 1974.
8. **Thorne, G. M., Deneke, C. F., and Gorbach, S. L.,** Hemagglutination and adhesiveness of toxigenic *Escherichia coli* isolated from humans, *Infect. Immun.*, 23, 690, 1979.
9. **Cantey, J. R. and Hosterman, D. S.,** Characterization of colonization of the rabbit gastrointestinal tract by *Escherichia coli* RDEC-1, *Infect. Immun.*, 26, 1099, 1979.
10. **Deneke, C. F., Thorne, G. M., and Gorbach, S. L.,** Attachment pili from enterotoxigenic *Escherichia coli* pathogenic for humans, *Infect. Immun.*, 26, 362, 1979.
11. **Deneke, C. F., Thorne, G. M., and Gorbach, S. L.,** Serogroups of attachment pili of enterotoxigenic *Escherichia coli* isolated from humans, *Infect. Immun.*, 32, 1254, 1981.
12. **Evans, D. G., Evans, D. J., Jr., Clegg, S., and Pauley, J. A.,** Purification and characterization of the CFA/I antigen of enterotoxigenic *Escherichia coli*, *Infect. Immun.*, 25, 738, 1979.

13. **Wevers, P., Picken, R., Schmidt, G., Jann, B., Jann, K., Golecki, J. R., and Kist, M.,** Characterization of pili associated with *Escherichia coli* 018ac, *Infect. Immun.*, 29, 685, 1980.
14. **Stirm, S., Ørskov, F., and Birch-Andersen, A.,** Episome-carried surface antigen K88 of *Escherichia coli*. III. Morphology, *J. Bacteriol.*, 93, 740, 1967.
15. **Stirm, S., Ørskov, F., and Mansa, B.,** Episome-carried surface antigen K88 of *Escherichia coli*. II. Isolation and chemical analysis, *J. Bacteriol.*, 93, 731, 1967.
16. **Isaacson, R. E.,** K99 surface antigen of *Escherichia coli:* Purification and partial characterization, *Infect. Immun.*, 15, 272, 1977.

Chapter 10

RELATION OF PILI TO OTHER VIRULENCE FACTORS

Ahmed Faris, Mats Lindahl, and Torkel Wadström

TABLE OF CONTENTS

I. INTRODUCTION

Cholera has become the prototype for a group of intestinal infections called enterotoxic enteropathies.[1] The mechanism of action of cholera toxin has been studied in great detail and encouraged a search for toxins in other intestinal infections (Table 1). After the discovery over 20 years ago that certain culture from *E. coli* strains cause fluid accumulation in rabbit intestinal loops, research has steadily been increasing on enterotoxigenic *E. coli* (ETEC). Smith in a series of publications on ETEC infections in animals established the concept that ETEC organisms produce (1) a heat labile (LT) and (2) heat stable (ST) enterotoxins, as well as (3) hemagglutinating surface antigens such as K88 and K99. In a series of animal experiments in piglets and lambs it became well established that either LT, ST, or both, together with the ETEC surface antigens K88 or K99 were necessary for reproducible diarrheal disease in young animals.[2]

These classical studies also showed that LT, ST, K88, and K99 virulence factors were plasmid mediated. These findings initiated search for ETEC as an etiology of diarrhea in children and adults (e.g., travelers' diarrhea). ETEC strains isolated from humans were found to harbor similar ent-plasmids for LT and ST production. The first study published in 1975 showed that human ETEC strains also possess a surface hemagglutinin claimed to be the adherence factor (adhesin, colonization factor antigen).[3] This observation encouraged epidemiological studies on human ETEC in various parts of the world. Studies on 106 ETEC strains from various countries showed that similar O:H serotypes were found repeatedly (including 06, H16, 08:H9, 015:H10, 025:H42, 078:H11, and 078:H12) in isolates from diarrheal patients.[4]

Surveys in Brazil and Ethiopia and on Cyprus showed that other enterobacteria such as klebsiella and enterobacter species produce enterotoxin(s).[5] A preliminary study on such non-*E. coli* strains have shown that an LT-like enterotoxin activates the cAMP system similar to cholera toxin and *E. coli* LT but no plasmid analysis has yet been reported on such non-*E. coli* ent^+ strains.[6]

II. VIRULENCE FACTORS IN ETEC AND ENTEROPATHOGENIC *E. COLI* (EPEC)

The discovery of human ETEC strains stimulated many studies on possible roles of LT and ST enterotoxins in the so-called enteropathogenic *E. coli* (EPEC) serotypes of *E. coli*. Such strains were reported as the etiology of severe outbreaks of infantile diarrhea already in the 1940s. The negative findings for LT and ST in several recent studies give no clear-cut evidence against the possibility of enterotoxin-causing diarrhea since rapid disappearance of LT and ST toxin production of many ETEC strains were reported and explained by loss of ent-plasmids upon storage of *E. coli* and other enterotoxic enterobacteria (ETB).[7]

Recently Rowe and co-workers reported on a cytotoxin for Vero cells as a possible virulence factor in EPEC of the classical 0111 and 026 serotypes.[8,9] A very interesting French study indicated that EPEC strains may produce a defective LT-like toxin composed of single A subunits.[10] The authors postulate that cell invasive properties for human enterocytes may explain how this toxin is delivered to its intracellular target in gut mucosal cells. Studies on hemagglutination patterns of EPEC strains showed several complex HA patterns which suggest that such strains may produce a number of different surface proteins which act as colonization factors in the small bowel.[11] Studies performed in this laboratory on *E. coli* isolates from diarrheal cases in Ethiopian children, reveal that there are certain ETEC strains which lack colonization factor antigen 1 (CFA/I) are either LT or ST positives. However, these strains possess type 1 pili. Adhesion studies with tissue culture cells 407, indicate that CFA/I^- isolates attach poorly to 407 cells in comparison to their isogenic positive parent

Table 1
VIRULENCE FACTORS IN ENTEROTOXIGENIC (ETEC), ENTEROPATHOGENIC (EPEC), AND ENTEROINVASIVE (EIEC) *E. COLI*

		Pili	Toxins	Tissue invasion
ETEC	Animal strains	K 88, K 99, 987P	LT, ST, HLY	No
	Human strains	CFA/I, CFA/II, ETC	LT, ST	No
EPEC	Animal strains	?	?[a]	?
	Human strains	MS, MR	L (A subunit) cytotoxin[b]	?
EIEC		?	?	Yes[d]

[a] Probably toxins are involved, e.g., edema disease in piglets (Wadström, 1977).
[b] Scheftel et al., *FEMS Microbiol. Lett.*, 9, 125, 1980.
[c] Cytotoxic factor for Vero cells (Scotland et al., 1980). However, these studies showed that 25 EPEC strains out 253 strains produced a toxin (VT) which caused morphological changes in Vero cells.
[d] Certain EIEC strains may possess plasmid-mediated invasive properties (Col V), M Williams, 81.

Table 2
ATTACHMENT OF ENTEROTOXIGENIC *E. COLI* TO HUMAN INTESTINAL CELLS

		Mean number of bacteria attached to 30 cells	
Strain	Pilus antigen	Isolated cells[a]	Intestinal cell line[b]
1628-15	CFA/I$^+$	30	20
1628-14	CFA/I$^-$	5	8
1049a-13	CFA/I$^+$	29	19
1049a-7	CFA/I$^-$	8	6
1058a-2	CFA/I$^+$	28	13
1058a-4	CFA/I$^-$	5	6

[a] Wadström, et al., 5th US-Japan Joint Cholera Conference, Bethesda, Md., 1980, 459.
[b] A human embryonic intestinal cell line (407) available from ATCC.

strains and they have low hydrophobic properties (Tables 2 and 3). We have also failed to detect any linkage between the presence of antibiotic resistance in these strains and colonization factor antigen (CFA/I, II) (Table 4). Similarly, two CFA/II-like ($ST^-LT^-MRHA^+_{bov}$) strains from Dr. Levine (University of Maryland, Baltimore) possess pili though they are toxin negative, but less hydrophobic than the true CFA/II (C922a-1) strain (Table 5).

III. ENTEROTOXINS AND PILI IN *YERSINIA ENTEROCOLITICA*

Several studies in the last few years showed that the 03, 08, and 09 serotypes of *Y. enterocolitica* pathogenic for humans and swine produce an enterotoxin with similar properties as *E. coli* ST.[12] This toxin is produced in low amounts in vitro and its role in disease has been questioned. Recent studies have shown that yersinia ST is under chromosomal control contrary to *E. coli* ST and that cell invasive properties have been suggested to be the main virulence factors in human infections and that they are under plasmid control.[26]

Table 3
RELATIVE SURFACE HYDROPHOBICITY OF SOME SPECIES OF ENTEROBACTERIACEAE AS MEASURED BY "SALTING-OUT" TECHNIQUE

Species[a]	Strain	Pili/HA[b]	HIC	Bacterial aggregation [$(NH_4)_2SO_4$]
Escherichia coli				
(ETEC)	1628-15	CFA/I /MRH	+	0.05 M
	C922a-1	CFA/II/MRB	+	0.09 M
	1593	K 99 /MRHS	+	1.3 M
	Bd3612	K 88 /MRG	+	1.3 M
Shigella	F3 103	Type 1/MSGC	+	0.1 M
flexneri	F5 12	Type 1/MSGC	+	0.3 M
	F2b1			
Salmonella	16-79	Type 1/MSG	(+)	1.5 M
typhimurium				
Klebsiella	75-1063	Type 1/MSGC	+	0.8 M
pnuemoniae	75-1416			
		Type 1/MSGC	+	0.8 M
Yersinia		Type 1/MRH	+	<0.02 M
enterocolitica				

[a] All strains were grown at 37°C except yersinia strain was grown at 20°C.
[b] HA = Hemagglutination; MRH = Mannose resistant HA of human red cells; MRB = Mannose resistant HA of bovine red cells; MSGC = Mannose sensitive HA of guinea pig and chicken erythrocytes.

Table 4
ABSENCE OF CORRELATION WITH ANTIBIOTIC RESISTANCE OF CFA PILI IN ENTEROTOXIGENIC *E. COLI* ISOLATED FROM ETHIOPIAN CHILDREN

	Antibiotic resistance to[b]		
Pili[a]	Ampicillin	Tetracycline	Chloramphenicol
CFA/I^+ (5)	R	R	R
CFA/I^- (5)	R	R	R
CFA/II^+ (5)	I	R	R
CFA/II^- (5)	I	R	R

[a] Five CFA and CFA variants were investigated by agarose gel electrophoresis. Our studies show contrary findings from those of Stieglitz et al., 1980. They showed linkage between ST and ampicillin resistance.
[b] R = Resistant; I = Intermediately resistant.

We have recently studied cell surface properties of *Y. enterocolitica* strains isolated from infections and found that hydrophobic cell surface properties as well as the enterotoxin are produced at a low temperature (<30°C) in certain culture media (Table 3). Recent reports on the autoagglutination properties of such enteropathogenic yersinia strains may well be a similar observation; i.e., that pronounced surface hydrophobicity causes autoagglutination at 35°C. We have used baterial autoagglutination as the principle for a simple test to measure

Table 5
HYDROPHOBIC PROPERTIES OF CFA/II AND CFA/II-LIKE *E. COLI* STRAINS AS DETERMINED BY THE SALT AGGREGATION TEST

Strain	Pili	Enterotoxin	Hemagglutination	Salt aggregation
				[$(NH_4(_2SO_4)$]
C922a-1	CFA/II	ST^+ LT^+	$MRHA_{bov}$	0.09 M
Slenne[a]	CFA-like	ST^- LT^-	$MRHA_{bov}$	0.4 M
Kyriakod[a]	CFA-like	ST^- LT^-	$MRHA_{bov}$	0.4 M

[a] These two strains had stronger hemagglutination reactions than the true CFA/II strains (C922a-1. (Kindly supplied by M. Levine, University of Maryland, Baltimore.)

surface properties, "the salt aggregation test".[13] Preliminary electron microscopy (EM) studies showed that autoaggregating strains of yersinia usually possess pili. Such strains also aggregate at low salt concentrations in the standard salt aggregation test and bind to hydrophobic gels under standard conditions.[30] Further studies with intestinal cells and brush border preparations will reveal if these surface factors are specific cell adhesins and if they are necessary for cell invasion of enterocytes. Preliminary studies show that piliated yersinia strains bind to a human embryonic intestinal cell line.

IV. ENTEROTOXINS AND PILI IN OTHER ENTEROPATHOGENS

A classical EM study on vibrio and aeromonas strains showed that both these organisms can possess surface appendages of pili-like nature.[14] However more recent studies in other laboratories have questioned the possible role of surface pili for intestinal colonization of *V. cholerae* but suggest a possible role of cell surface hemagglutinins such as "cholera lectin" as cell adhesins.[27] A recent Canadian study revealed cell surface pili of the neoenteropathogen *Aeromonas hydrophila* which could be defined as cell surface lectins which can be inhibitable by specific monosaccharides.[15]

Recent studies in our laboratory showed that serial passages of vibrio strains in broth cultures enhance hemagglutination as well as hydrophobic surface properties (Tables 6 and 7). An extensive study on cell surface and extracellular hemagglutinins of *V. cholerae* also showed that these proteins bind to hydrophobic gels such as Phenyl-Sepharose.[28] Such factors may well be important components for future cholera vaccines and should be further investigated.

Extensive studies in several laboratories in recent years on the role of enterotoxins in salmonella infections have not given conclusive evidence for either LT or ST-like enterotoxins despite positive *in vivo* tests in animal intestinal models or cell culture assays such as the CHO cell test.[16] Genetic studies with DNA probes for both LT and ST genes in enterotoxin producing salmonella strains did not support the view that salmonella belonging to the diarrheal diseases to be classified as enterotoxic enteropathies.[26]

Salmonella typhimurium strains reported as a species which produce enterotoxin-like activity was reported long ago to possess surface pili similar to type 1 but with no hemagglutinating (HA) properties and thus named type 2 in the scheme of Old and Duguid.[17]

Serial passages of *Salmonella typhimurium* in broth cultures confirmed that type 2 fimbriae as classical type 1 fimbrie of *E. coli* cause formation of surface pellicles. Such organisms as well as piliated *Shigella flexneri* strains show pronounced hydrophobicity in hydrophobic interaction chromatography (HIC) and in the salt aggregation test (Table 3). Recent studies

Table 6
HEMAGGLUTINATION AND HYDROPHOBIC PROPERTIES OF SOME VIBRIO STRAINS ISOLATED FROM HUMAN DIARRHEA

	Strains	Hemagglutination[a]	HIC	Salt aggregation
Vibrio cholerae	RHV/1191-2B	+ + +	+	[$(NH_4)_2SO_4$]
	W-603-B	+ + +	+	<0.02 M
	VC 454-80	+ + +	+	<0.02 M
	VC 537-80	+ + +	+	<0.02 M
NAG Vibrio	VC 053	+ + + +	+	<0.02 M

[a] All strains showed strong MSG, MSC, and MRH reactions after four passages in nutrient broth.

Table 7
HEMAGGLUTINATION AND HYDROPHOBIC PROPERTIES OF *VIBRIO CHOLERAE* AFTER SERIAL PASSAGES IN BROTH CULTURES

Growth conditions[a]	Hemagglutination[b] Guinea pig	Chicken	Human	HIC	Salt aggregation [$(NH_4)_2SO_4$]
Passage 1	—	—	—	—	>2 M
2	—	—	—	—	>2 M
3	(MSG)	(MSC)	(MRH)	(+)	<0.02 M
4	MSG	MSC	MRH	+	<0.02 M

[a] All strains were grown in Tryptone-water (TW) broth, Nutrient Broth (NB) nr. 2 (Oxoid). The passages were performed by adding 1 drop of 48 hr culture into fresh broth and incubated at 37°C, 48 hr for each passage. Growth in the two media gave similar results.

indicate that *Salmonella typhimurium* may produce a hemagglutinin which may be necessary for invasion of Hela cells.[18]

Serious attempts to express pili on shigella strains of other shigella species other than *S. flexneri* has been unsuccessful. Thus it is possible that cell surface pili are not necessary for these shigella species to cause contact with mucus layer or intestinal enterocytes before a local mucosal invasion. However recent findings indicate that there are complicated metabolic regulations for different pili biosynthesis, e.g., glucose and alanine suppression.[19] It is impossible at present to totally exclude the role of pili or other cell surface proteins as a prerequisite of adhesion for these shigella species to enterocytes before the onset of shigellosis.

V. ROLE OF PILI AND SURFACE CAPSULE IN INTESTINAL COLONIZATION

Studies on cell surfaces of human intestinal anaerobes such as *Bacteroides fragilis* have shown that capsules may be involved in cell adhesion.[20]

However, no systematic study on possible adhesion of intestinal anaerobes to enterocytes or brush border preparations have yet been reported and no EM studies have yet revealed any pili structures on the surface. Preliminary studies showed that *Bacterioides fragilis* strains easily aggregate after cultivation in conventional anaerobic broth culture; an indication of cell surface hydrophobic properties.[29]

Table 8
INFLUENCE OF GROWTH MEDIUM ON THE SURFACE PROPERTIES OF RDEC-1 STRAIN

Medium	Hemagglutination[a]	HIC	Salt aggregation $[(NH_4)_2SO_4]$
Penassay broth[b]	Nil	+	0.6 M
Trypticase soy broth	Nil	+	>2 M

[a] No hemagglutination of bovine, human, chicken, or guinea pig erythrocytes in standard HA-assay.

[b] Bacto, Antibiotic medium 3, Penassay broth and TSB from Difco Labs.

Capsule production as a possible virulence factor involved in cell invasion has been described in *Vibrio parahemolyticus*.[21] Both capsule and pili formation by an *E. coli* rabbit pathogen (strain RDEC-1) have been suggested as important surface factors for intestinal colonization.[22,23]

We have recently confirmed that culture conditions markedly influence cell surface properties of the RDEC-1 strain (Table 8). Adhesion studies with this strain cultured under conditions to enhance or suppress surface hydrophobicity should reveal if pili capsule or both may be necessary for adhesions to the rabbit intestinal mucosa. Recent observations that pili can be expressed on the surface of capsulated *E. coli* as well as on rough strains lacking a complete lipopolysaccharide O-antigen suggest that cell adhesion to enterocytes and other cells may be a complex process.[30]

VI. DISCUSSION AND CONCLUSIONS

Successful recent development of LT toxoid as well as pili vaccines for prevention of ETEC infections in animals have confirmed that enterotoxins as well as pili are key virulence factors in enterotoxic enteropathies caused by ETEC.[24] However, recent observations on selection of ETEC with pili such as the nonhemagglutinating 987 and possibly other surface antigens in porcine ETEC strains after vaccinations suggest that there is a vast variation in the pathogenesis of ETEC and intestinal infections in animals and humans. There is a need, therefore, for new tests for assaying and detecting nonhemagglutinating cell surface factors which may be potential colonization factors (adhesins). Such studies are important for introduction of new vaccines. New vaccines will most probably cause selective pressure for previously undiscovered adhesins and will aim to expose uncharacterized new adhesins which would be incorporated in future vaccines. The pronounced surface hydrophobicity of all virulence factors we know today (K88, K99, 987 p, CFA/I, CFA/II etc.) as well as *E. coli* LT and ST suggest that the hydrophobic principle may be one way of combating infections at the mucosal surface. Further characterization of specific receptors in the mucus layer and on enterocytes may also lead to new preventive methods (''receptor therapy'') as suggested by Keutsch.[25]

ACKNOWLEDGMENTS

This study was supported by grants from the Swedish Medical Research Council (16 × 04723) and the Swedish Board for Technical Development.

REFERENCES

1. **DuPont, H. L. and Pickering, L. K.,** Infection of the gastro-intestinal tract: microbiology, pathophysiology, and clinical features, in *Current Topics in Infectious Diseases*, Greenough, W. B. and Merigan, T. C., Eds., Plenum Press, New York, 1980.
2. **Smith, H. W.,** Neonatal *Escherichia coli* infections in domestic mammals: transmissibility of pathogenic characteristics, in *Ciba Foundation Symposium No. 42, Acute Diarrhea in Childhood*, Elsevier, Amsterdam, 1975.
3. **Evans, D. G., Silver, R. P., Evans, D. J., Jr., Chase, D. G., and Gorbach, S. L.,** Plasmid-controlled colonization factor associated with virulence in *Escherichia coli* enterotoxigenic for humans, *Infect. Immun.*, 12, 656, 1975.
4. **Ørskov, F., Ørskov, I., Evans, D. J., Jr., Sack, R.B., Sack, D. A., and Wadstrom, T.,** Special *Escherichia coli* serotypes among enterotoxigenic strains from diarrhea in adults and children, *Med. Microbiol. Immunol.*, 162, 73, 1976.
5. **Bäck, E., Möllby, R., Kaijser, B., Stintzing, B., Wadström, T., and Habte, D.,** Enterotoxigenic *Escherichia coli* and other gram-negative bacteria of infantile diarrhea: surface antigens, hemagglutinins, colonization factor antigen, and loss of enterotoxigenicity, *J. Infect. Dis.*, 142, 318, 1980.
6. **Dorner, F., Ehn, G., Plot, R., Scheirer, W., and Wsserbauer, E.,** Comparative effects of enterotoxins from diverse strains of *Escherichia coli, Klebsiella pneumoniae, Enterobacter cloacae,* and *Vibrio cholera* in tissue culture assays and in adenylate cyclase systems, *Abstr. XII Int. Congr. Microbiol.* 1978, 17.
7. **Wadström, T.,** Relative importance of enterotoxigenic and invasive enteropathogenic bacteria in infantile diarrhea, *Zbl. Bakt. Hyg. I Abt. Orig. A*, 242, 52, 1978.
8. **Wade, W. G., Thom, B. T., and Evans, N.,** Cytotoxic enteropathogenic *Escherichia coli, Lancet*, 1, 1235, 1979.
9. **Scotland, S., Day, N. P., and Rowe, B.,** Production of a cytotoxin affecting Vero cells by strains of *Escherichia coli* belonging to traditional enteropathogenic serogroups, *FEMS Microbiol. Lett.*, 7, 15, 1980.
10. **Scheftel, J. M., Martin, C., Bober, C., and Monteil, H.,** Isolation of an enterotoxic factor elaborated by human enteropathogenic *Escherichia coli, FEMS Microbiol. Lett.*, 9, 125, 1980.
11. **Evans, D. J., Evans, D. J., Jr., Young, L. S., and Pitt, J.,** Hemagglutination typing of *Escherichia coli:* definition of seven hemagglutination types, *J. Clin. Microbiol.*, 12, 235, 1980.
12. **Olsson, E., Krovacek, K., Hurvell, B., and Wadström, T.,** Enterotoxigenic of porcine and human isolates of *Yersinia enterocolitica* in Sweden, *Curr. Microbiol.*, 3, 267, 1980.
13. **Lindahl, M., Faris, A., Wadstrom, T., and Hjerten, S.,** A new test based on "salting out" to measure relative surface hydrophobicity of bacterial cells, *Biochim. Biophys. Acta*, 677, 471, 1981.
14. **Tweedy, J. M., Park, R. W. A., and Hodgkiss, W.,** Evidence for the presence of fimbriae on vibrio species, *J. Gen. Microbiol.*, 51, 235, 1968.
15. **Atkinson, H. M. and Trust, T. J.,** Haemagglutination and adherence ability of *Aeromonas hydrophila, Infect. Immun.*, 27, 938, 1980.
16. **Peterson, J. W.,** *Salmonella* toxin, *Pharmacol. Ther.*, 11, 719, 1980.
17. **Duguid, J. P. and Old, D. C.,** Adhesive properties of enterobacteriaceae, in *Bacterial Adherence*, Vol. 6, Chapman and Hall, London, 1980.
18. **Jones, G. W., Richardson, L. A., and vanden-Bosch, J. L.,** *Microbial Adhesion to Surfaces*, Berkeley, R. C. W., Lynch, J. M., Melling, J., Rutter, P. R., and Vincent, B., Eds., Elois Horwood Ltd. Publisher, Chichester, England, 1980, 211.
19. **deGraaf, F., Wientjes, K., and Klaasen-Boor, P.,** Production of K99 antigens by enterotoxigenic *Escherichia coli* of serogroups 08, 09, 020, and 0101 grown at different conditions, *Infect. Immun.*, 27, 216, 1980.
20. **Onderdonk, A. B., Moon, N. E., Kasper, D. L., and Bartlett, J. G.,** Adherence of *Bacteroides fragilis* in vivo, *Infect. Immun.*, 19, 1083, 1978.
21. **Carruthers, M. M. and Anderson, B.,** Inhibition by polyanions of adherence by Kanagawa-positive *Vibrio parahaemolyticus:* a physicochemical effect, *J. Infect. Dis.*, 140, 119, 1979.
22. **Boedeker, E. C., Cheney, C. P., Schad, P. A., and Formal, S. B.,** Hydrophobic bacterial properties correlates with the adherence of enteropathogens to intestinal brush border membranes (BBM), *Gastroenterology*, 76, 1105, 1979.
23. **Cantey, J. R., Lushbaugh, W. S., and Inman, L. R.,** Attachment of bacteria to intestinal epithelial cells in diarrhea caused by *Escherichia coli* strain RDEC-1 in the rabbit: stages and role of capsule, *J. Infect. Dis.*, 143, 219, 1981.
24. **Moon, H. W.,** Protection against enteric colibacillosis in pigs suckling orally vaccinated dams: evidence for pili as protective antigens, *Am. J. Vet. Res.*, 42, 173, 1981.
25. **Keusch, G. T.,** Specific membrane receptors: pathogenetic and therapeutic implications in infectious diseases, *Rev. Infec. Dis.*, 1, 517, 1979.
26. **Falkow, S.,** personal communication.

27. **Finkelstein, R.,** personal communication.
28. **Finkelstein, R. and Clements, X.,** in press.
29. **Ryden, A. D.,** personal communication.
30. **Wadstrom, T., Faris, A., Lindahl, M., and Hjerten, S.,** Hydrophobic properties of pili, in *Attachment of Organisms to the Gut Mucosa,* Vol. I, Boedeker, E. C., Ed., CRC Press, Boca Raton, Fla., 1984.

Chapter 11

PILI AS ADHERENCE FACTORS IN *ESCHERICHIA COLI* STRAIN RDEC-1

Edgar C. Boedeker and Christopher P. Cheney

TABLE OF CONTENTS

I. INTRODUCTION

The diarrheal disease manifestations, characteristics of gut colonization, immunology, and pathogenesis of gastrointestinal infection of the rabbit with the enteroadherent *E. coli* strain RDEC-1 have been described by Cantey in Chapter 5 of this volume.[1] Briefly, within 6 to 10 days of oral inoculation of postweaning rabbits with as few as 10^2 organisms, RDEC-1 *E. coli* are found closely adherent to the apical membranes of intestinal epithelial cells of the ileum and cecum.[2,3] These cells are not invaded but there is the close (11 nm) apposition of the bacteria to pedestal-like extensions of the apical cell membranes which are lacking characteristic microvillar architecture.[4] This is a late stage of adherence which is observed at the time of clinically observed diarrhea. Earlier stages in this associative process, in which RDEC-1 organisms interact with the apical portions of intestinal epithelial cells without disrupting their apical cytoskeleton have also been described during in vivo infection by Cantey.[5]

In order to study the determinants, on both the RDEC-1 organisms and on the intestinal surface, which are responsible for the mucosal attachment phenomena observed in vivo, our approach has been to establish a reproducible and quantifiable assay for in vitro attachment.[6] We have sought to provide evidence, chiefly from considerations of corresponding species, age, and tissue specificity for the two phenomena, that the in vitro interaction is a reflection of the in vivo process.[7] Elsewhere in this volume, the age-related appearance of host receptors for RDEC-1 organisms on intestinal epithelial cell brush border membranes of rabbits is discussed.[8] The current chapter describes our approach to the definition of the bacterial determinants of the adherence process.

II. METHODS OF STUDY

Some of our earliest experiments indicated that a rapid and vigorous interaction takes place in vitro between isolated rabbit intestinal brush border (microvillus) membranes and RDEC-1 organisms (Figure 1). When organisms grown overnight in Penassay broth were added to freshly prepared suspensions of rabbit ileal brush borders similar agglutination of intact brush borders and RDEC-1 took place. Rapid, visually apparent aggregation took place within minutes at 37°C. When these aggregates were observed under phase microscopy, large clumps of coagglutinated bacteria and BBMS were seen (Figure 2A), but quantitation of this process was difficult. Brief vortexing of the aggregates proved sufficient to disrupt the large clumps (Figure 2B) without eliminating the attachment of bacteria to individual brush borders (BBs) so that the number of bacteria per BB could be enumerated. For our standard adherence assay we chose 40 brush borders as a convenient number to be quantitated and Figure 3 shows the distribution of adherent *E. coli* on rabbit brush borders. The percentage of brush borders with a given range of adhering bacteria is plotted against the number of adherent bacteria per brush border (from 0 to 34). The results for RDEC-1 and two control *E. coli* are compared. H.S. is a nonpathogenic *E. coli* strain isolated from human stool. 640 is an *E. coli* strain isolated from the stools of a healthy rabbit. Both strains 640 and H.S. express type 1 pili. More than half of the brush borders incubated with strain 640 have no adherent organisms. With H.S., 95% of the brush borders had fewer than nine adherent organisms. In contrast, over half of the brush borders incubated with RDEC-1 had greater than 10 adherent organisms. Using this basic assay, we were able to define the following characteristics of RDEC-1 adherence to rabbit intestinal brush borders:[6] (1) The reaction is rapid and goes to completion in 15 min at 37°C. (2) The reaction is temperature sensitive and its rate is measurably inhibited at 4°C. (3) The optimum pH for interaction is between 6.5 and 7.0. (4) The reaction is not influenced by the addition of mannose or other membrane sugars in concentrations as high as 20 mg/mℓ (125 m*M*). (5) Within the rabbit intestine,

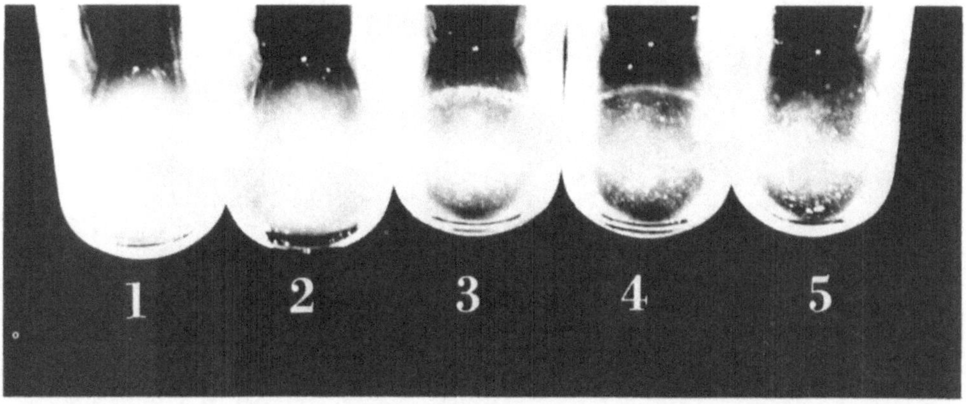

FIGURE 1. The coagglutination of RDEC-1 organisms and rabbit ileal brush border (microvillus) membranes is shown. Each tube contains 10^8 RDEC-1 organisms grown overnight in Penassay broth and suspended in phosphate buffered saline. Decreasing amounts of brush border membrane suspensions, expressed as μg membrane protein, were added to tubes 1 to 5. (Tube 1, 200 μg; tube 2, 100 μg; tube 3, 50 μg; tube 4 20 μg; tube 5, 10 μg). Incubations for 15 min at 37°C.

there is a regional specificity for the ileum with maximal adherence to jejunal brush borders only half that for ileum. (This finding correlates with the distal colonization of rabbit intestine observed in vivo by Cantey for RDEC-1.[2,3]

To further establish the relation of the in vitro adherence phenomenon to the in vivo process, we compared in vitro brush border adherence with in vivo colonization and infectivity in three species.[7] Table 1 compares the adherence of RDEC-1 to brush borders from the rabbit, rat, guinea pig, and human small intestine. No significant adherence of RDEC-1 to any of the brush borders from species other than the rabbit occurred. Although infectivity and colonization studies were not done in humans, neither the rat nor guinea pig could be infected or colonized with RDEC-1 following intragastric inoculation.[7]

The species specificity of RDEC-1 adherence, as well as the mucosal localization of adherence, was shown by another method as well. Figure 4 shows the results of overlaying RDEC-1 *E. coli* on 2 μm cryostat sections of ileal mucosa from the rabbit, as well as from the rat, guinea pig, and human intestine. After incubation at 37°C, the sections were washed and stained with fluorescein using a double-antibody technique employing anti-RDEC-1 serum. The outlines of the intestinal villi can be appreciated. RDEC-1 is localized specifically at the mucosal surface of the rabbit ileum, whereas it is randomly and sparsely distributed over the lamina propria of the other three species.[7]

III. EVIDENCE FOR A ROLE FOR PILI (AF/R1) IN RDEC-1 ADHERENCE

All of the experiments described thus far utilized cultures of RDEC-1 grown overnight in Penassay broth. Figure 5 shows the typical electron microscopic appearance, as demonstrated by platinum shadow casting of the majority of organisms grown in this way. The majority of the organisms are heavily piliated.

The remaining experiments to be described were designed to determine whether these pili determine the adherence of RDEC-1 to rabbit intestinal mucosa. The following experimental approaches were utilized to determine the importance of pili to RDEC-1 adherence: (A) Suppression of pilus expression was achieved.[9] (B) RDEC-1 (AF/R1) pili were genetically transferred to other organisms.[9] (C) Isolated RDEC-1 (AF/R1) pili were prepared and their adherence to intestinal mucosa was determined.[10]

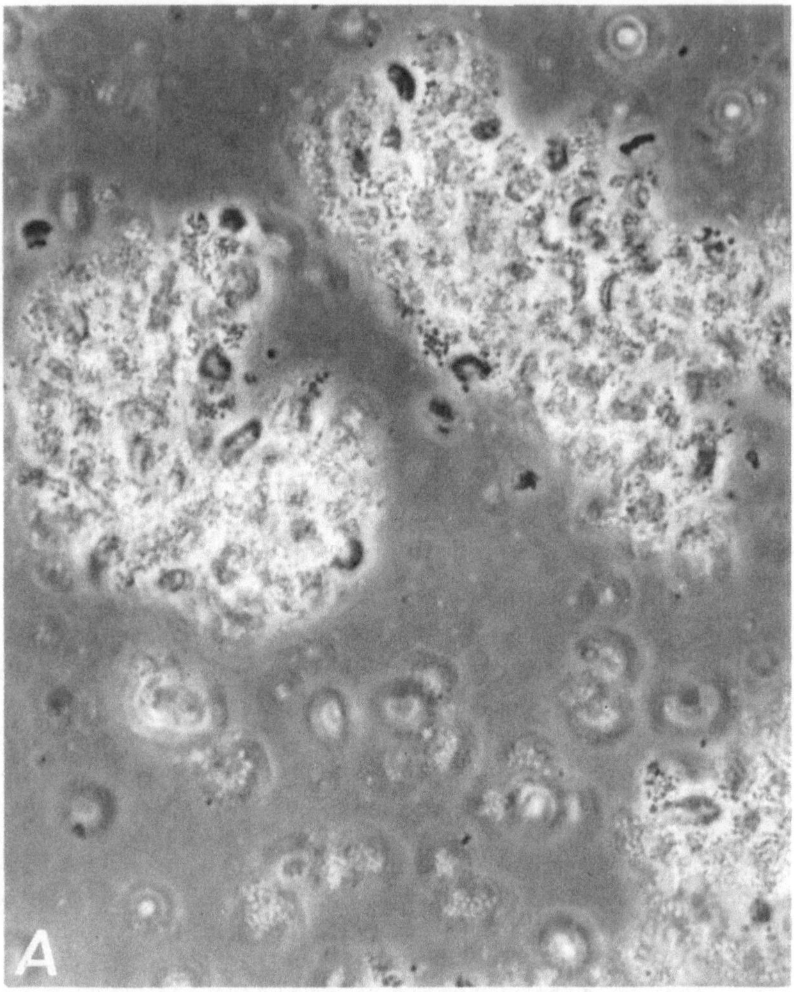

FIGURE 2. Phase contrast photo micrographs (magnification × 600) of the interaction of RDEC-1 with rabbit ileal brush borders. 10^8 bacteria were incubated with 10^6 brush borders in 100 μℓ phosphate buffered saline for 30 min at 23°C. (A) Coaggregates of RDEC-1 and brush borders. (B) Vortexed dispersion of coaggregates. Organisms can be seen adhering to individual brush borders.

A. Suppression of Pilus Expression

The first two lines of Table 2 indicate the difference in RDEC-1 adherence which was observed when the organisms were grown overnight in Brain Heart Infusion (BHI) medium as opposed to the standard Penassay Broth (PAB). The observation was fortuitous in that the relatively more enriched Brain Heart Infusion medium was utilized in an attempt to increase the yield of organisms in overnight culture. Increased yield was obtained; however, it was consistently found that organisms grown in Brain Heart Infusion medium did not adhere to rabbit brush borders in vitro. Systematic electron microscopic evaluation of the BHI and PAB grown organisms revealed a striking absence of pili in the BHI organisms. The suppression of pilus expression under these conditions was not absolute but it was nearly so. As indicated in Table 2, only a few percent (+/−) of the organisms grown overnight on BHI were piliated as compared to nearly 100% (+/+) of the PAB grown organisms. Brush border adherence was also quantitated for each of the strains under different growth conditions and there is an apparent correlation of pilus expression and mucosal adherence.[9]

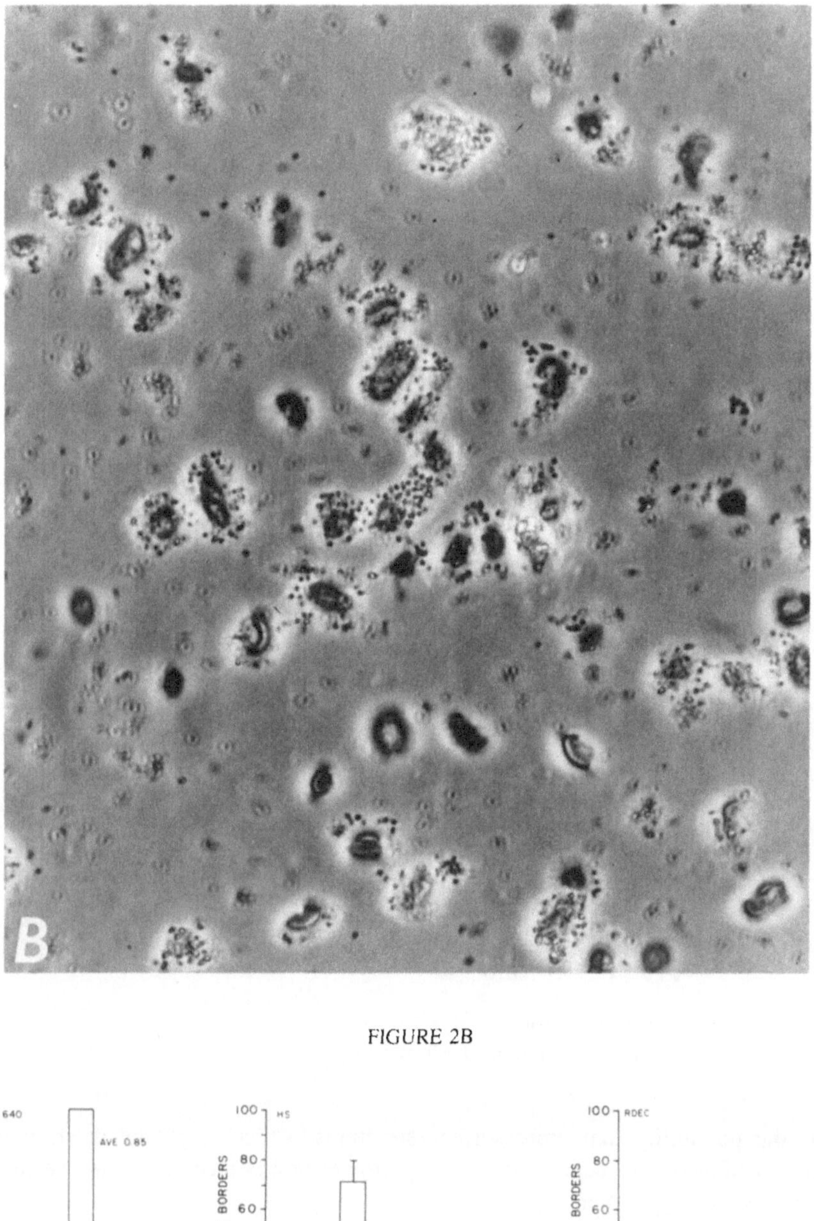

FIGURE 2B

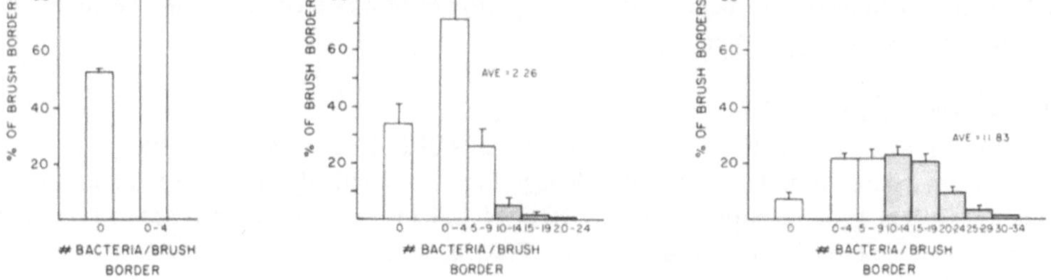

FIGURE 3. Distribution of adherent bacteria on rabbit ileal brush borders for three *E. coli* strains: 640, left panel; HS, center panel; RDEC-1 right panel (see text for descriptions of the strains). Following incubation of 10^8 bacteria with 10^6 brush borders the number of bacteria adhering to each of 40 brush borders was determined. The ranges of bacteria/brush border are shown on the abscissa. The percentage of brush borders with a given range of adherent organisms is plotted on the ordinate. Stippled bars indicate brush borders with 10 or more adherent organisms.

Table 1
SPECIES SPECIFICITY OF IN VITRO BRUSH BORDER ADHERENCE AND IN VIVO ILEAL COLONIZATION BY RDEC-1

Species	No. of adherent organisms[b] per ileal brush border	Mean $\log_{10}$ RDEC-1 *E. coli*[c] growth per gram ileal tissue
Rat	<1	0
Guinea pig	<1	1.4
Rabbit	10—12	8.6
Human	<1	N.D.[a]

[a] N.D. = not done.
[b] No. of adherent organisms determined by standard adherence assay (see text).
[c] MaConkey agar plates containing nalidixic acid were used to quantitate *E. coli* growth following oral inoculation of 10^9 nalidixic acid resistant RDEC-1 organisms.

Table 2
EFFECTS OF GROWTH MEDIUM ON BACTERIAL PILIATION AND ADHERENCE

Organism	Growth medium	Piliation	Brush border adherence
RDEC-1	PAB[a]	+ + +	Yes
RDEC-1	BHI[b]	+/−	No
M4243	PAB	−	No
M4243	BHI	−	No
D-1-5	PAB	+ + +	Yes
D-1-5	BHI	+/−	No
D-1-2	PAB	−	No
D-1-2	BHI	−	No

[a] Penassay broth.
[b] Brain Heart Infusion medium.

However, the possibility that other surface structures, which might be determinants of adherence, could also have been suppressed by growth in this medium cannot be ruled out by this single set of experiments.

B. Genetic Transfer of Pili

For this reason, an attempt was made to transfer RDEC-1 adherence properties to another class of organisms, in this case the avirulent *flexneri 2a* strain M4243.[9] Based on an initial assumption that genetic information for RDEC-1 pili resided on the chromosome, a high frequency recombinant form of adherent RDEC-1 was prepared and conjugal mating with a nonadherent shigella was attempted. Appropriate markers were selected to eliminate parental strains and daughter strains were selected and screened for adherence. An adherent daughter strain (D-1-5) and a nonadherent daughter strain (D-1-2) were selected for further study. Table 2 indicates the adherence to rabbit brush borders of the two daughter strains. Both D-1-2 and the shigella parent M4243 failed to adhere to brush borders. D-1-5 had an adherence pattern similar to that of the parent RDEC-1. Electron microscopic examination of the adherent and nonadherent strains revealed the presence of pili on D-1-5 but not on

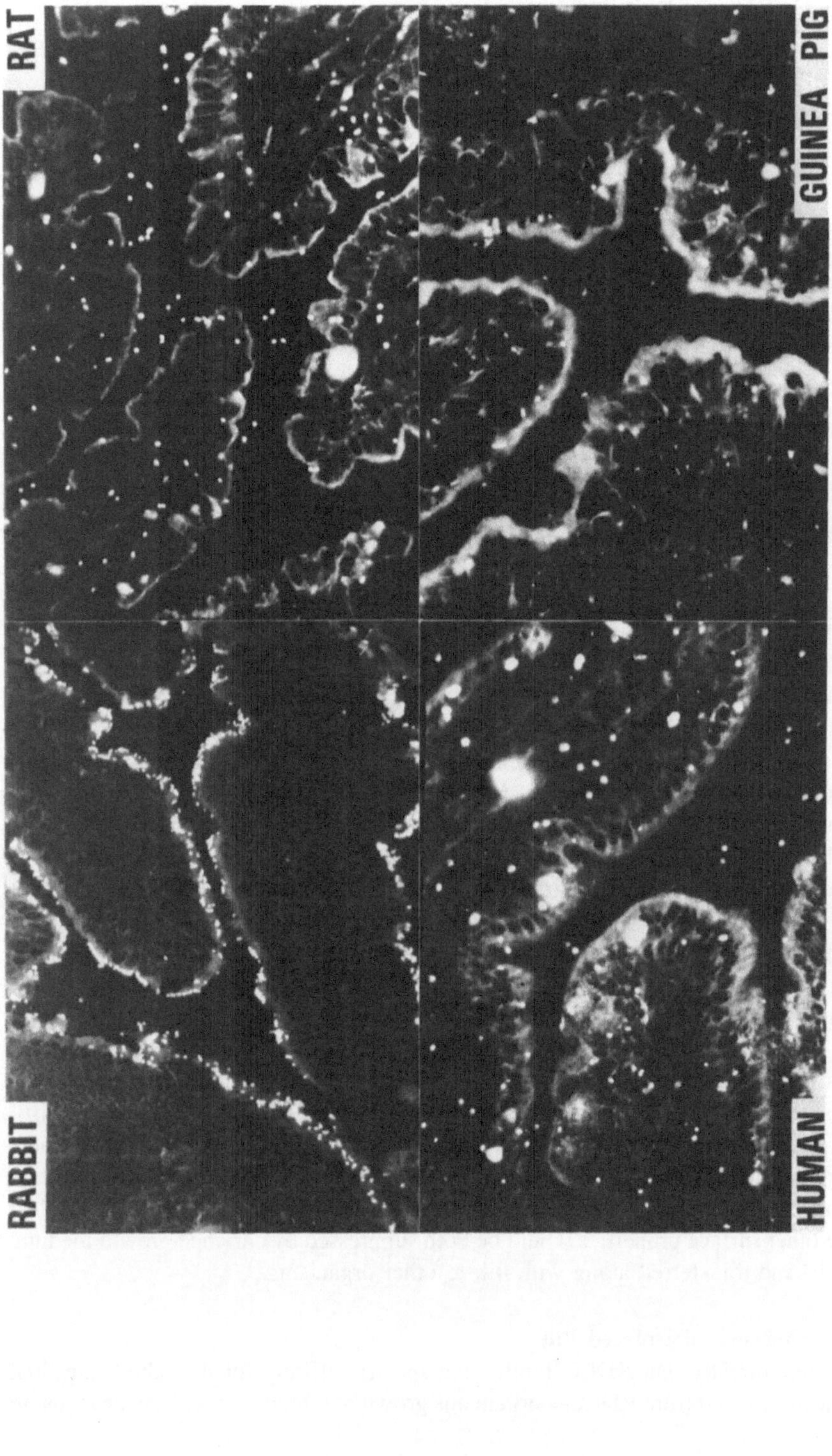

FIGURE 4. In vitro immunofluorescent localization of RDEC-1 organisms adhering specifically to the mucosal surface of a frozen section of rabbit ileum, but nonspecifically distributed over frozen sections of rat, guinea pig, and human ileal tissue.

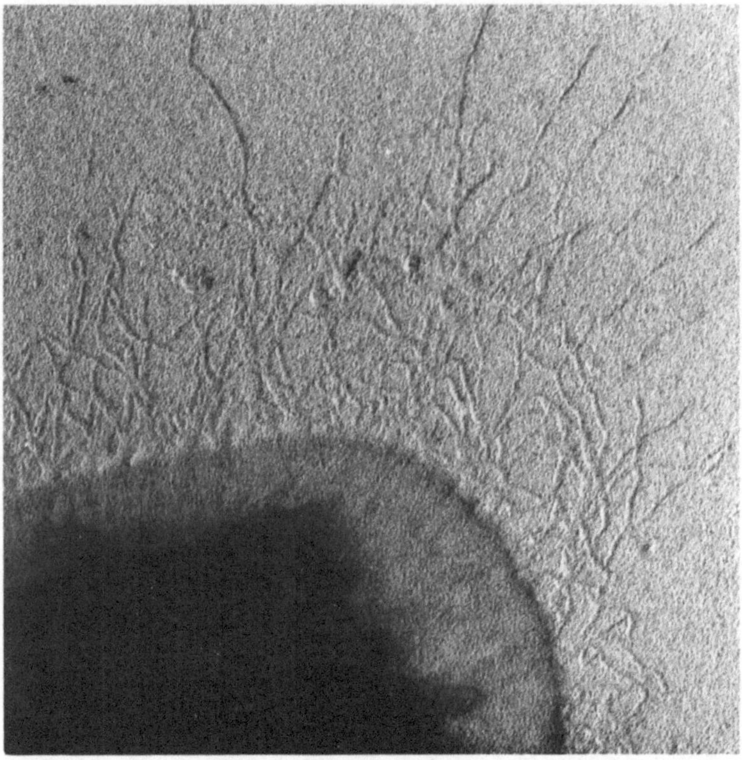

FIGURE 5. Platinum shadow caset electron micrograph of the surface of a piliated RDEC-1 organism from overnight growth in Penassay broth (Magnification × 120,000).

D-1-2. Moreover, the daughter strains responded to growth in BHI medium in a manner similar to the parent RDEC-1 strain with respect to both adherence and piliation. These results are summarized in Table 2. The parent RDEC-1 is adherent and piliated when grown in PAB but not in BHI. The other parent M4243 is neither piliated or adherent in either growth medium. D-1-5 is piliated and adherent to brush borders when grown in PAB but not BHI. D-1-2 is neither piliated nor adherent.

These experiments suggest that the transfer of RDEC-1 pili correlated with the transfer of adherence properties. Although the experimental design was intended to promote the transfer of chromosomal DNA, (since RDEC-1 had previously been reported to be capable of expressing type 1 pili,[11] and the genetic locus for type 1 pili is chromosomal), the transfer of pili and adherence, in fact, correlated with the transfer of a specific plasmid from RDEC-1 to M4243 (Figure 6). Moreover, type 1 pili did not seem to be involved since the hemagglutination patterns of piliated, adherent RDEC-1 organisms did not demonstrate mannose sensitive hemagglutination of guinea pig erythrocytes (a hallmark of type 1 pili). Furthermore, organisms such as 640 or H.S., known to possess type 1 pili, did not adhere to rabbit brush borders (Figure 3). Again, although other surface properties might have been transferred together with the genetic information for specific RDEC-1 pili, it is much less likely that the other surface properties would be both suppressed by enriched growth medium along with pili, and transferred along with pili to other organisms.

C. Mucosal Adherence of Isolated Pili

To demonstrate directly that RDEC-1 pili have specific affinity for the rabbit intestinal mucosa, pili were isolated from RDEC-1 organisms grown overnight in Penassay broth using

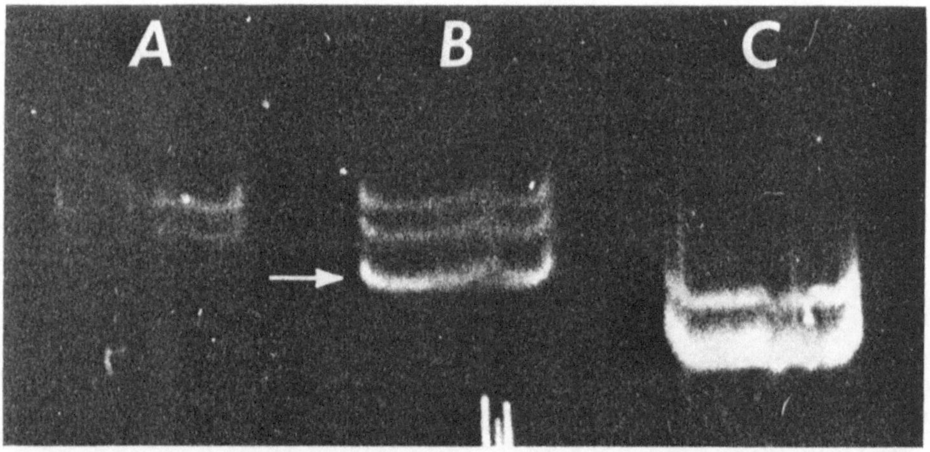

FIGURE 6. Agarose gel electrophoresis of plasmid DNA from: (A) the nonpiliated, nonadherent, shigella recipient strain M4243; (B) the piliated, adherent shigella exconjugate strain, D-1-5; (C) RDEC-1. The arrow indicates a plasmid in D-1-5, not present in the M4243 parent, but apparently obtained from RDEC-1.

a modification of Isaacson's method.[12] The pilus preparation is shown in Figure 7. These pili migrated as a single band of molecular weight approximately 18,000 following their dissociation into subunits by heating in the presence of SDS and betamercaptoethanol at 100°C for 3 min. This pilus preparation was then used to prepare antipilus antisera for use in immunofluorescent localization. Pili were suspended and overlain over frozen sections of rabbit. rat, guinea pig, and human intestine, in a manner exactly analogous to that illustrated in Figure 4 for the whole RDEC-1 organisms. Figure 8 shows that pili distribute specifically at the mucosal surface of the rabbit ileum as shown by a pattern of linear fluorescence. This did not occur with the rat, guinea pig, or human mucosa. Thus, isolated RDEC-1 pili adhere with the same distribution and species specificity as whole organisms to the intestinal mucosa of the rabbit, indicating that specific mucosal adhesive properties reside in these pili.[10]

IV. SUMMARY

In summary, we have provided evidence that pili are adherence factors for RDEC-1 *E. coli*, since (1) suppression of pilus expression in BHI medium also results in loss of RDEC-1 adherence to rabbit brush border membranes; (2) transfer of RDEC-1 pili to M4243 *S. flexneri* correlated with the transfer of brush border adherence; (3) isolated RDEC-1 pili adhere to the intestinal mucosa in a species specific manner with a distribution paralleling that of the whole RDEC-1 organisms. Thus, we feel that RDEC-1 represents another example of an *E. coli* enteropathogen whose virulence depends on mucosal attachment to the small intestine of its host, a process which is mediated by specific pili. In this respect, RDEC-1 seems analogous to the K88 and K99 strains of *E. coli* discussed in Chapter 3[13] and Chapter 4[14] of this volume.

V. DISCUSSION

In Chapter 20 of this volume,[15] Cantey provides elegant evidence using high voltage electron microscopy, ruthenium red, and specific antibody staining, that the bacterial surface polysaccharide is expressed in vivo during the attachment process.[5] Similar direct evidence for the presence of pili during in vivo attachment is lacking, i.e., it has not yet been technically possible to document the expression of pili in vivo at the time of mucosal attachment,

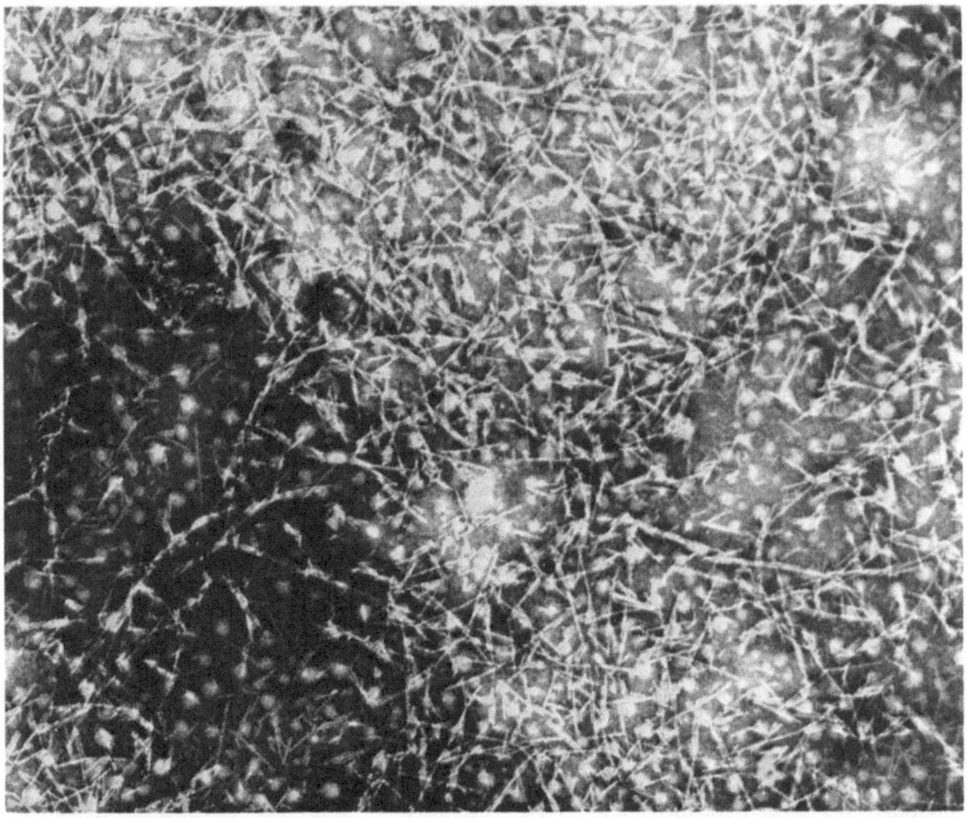

FIGURE 7. Electron micrograph examination (Magnification × 100,000) of a negatively stained preparation of purified RDEC-1 (AF/R1) pili.

although Cantey, et al.[5] did see some negatively stained pilus-like structures. However, the presence of the bacterial surface polysaccharide during the adherence process does not represent firm evidence that this structure is responsible for adherence. In a strict sense it only demonstrates that the surface polysaccharide does not prevent adherence. Conceivably, surface polysaccharide and pili could each be responsible for separate stages (early and late) seen in the in vivo studies of Cantey. We favor the hypothesis, based on our in vitro studies, that pili are responsible for the early stage association between RDEC-1 and the mucosal surface and may also be responsible for the characteristic close late stage of association seen between RDEC-1 and the apical plasma membrane of rabbit intestinal epithelial cells.

Cantey and Inman[16,17] have also provided very recent evidence that the earliest documentable association between RDEC-1 organisms and the intestine, at a stage prior to the development of diarrhea, is with the specialized epithelial M cells overlying the lymphoid follicles of the Peyer's patch. This association can be seen as early as 2 days postinoculation and these authors postulate that this association may be required for the subsequent association with, or attachment to, the absorptive epithelial cell. The surface of absorptive epithelial cells may be less accessible to luminal organisms, because of well-developed mucus and glycocalyx layers, than are the specialized M cells. The M cell appears to readily sample luminal antigens and usually appears to be relatively free of mucus and to have only a rudimentary glycocalyx on scanning electron microscopic studies. The interaction of RDEC-1 with the absorptive epithelial cell surface may be independent of previous interaction of RDEC-1 organisms with the Peyer's patch epithelium. Certainly our in vitro experiments

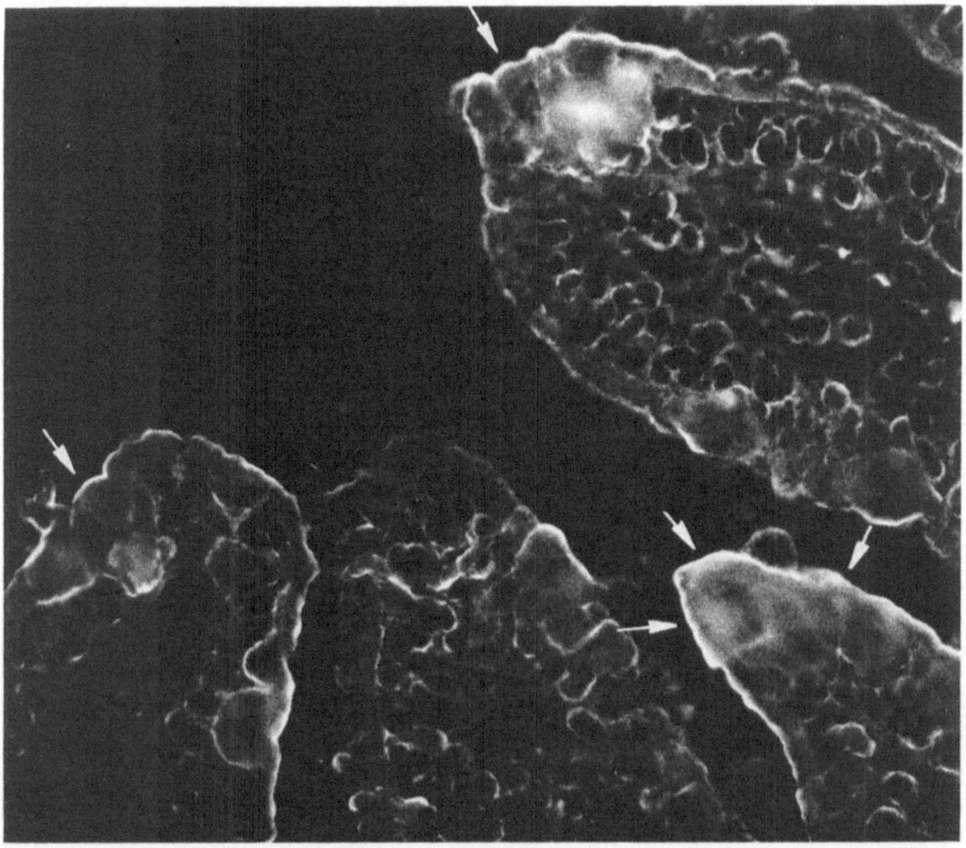

FIGURE 8. Photomicrograph of a frozen section of rabbit ileum demonstrating the tips of several villi. The preparation has been overlain with RDEC-1 (AF/R1) pili and stained with anti-RDEC-1 (AF/R1) pilus serum followed by fluorescein conjugated anti-immunoglobulin. Arrows indicate areas of linear fluorescence indicating RDEC-1 (AF/R1) pilus binding at the mucosal surface.

suggest that piliated RDEC-1 organisms, grown in vitro and never exposed to the host immune system, are capable of interacting in a highly specific way with the absorptive epithelial cell surface. This capacity for mucosal adherence of RDEC-1 appears, on the basis of our in vitro evidence, to be conferred by specific pili, distinct from type 1. We have designated these pili AF/R1 since they are the first pilus structures demonstrated to serve as specific adherence factors (AF) for rabbit (R) intestine.[10] Whether the bacterial determinants for association of RDEC-1 with the Peyer's patch epithelium are the same as those for the absorptive epithelium, and the role of AF/R1 in initiating a mucosal immune response, remain to be determined in future studies.

REFERENCES

1. **Cantey, J. R.,** The rabbit model of *Escherichia coli* (strain RDEC-1) diarrhea, in *Attachment of Microorganisms to the Gut Mucosa, Vol. 1,* Boedeker, E. C., Ed., CRC Press, Boca Raton, Fla., 1984, chap.5.
2. **Cantey, J. R. and Blake, R. K.,** Diarrhea due to *Escherichia coli* in the rabbit: a novel mechanism, *J. Infect. Dis.,* 135, 454, 1977.
3. **Cantey, J. R. and Hosterman, D. S.,** Characterization of colonization of the rabbit gastrointestinal tract by *Escherichia coli* RDEC-1, *Infect. Immun.,* 26, 1099, 1979.

4. **Takeuchi, A. K., Inman, L. R., O'Hanley, P. D., Cantey, J. R., and Lushbaugh, W. B.,** Scanning and transmission electron microscopic study of *Escherichia coli* 015 (RDEC-1) enteric infection in rabbits, *Infect. Immun.*, 19, 686, 1978.
5. **Cantey, J. R., Lushbaugh, W. B., and Inman, L. R.,** Attachment of bacteria to intestinal epithelial cells in diarrhea caused by *Escherichia coli* strain RDEC-1 in the rabbit: stages and role of the capsule, *J. Infect. Dis.*, 143, 219, 1981.
6. **Cheney, C. P., Boedeker, E. C., and Formal, S. B.,** Quantitation of the adherence of an enteropathogenic *Escherichia coli* to isolated rabbit intestinal brush borders, *Infect. Immun.*, 28, 1019, 1980.
7. **Cheney, C. P., Schad, P. A., Formal, S. B., and Boedeker, E. C.,** Species specificity of *in vitro* Escherichia coli adherence to host intestinal cell membranes and its correlation with *in vitro* colonization and infectivity, *Infect. Immun.*, 28, 1019, 1980.
8. **Cheney, C. P. and Boedeker, E. C.,** Appearance of host intestinal receptors for pathogenic *E. coli* with age, in *Attachment of Organisms to the Gut Mucosa*, Vol. 2, Boedeker, E. C., Ed., CRC Press, Boca Raton, Fla., 1984, chap. 16.
9. **Cheney, C. P., Schad, P. A., Formal, S. B., and Boedeker, E. C.,** The genetic transfer of mucosal adherence factors from the *E. coli* strain RDEC-1 into a *Shigella flexneri* strain, *J. Infect. Dis.*, 1983, in press.
10. **Berendson, R., Cheney, C. P., Schad, P. A., and Boedeker, E. C.,** Isolated *E. coli* pili adhere specificity to the intestinal mucosal surface: evidence that pili are an *E. coli* adherence factor, *Gastroenterology*, 1983, in press.
11. **Fusco, P., To, A. C.-C, To, S. C. Om, and Brinton, C. C., Jr.,** The purification and characterization of four types of *E. coli* pili and the specificity of *E. coli* pili for immunity, colonization and adhesion, XIIIth Joint Conference on Cholera DHEW Publication No (NIH) 78-1590, 30, 1978.
12. **Isaacson, R. E.,** K99 surface antigen of *Escherichia coli:* purification and partial characterization, *Infect. Immun.*, 15, 272, 1977.
13. **Sellwood, R.,** The K88 adherence system in swine in *Attachment of Organisms to the Gut Mucosa*, Vol. 1, Boedeker, E. C., Ed., CRC Press, Boca Raton, Fla., 1984, chap. 3.
14. **Moon, H. W. and Runnels, P. L.,** The K99 adherence system in cattle, in *Attachment of Organisms to the Gut Mucosa*, Vol. 1, Boedeker, E. C., Ed., CRC Press, Boca Raton, Fla., 1984, chap. 4.
15. **Cantey, J. R. and Inman, L. R.,** The role of negatively charged surface polysaccharide in adherence of *Escherichia coli* strain RDEC-1 to gut epithelial cells in the rabbit, in *Attachment of Organisms to the Gut Mucosa*, Vol. 1, Boedeker, E. C., Ed., CRC Press, Boca Raton, Fla., 1984, chap. 20.
16. **Cantey, J. R. and Inman, L. R.,** Diarrhea due to *Escherichia coli* strain RDEC-1 in the rabbit: the Peyer's patch as the initial site of attachment and colonization, *J. Infect. Dis.*, 143, 440, 1981.
17. **Inman, J. R. and Cantey, J. R.,** Specific adherence of *Escherichia coli* (strain RDEC-1) to membranous (M) cells of the Peyer's patch in *Escherichia coli* diarrhea in the rabbit, *J. Clin. Invest.*, 71, 1, 1983.

Chapter 12

HYDROPHOBIC PROPERTIES OF PILI

Torkel Wadström, Ahmed Faris, Mats Lindahl, and Stellan Hjertén

TABLE OF CONTENTS

I. INTRODUCTION

More than two decades ago, studies on the piliated phase of *Escherichia coli* showed that such organisms possess a reduced surface charge. More recently, Jones[1] in the first extensive review on bacterial cell interactions claimed that in general this reduction in surface charge minimizes the repulsive forces between bacteria and animal cell surfaces.[1] Electrophoretic studies on piliated and nonpiliated phases have confirmed this proposal.

Purification of pili from different organisms has shown that in general these surface protein antigens possess a relatively high percentage of hydrophobic amino acids (4 to 40%). It has therefore been suggested that the hair-like structures on the cell surface are formed by hydrophobic interaction between the hydrophobic pili subunits.[2]

Early observations by Brinton[3] suggested that pili stick to particles such as latex beads with pronounced hydrophobic (syn. amphiphilic, lipophilic) properties. This original observation indicated that hydrophobic residues were exposed on the *surface* of the pili.[3] Furthermore, the pronounced tendency of pili to clump alongside on heavily piliated strains further indicated that hydrophobic forces and exclusion of water caused this phenomenon.[4]

Several methods have been developed to measure cell surface hydrophobicity (Table 1). We developed a simple, small-scale chromatographic procedure (Hydrophobic Interaction Chromatography, HIC) on commercial Octyl® and Phenyl® Sepharose® (amphiphilic gels) to investigate bacterial surface hydrophobicity in Pasteur pipettes. Suspending fimbriated bacteria in salt solutions of different ionic strength to measure binding to these HIC gels gave a simple semiquantitative method to define bacterial cell surface hydrophobicity. A great number of strains of different Gram-negative species have been investigated. A good correlation between binding to amphiphilic gels and expression of surface pili has been demonstrated by electron microscopy and hemagglutination tests. Cell surface exposure of hydrophilic parts of lipopolysaccharides (LPS) of nonpiliated organisms prevents interaction of such strains with these HIC gels even at high salt concentrations, while rough strains with an incomplete LPS structure bind only at very high salt concentrations ($>1\ M$ ammonium sulfate)[5] (Table 2).

Recently, other pathogenic bacteria which colonize different surfaces were investigated by the standard HIC procedure. Thus, virulent group A streptococcus strains with surface M-protein showed more pronounced surface hydrophobicity than fimbriated strains of *E. coli*, while nonvirulent M-negative strains did not bind to Octyl Sepharose®.[6] *Staphylococcus aureus* and *S. saprophyticus*, which preferentially bind to urinary tract epithelium, also showed pronounced surface hydrophobicity.[7]

II. ALTERNATIVE METHODS TO MEASURE CELL SURFACE HYDROPHOBICITY

Magnusson and colleagues developed a standard phase partition system to study cell surface properties of S and R forms of salmonella strains.[8] Recent studies in both laboratories using this method and HIC have shown that the data obtained are not easily comparable. We believe that the two methods "sense" surface hydrophobicity differently, i.e., the fatty acid ligands in the partition experiments penetrate the pilus surface deeper than in the chromatographic method. Comparative studies with low molecular weight hydrophobic compounds, such as ^3H palmitic acid or certain hydrophobic fluorescent dyes such as ANS (aniline naphthalene sulfonic acid), would also be interesting to utilize. New methods such as measuring binding to hydrocarbons should also be employed as well as surface tension methods.[9]

Autoagglutination of enteric organisms in slide agglutination tests in the diagnostic laboratory has mainly been attributed to rough (R) forms of salmonella and other enteric

Table 1
METHODS TO MEASURE SURFACE HYDROPHOBICITY OF CELLS

Phase partition chromatography
Hydrophobic interaction chromatography
Wide angle diffraction
Binding to hydrocarbons
Hydrophobic probes (^{14}C-dodecanoic acid)
"The salt aggregation test"

Table 2
HEMAGGLUTINATING AND HYDROPHOBIC PROPERTIES OF *E. COLI* ROUGH STRAINS WITH AND WITHOUT FIMBRIAE

Type of strains	Strain	Hemagglutination (HA)			HIC	Salt aggregation [$(NH_4)_2SO_4$]
Rough F^{+}[a] (Fl)[b]	R 337	MSG[c]	MSC[d]	MRH[e]	+	0.3 M
Rough F^{+} (Fl)	R 265	MSG	MSC	MRH	+	0.2 M
Rough K^{+}	R 135	MSG	MSC	MRH	+	0.2 M
Rough	R 346	—	—	—	(+)	>2 M
Rough F^{-} K^{-}	R 125	MSG	MSC	—	+	<0.08 M

Note: All strains were grown in CFA-broth.[12] Strains were kindly donated by Dr. H. Tshäpe, Wernigerode, DDR.

[a] F^{+} = fimbriated (piliated).
[b] Fl = type 1 pili.
[c] MSG = mannose-sensitive guinea pig.
[d] MSC = mannose-sensitive chicken.
[e] MRH = mannose-resistant human.

organisms. However, more recent studies on autoagglutinating strains of *Yersinia enterocolitica* and comparative studies by HIC and the newly developed salt aggregation test suggest that surface fimbriae are often involved in this phenomenon (which can prevent interpretation of simple diagnostic tests such as slide agglutination to determine O and H bacterial surface antigens). Systematic studies based upon the principle that salting out occurs at low salt concentrations of proteins with hydrophobic properties such as intrinsic membrane proteins are highly desirable. We have shown that P^{+} *E. coli* from various sources precipitate at low salt concentrations and that different salts differ in their efficiency to cause precipitation. (The salts can be arranged in the so-called Hofmeister series.[10])

This new test has permitted very reproducible semiquantitative measurements of surface hydrophobicity of piliated strains of different species and from different sources.[11] This method is also suitable for rapid screening of a large number of clinical isolates. Preliminary findings indicate that nonhemagglutinating pili can be detected by this method.[24] Other methods such as electron microscopy or the other methods developed to measure surface hydrophobicity seem less suited for this particular purpose. A set of up to 14 different erythrocyte species in screening strains for hemagglutinating properties also seemed to be less efficient than this new "salting out" method. Investigations on nonhemagglutinating pili such as 987p also showed that this simple method can be used to define optimal production of cell surface fimbriae of great importance in vaccine production.

Table 3
BINDING OF ENTEROTOXIGENIC *E. COLI* WITH CFA/I PILI TO DIFFERENT AMPHIPHILIC GELS

Gel	Ligand		Binding of CFA/I strain[a] 0.15 M NaCl	Binding of CFA/I strain[a] 4.0 M NaCl[b]
Sepharose® CL-4B	Octyl	(C 8)	−	+
	Palmitoyl	(C15)	+	+
	Stearoyl	(C17)	+	+

[a] Bacterial suspension (100 μℓ = 1 × 10^8 bacteria) in saline was applied to each column (1 mℓ gel bed in a Pasteur pipette). + = more than 90% adsorption; − = less than 10% adsorption of bacteria applied.

[b] Salt solutions were buffered with 0.01 *M* sodium phosphate buffer, pH 6.8. Note that gels with longer aliphatic chains (hydrophobic ligands) have more pronounced hydrophobic properties and bind enterotoxigenic *E. coli* (ETEC) bacteria at lower salt concentration. However, all of these gels bind the ETEC strain in the standard HIC-procedure (i.e., in 1 *M* $(NH_4)_2SO_4$) or 4 *M* NaCl).

III. PREVENTION OF ETEC DIARRHEA IN THE INFANT RABBIT BY HYDROPHOBIC GELS

Classical studies on the pathogenesis of porcine *E. coli* diarrhea by Smith, Jones, and Rutter showed that surface proteins (K88) are necessary to cause colonization of the small intestine.[1] More recently, other surface protein antigens have also been characterized in porcine *E. coli* diarrhea (K99, 987p) and in human enterotoxigenic *E. coli* (ETEC) infections (CFA/I, CFA/II). Binding of piliated ETEC bacteria (K88, CFA/I, etc.) was found to occur only at high nonphysiological salt concentrations which amplifies hydrophobic interactions between hydrophobic (amphiphilic) molecules and the nonpolar ligands.

By substituting stronger hydrophobic ligands on different gel surfaces such as dextran, agarose, cellulose, and amphiphilic gels were developed to bind fimbriated ETEC at physiological salt concentrations. One such gel was then tried in the infant rabbit model, i.e., in 3- to 6-day-old rabbits.[12] One batch of palmitoyl agar with good in vitro binding capacity for CFA/I ETEC was selected for these experiments (Table 3). Animals were divided into groups A, B, and C (Figure 1), always with animals of the same litter in the control group (A) and the experimental groups (B or C). Gel was preswelled to prevent adsorption of diarrheal gut fluid. A fairly large amount of gel (0.25 g dry weight) was given in order to allow adsorption of enough of the very high inoculum dose (10^9 organisms orally) to prevent intestinal colonization and onset of diarrhea. This high dose of organisms was found necessary to give diarrheal symptoms in all animals in the control group (Figure 1).

Gels given 15 min before, 6 hr and 12 hr after the oral dose of 10^9 ETEC organisms were found to prevent diarrhea effectively in the animals (Figure 2). High mortality rates in the untreated control groups in some litters receiving CFA/I strains (and also CFA/II strains, data not shown) could not be effectively prevented by oral and subcutaneous rehydration every 3 to 4 hr. Furthermore, analyses of small intestinal luminal contents and the intestinal walls of animals given CFA/I positive and negative variants did not reveal a true intestinal gut wall colonization by CFA positive organisms in this rabbit model.

The results of these experiments with one CFA/I ETEC strain are summarized in Figure 1. Eighty-eight percent of animals in the untreated control group got diarrhea as defined by repeated unformed stools visible as patches on the filter paper covering the bottom of each cage. Addition of a small amount of carbon to the inoculum did not affect the clinical diarrhea and was found to be a very practical marker to semiquantitate the amount of diarrhea

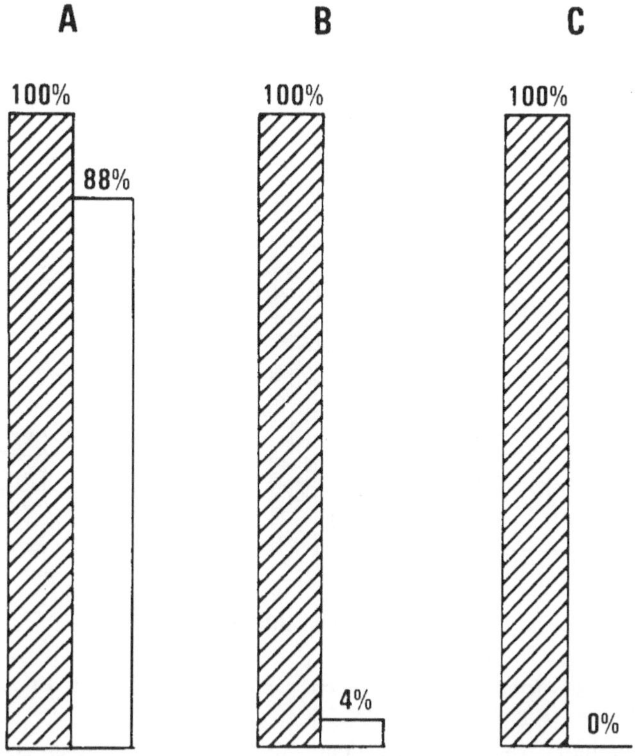

FIGURE 1. Prevention of diarrhea in infant rabbits by a hydrophobic gel (palmitoyl agar). Hatched bars: percentage of animals given 10^9 ETEC organisms. Open bars: percentage of animals with diarrhea. (A) Control group, no gel given. (B) Gel group, 0.25 g preswelled gel was given prior to bacterial challenge. (C) Gel group, 0.25 g preswelled gel was given 6 hr after bacterial challenge.

stool for each animal during an experiment. Repeated changes of filter paper revealed an acute onset usually about 12 to 14 hr after oral challenge with ETEC organisms also with CFA negative organisms (Figure 1).

However, analysis of the results with CFA/I+ and CFA/I− ETEC strains showed a much higher attack rate in animals receiving CFA/I+ ETEC with maximal diarrheal symptoms 18 hr after challenge (Figure 2). Clinical diarrhea usually declined within the next 10 hr and most animals in the control group then usually recovered spontaneously (c.f. symptoms for 2 to 7 days of cholera-like disease of CFA/I ETEC in human volunteers).[13]

IV. DISCUSSION AND CONCLUSIONS

Reports in old literature show that different absorbtive agents such as charcoal, colloidal clays, kaolin, and pectin were used in India and China many hundreds of years ago.[14] Such agents have also been used in the treatment of severe infantile diarrhea in the preantibiotic era during this century in different parts of the world. A study in Dacca showed that charcoal as such was ineffective in the treatment of cholera patients, while adsorption of cholera toxin on GM_1 charcoal drastically decreased the duration of symptoms in patients with severe cholera.[15]

Recent studies at our institute showed that *E. coli* LT and ST toxins bound to certain amphiphilic gels in physiological buffers. Further experiments in other models are necessary to determine if toxin adsorption is important to explain the effect of amphiphilic gel given

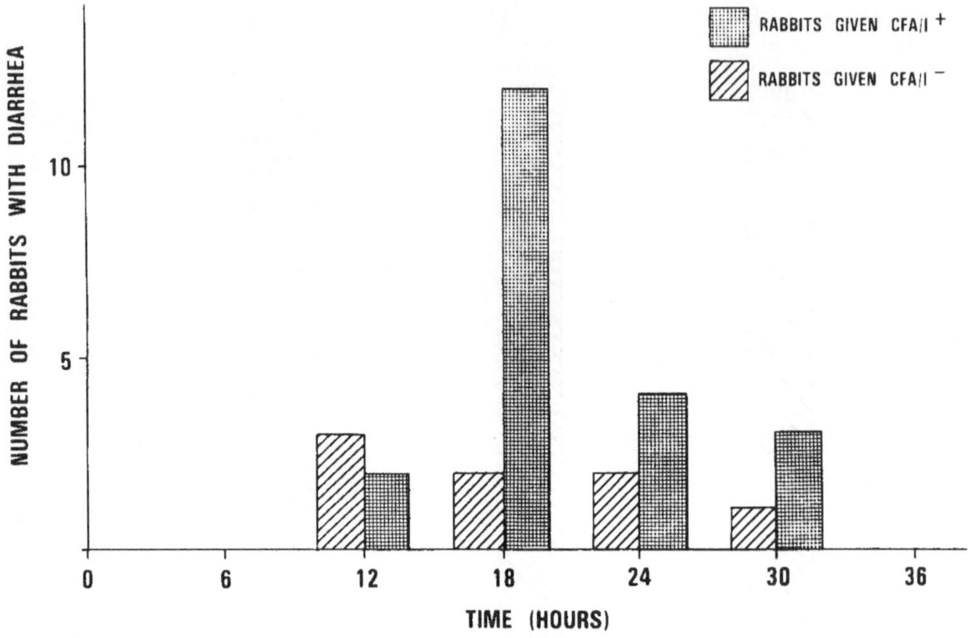

FIGURE 2. Onset of diarrhea in infant rabbits receiving 10^9 CFA/I$^+$ or CFA/I$^-$ ETEC organisms.

before onset of symptoms (as a prophylacticum) or after onset, as treatment. Moreover, in vitro experiments showed that other fimbriated and nonfimbriated intestinal pathogens such as vibrios and aeromonas organisms bind to amphiphilic gels, while nonfimbriated *E. coli* or *E. coli* with type 1 fimbriae show poor binding properties.

In conclusion, we believe that the application of amphiphilic gels utilizing the hydrophobic surface properties of intestinal pathogens is worthwhile to further investigate as a possible new method to prevent and treat intestinal infections such as ETEC and cholera primarily localized to the small intestine. Recent reports on fimbriated *Yersinia enterocolitica* suggest that investigations of other pathogens may be worthwhile, despite the fact that shigella (type 1) and salmonella (type 2, i.e., type 1-like) seem to show fimbriation of only certain species.[16] However, not yet described nonhemagglutinating surface protein antigens may be revealed by new methods such as HIC or salting out experiments.[24] Such studies are already in progress including also studies of biospecific (lectin-like) interactions caused by the hemagglutinating properties reported for most fimbrial antigens.

The discovery that GM_2 glycoconjugates such as gangliosides and glycopeptides with similar sugar sequences may be specific cell receptors opens alternative methods to prevent mucosal colonization by pathogens. However, the great diversity to be expected for different antigenic types of fimbriae (such as F7 to F12 in UTI *E. coli*) indicates that the lectin approach is most probably much more difficult.[17] Nevertheless, the binding of d-mannose to type 1 fimbriae and CFA/I to GM_2 ganglioside suggests that combined treatment may be a possibility to prevent small bowel colonization by ETEC lacking CFA/I, CFA/II, or other MRHA hemagglutinins. Levine and others have estimated that the majority of such strains produce type 1 fimbriae.[18] Several plant lectins such as concanavalin A (con A) and wheat germ agglutinin (WGA) possess hydrophobic surface properties.[19] We propose that bacterial cell surface fimbriae have a similar nature and that hydrophobic interactions are probably of major importance to permit strong binding after an initial lectin-like bacterial eukaryotic cell interaction. However, specific charge interactions may also be involved in the specific binding of fimbriated enteric organisms to the mucus surface layer as well as intestinal cell surfaces.

Ion-exchanging gels were found to bind salmonella organisms and a similar resin (cholestyramine) was used as an adsorbent of *E. coli* enterotoxins in animal models with limited success.[20,21]

Recent studies in our laboratory have shown that both *E. coli* LT and ST enterotoxins have pronounced hydrophobic properties which can be used for purification of both toxins.[25] In vitro experiments have also shown that these toxins bind to certain hydrophobic gels depending on the choice of amphilic ligand and the degree of substitution of the gel. These two parameters are probably also important to determine the binding of other constituents of the intestinal fluid, such as bile acids and components on the mucosal surface such as mucus components.

The batch of palmitoyl agar used in this study showed a very poor toxin binding capacity, while gels with other ligand and another gel matrix showed high toxin-binding capacity for both *E. coli* LT and ST (unpublished). The recent discovery that the toxin(s) produced by *Clostridium difficile* binds to cholestyramine gel as well as to sterols suggests that also other toxins may bind to hydrophobic gels.[22] The hypothesis that membrane-binding toxins as well as several hormones show hydrophobic surface properties will soon be further explored.[4,22,23]

ACKNOWLEDGMENTS

We would like to acknowledge K. Lövgren, B. Ågerup, and T. de Belder for a stimulating collaboration. This study was supported by grants from the Swedish Medical Research Council (16X 04723) and a grant from the Swedish Board for Technical Development.

REFERENCES

1. **Jones, G. W.,** The attachment of bacteria to the surfaces of animal cell, in *Microbial Interactions*, Receptors and Recognition, Series B, Vol. 3, Reissig, J. L., Ed., Chapman and Hall, London, 1977, 139.
2. **Buchanan, T. M., Chen, K. C. S., Jones, R. B., Hilderbrandt, J. F. H., Pearce, W. A., Hermodson, M. A., Newland, J. C., and Luchtel, D. L.,** Pili and principal outer membrane of *Neisseria gonorrhoeae:* immunochemical, structural, and pathogenic aspects, in *Proc. on Immunobiology of Neisseria gonorrhoea*, Brooks, G. F., Gotschlich, E. C., Holmes, K. K., Sawyer, W. D., and Young, F. E., Eds., ASM, Washington, D.C., 1978, 145.
3. **Brinton, C. C., Bryan, J., Dillon, JO-A., Guerina, N., Jacobson, L. J., Labik, A., Lee, S., Levine, A., Lim, S., McMichael, J., Polen, S., Rogers, K., To, A. C.-C., and To, S. C.-M.,** Uses of pili in gonorrhea control: role of bacterial pili in disease, purification and properties of gonococcal pili, and progress in the development of a gonococcal pilus vaccine for gonorrhoea, in *Proc. on Immunobiology of Neisseria gonorrhoeae*, Brooks, G. F., Gotschlich, E. C., Holmes, K. K., Sawyer, W. D., and Young, F. E., Eds., ASM, Washington, D.C., 1978, 155.
4. **Wadström, T., Smyth, C. J., Faris, A., Jonsson, P., and Freer, J. F.,** Hydrophobic absorptive and haemagglutination properties of enterotoxigenic *Escherichia coli* with different colonization factors: K88 antigen, K99 antigen, colonization factor antigen and adherence factor, in *Proc. 2nd. Int. Sympos. Neonatal Diarrhoea*, Acres, S., Ed., Stuart Brandle Publishing Service Ltd., Edmonton, Canada, 1979, 29.
5. **Faris, A., Wadstöm, T., and Freer, J. H.,** Hydrophobic adsorptive and hemagglutinating properties of *Escherichia coli* possessing colonization factor antigens (CFA/I or CFA/II), type 1 pili, or other pili, *Curr. Microbiol.*, 2, 67, 1981.
6. **Tylewska, S. K., Hjertén, S., and Wadström, T.,** Contribution of M protein to the hydrophobic surface properties of *Streptococcus pyogenes, FEMS Microbiol. Lett.*, 6, 249, 1979.
7. **Wadström, T., Faris, A., Lindahl, M., Hjertén, S., and Ågerup, B.,** A new principle for prevention of diarrhoea caused by enterotoxigenic *Escherichia coli* (ETEC) possessing colonization factor antigen (CFA/I), *Scand. J. Infect. Dis.*, 13, 129, 1981.

8. **Magnusson, K.-E., Stendahl, O., Tagesson, C., Edebo, L., and Johansson, G.**, The tendency of smooth and rough *Salmonella typhimurium* bacteria and lipopolysaccharide to hydrophobic and ionic interaction, as studied in aqueous polymer two-phase systems, *Acta Pathol. Microbiol. Scand.*, 85, 212, 1977.
9. **Rosenberg, M., Gutnick, D., and Rosenberg, E.**, Adherence of bacteria to hydrocarbons: a simple method for measuring cell-surface hydrophobicity, *FEMS Microbiol. Lett.*, 9, 29, 1980.
10. **Lindahl, M., Faris, A., Wadström, T., and Hjérten, S.**, A new test based on 'salting-out' to measure relative surface hydrophobicity of bacterial cells, *Biochim. Biophys. Acta*, 677, 471, 1981.
11. **Faris, A., Lindahl, M., and Wadström, T.**, Relation of pili to other virulence factors, in *Attachment of Organisms to the Gut Mucosa*, Vol. 2, Boedeker, E. C., Ed., CRC Press, Inc., Boca Raton, Fla., 1984.
12. **Evans, D. G., Silver, R. P., Evans, D. J., Jr., Chase, D. G., and Gorbach, S. L.**, Plasmid controlled colonization factor associated with virulence in *Escherichia coli* enterotoxigenic for humans, *Infect. Immun.*, 12, 656, 1975.
13. **Satterwhite, T. K., DuPont, H. L., Evans, D. G., and Evans, D. J;, Jr.**, Role of *Escherichia coli* colonisation factor antigen in acute diarrhoea, *Lancet*, 1, 181, 1978.
14. **Martin, G. J.**, *Ion Exchange and Adsorption Agents in Medicine. The Concept of Intestinal Bionomics*, Little, Brown and Co., Boston, 1955.
15. **Holmgren, J.**, Actions of cholera toxin and the prevention and treatment of cholera, *Nature (London)*, 292, 413, 1981.
16. **Duguid, J. P. and Old, D. C.**, Adhesive properties of enterobacteriaceae, in *Bacteria Adherence*, Receptor and Recognition Series B, Vol. 6, Beachey, E. H., Ed., Chapman and Hall, London, 1980, 185.
17. **Ørskov, I., Ørskov, F., and Birch-Andersen, A.**, Comparison of *Escherichia coli* fimbrial antigen F7 with type 1 fimbriae, *Infect. Immun.*, 27, 657, 1980.
18. **Levine, M. M.**, Adhesion of enterotoxigenic *Escherichia coli* in humans and animals, in Adhesion and Microorganism Pathogenicity, Ciba Foundation Symposium 80, Pitman Medical, 1981, 142.
19. **Wadström, T. and Trust, T.**, Bacterial surface lectins, in *Medical Microbiology*, Easmon, C. S. F. and Jeljaszewicz, I., Eds., 1984.
20. **Hall, A. N., Hogg, S. D., and Phillips, G. O.**, Gradient elution of *Salmonella typhimurium* and *Escherichia coli* strains from a DEAE-cellulose column, *J. Appl. Bacteriol.*, 41, 189, 1976.
21. **Mullan, N. A., Burgess, R. J., and Newsome, P. M.**, The ability of cholestyramine resin and other adsorbents to bind *Escherichia coli* enterotoxins, *J. Med. Microbiol.*, 12, 487, 1979.
22. **Kreutzer, E. W. and Milligan, F. D.**, Treatment of antibiotic associated pseudomembraneous colitis with cholestyramine resin, *Johns Hopkins Med. J.*, 143, 67, 1978.
23. **Chang, T.-W. and Gorbach, S. L.**, Inhibition of binding of *Clostridium difficile* toxin by steroids, *J. Infect. Dis.*, 142, 113, 1980.
24. **Ljungh, Å. and Wadström, T.**, manuscript in preparation.
25. **Rönnberg, B.**, manuscript in preparation.

Chapter 13

REGULATION OF PILUS EXPRESSION

Richard E. Isaacson

TABLE OF CONTENTS

I. ABSTRACT

The expression of *Escherichia coli* somatic pili is a regulated process. The pilus K99 was used as a prototype to study quantitative variation of pilus expression. The appearance of K99 on the extracellular surface of *E. coli* occurs only during the logarithmic phase of growth. However, net subunit synthesis occurs throughout the life cycle of the bacterial culture including the times when extracellular cell-bound K99 is not being produced. In minimal medium, glucose does not affect the appearance of K99 on the cell surface, contrary to the effect observed in a rich medium, yet net subunit synthesis is decreased by glucose. Expression of K99 on the cell surface is unaffected by chloramphenicol while subunit synthesis is totally inhibited. These results are consistent with the hypothesis that pilus outgrowth (appearance on the cell surface) and subunit synthesis are two different processes and are possibly controlled independently. Production of K99 by cells grown at 18°C is completely inhibited.

II. INTRODUCTION

Colonization of and adhesion to small intestines is a necessary prerequisite for enterotoxigenic *Escherichia coli* (ETEC) if they are to induce diarrheal disease. This process is mediated by bacterial somatic pili (adhesins) that form a bridge between bacterium and enterocyte microvilli. Expression of these pili is thus an attribute of virulence.

At least two regulatory processes can govern the expression of pili. The first is a reversible, qualitative, on-off process called phase-variation.[1] The second is a quantitative process (quantitative variation) which regulates the amount of pilus produced.[1,2] All pili appear to be regulated by the latter process and some by the former process.

The pilus K99 is a plasmid-encoded adhesin[3,4] that facilitates adhesion of ETEC to the small intestines of pigs,[5] calves,[3,4] and sheep[3,4] and is regulated by quantitative variation only. This adhesin has been used as a model for studying quantitative variation. Factors known to affect K99 quantitative variation include repression by L-alanine[6] and glucose[7,8] and derepression by cyclic AMP[8] and aeration.[8] Expression is also inhibited by growth of cells at 18°C.[4,9]

The studies described in the manuscript were designed to further define and understand the processes regulating K99 quantitative variation.

III. MATERIALS AND METHODS

A. Bacterial Strains and Growth Media

The K99 plasmid from *E. coli* strain 1474[10] was moved into *E. coli* K12 W3110 by conjugal mating. Transconjugantes were K99$^+$ and tetracycline resistant and contained two plasmids (52 and 58 megadaltons) as determined by agarose gel electrophoresis.[11] One transconjugante I110 along with the parent strain 1474 and the isogenic, plasmidless strains 1475 and W3110 were used in the experiments described.

Cells were grown in trypticase soy broth (rich medium) or Tris-salts[12] containing 0.15 glucose (unless specified) and 0.15% casein acid hydrolysate. To label cells, carrier free ^{35}[S] sulfuric acid was added (250 μci/mℓ medium).

B. K99 Assays

The ELAC assay was performed as previously described.[8] Nonequilibrium two-dimensional polyacrylamide gels (2-D gels) were described by O'Farrell.[13] Ampholytes used were pH 3.5 to 10 (Biolytes, BioRad). After electrophoresis, the gels were dried and exposed to Kodak® X-Omat AR X-ray film at −70°C. The number of counts in the spots of interest were measured in a Beckman LS7000 liquid scintillation spectrometer.

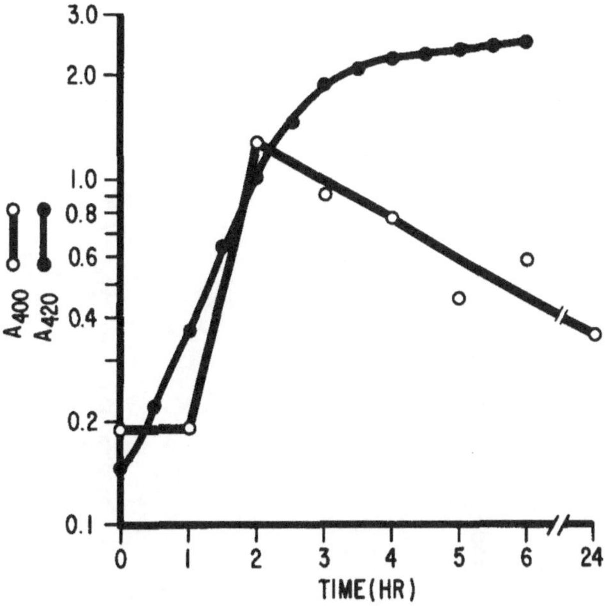

FIGURE 1. Kinetics of K99 production of *E. coli* 1474 in the rich medium trypticase soy broth. At timed intervals, samples of culture were removed; the turbidity measured (●) in a spectrophotometer at 420 nm and K99 was determined (○) by ELAC assay (A_{400}). (From Isaacson, R. E., *Infect. Immun.*, 30, 125, 1980. With permission.)

IV. RESULTS

A. Time of K99 Expression

Expression of K99 as detected by ELAC occurs in logarithmic phase of bacterial growth (Figure 1).[8] Extracellular, cell-bound K99 is not expressed until cells enter log phase and the amount of K99 per cell decreases in stationary phase. The data shown in Figure 1 were obtained when cells were grown in the rich medium trypticase soy broth. The same results were obtained when cells were grown in the minimal medium Tris-salts (not shown) except that the growth rate and rate of K99 expression were lower when compared to the rates obtained in rich medium.

Since the ELAC assay measured extracellular, cell-bound K99 and may not measure actual K99 subunit synthesis, two-dimensional gels were employed. Cells labeled with free ^{35}[S] were lysed and electrophoresed as described. K99 was located after autoradiography and quantitated by measurement of ^{35}[S] in the K99 spots. Figure 2 is an autoradiogram of a whole cell lysate of K99$^+$ cells run in 2-D gels. The arrows point to the K99 spots. Note that a comparison of the amount of K99 detected by ELAC and 2-D gels using cells grown in minimal medium is shown in Figure 3. Cells in log phase (between 2 and 4 hr) exhibit the characteristic increase in K99 measured by ELAC. However, during this same period, the amount of total K99 remains constant. The amount of K99 as measured by 2-D gels after 24 hr of incubation is approximately 0.2% of the total cellular proteins. Thus K99 synthesis occurs throughout the entire life cycle of *E. coli* while its appearance extracellularly occurs only in log phase.

B. Effect of Glucose on K99 Expression

Previous experiments showed that glucose caused a decrease in the amount of K99 produced when antibodies were used to demonstrate its presence. Cells growing in rich medium

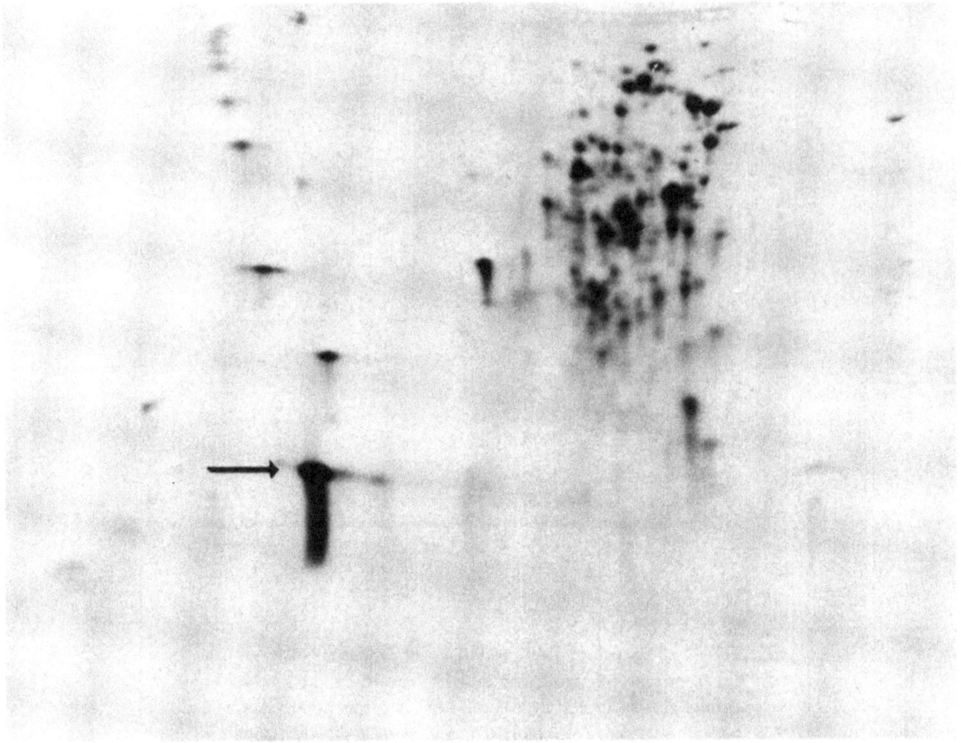

FIGURE 2. Autoradiogram of a 2-D gel of a whole cell lysate of *E. coli* strain I 110(K99⁺). The arrow indicates a K99 spot.

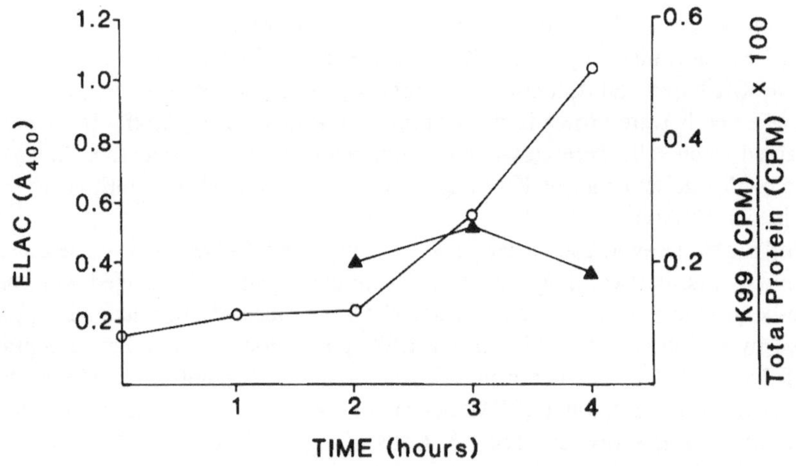

FIGURE 3. Kinetics of K99 production of *E. coli* I110 grown in the minimal medium Tris-salts. K99 was measured by ELAC (○) and 2-D gel (▲).

Table 1
EFFECT OF GLUCOSE ON K99 EXPRESSION

Glucose (%)	ELAC (A_{400})	2-D Gel (CPM)
0.15	0.59	3867
	1.01[b]	1.9[b]
2.0	0.58	2015

[a] Results are the mean of three separate experiments.
[b] Ratio of 0.15%:2.0 glucose for ELAC results and 2-D gel results.

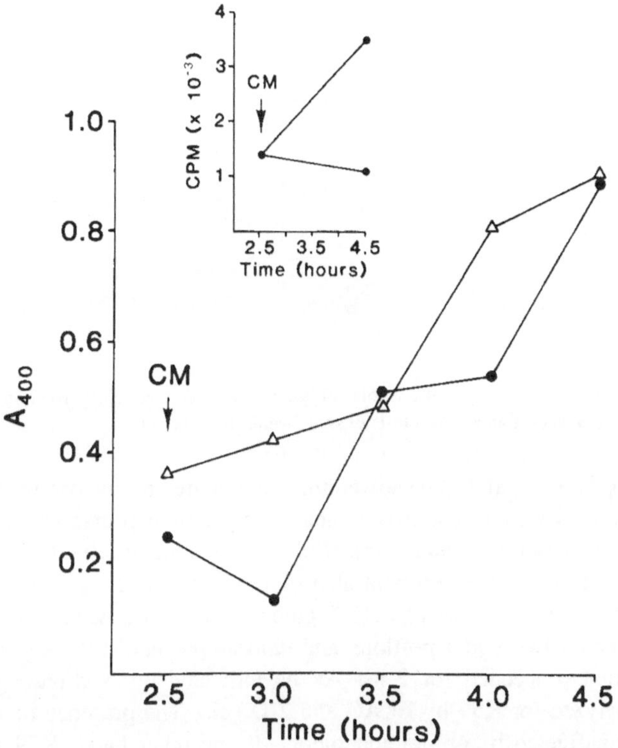

FIGURE 4. Effect of chloramphenicol on K99 production. *E. coli* strain I110 was grown in the minimal medium Tris-salts for 2.5 hr in the presence of ^{35}[S]. The culture was split and chloramphenicol (250 μg/mℓ) added to one portion. At the indicated time samples were removed and analyzed by ELAC or 2-D gel (insert) ●-● with chloramphenicol, Δ-Δ without chloramphenicol. The lower line in the insert is with chloramphenicol.

supplemented with 2% glucose produced small amounts of K99, while supplementation with 0.15% glucose resulted in good K99 production. Addition of cyclic AMP (0.5 m*M*) to cultures growing in the presence of 2% glucose alleviated the apparent repression of K99 expression. The effect of glucose on K99 expression using cells grown in minimal medium is shown in Table 1. No repression by glucose could be detected when the ELAC assay was used. This result is different from results obtained with rich medium. Net K99 synthesis, as measured by 2-D gels, was, however, reduced 1.9-fold in the presence of 2% glucose. Although this reduction was not particularly striking, in the three experiments used to obtain

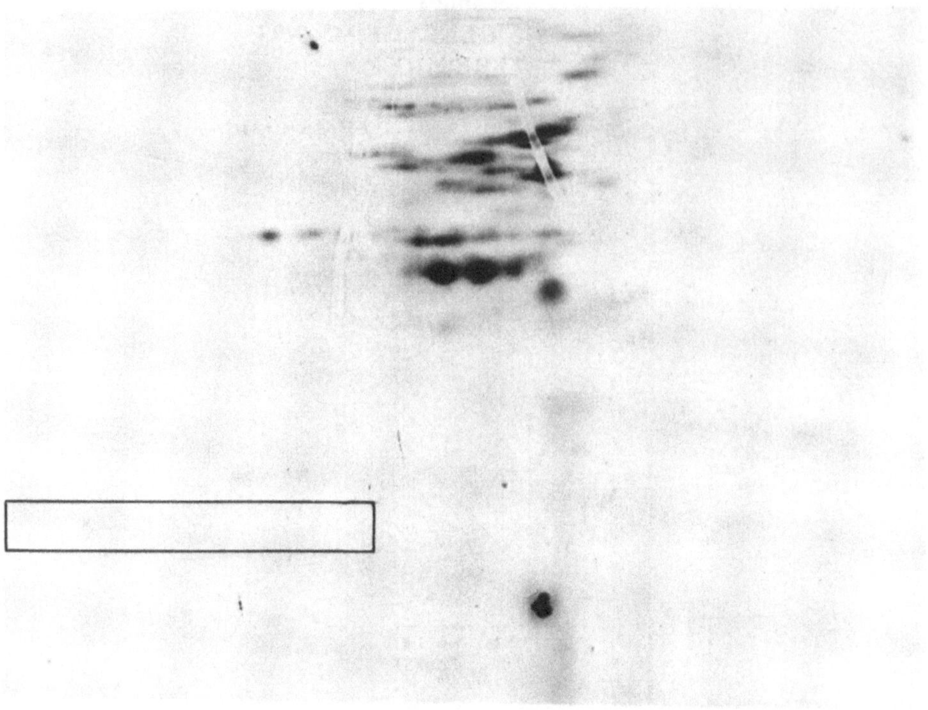

FIGURE 5. Autoradiogram of 2-D gel of a whole cel lysate of *E. coli* strain 1110 grown in the minimal medium Tris-salts at 18°C. The boxed-in area is where K99 spots would be expected.

the data shown in Table 1, all three showed the same degree of glucose-mediated repression. These results, along with the results obtained from the time course studies (Figure 3) lead to the conclusion that synthesis and extracellular appearance of K99 do not reflect the same phenomenon. The following experiment also supports this conclusion. Cells were grown in minimal medium in the presence of 0.15% glucose and ^{35}[S]. After 2.5 hr of growth, the culture was split into two equal portions and chloramphenicol (250 μg/mℓ) added to one portion. Incubation proceeded for 2 hr. At the indicated times (Figure 4) samples were removed and analyzed for K99 by ELAC and 2-D gels. The presence of K99 as measured by ELAC was unaffected by chloramphenicol. On the other hand, K99 subunit synthesis was completely inhibited by chloramphenicol (insert of Figure 4).

C. Effect of Temperature on K99 Expression

The effect of growth at 18°C on K99 expression was determined by the ELAC and 2-D gel analysis. K99$^+$ cells grown at 18°C do not agglutinate in K99-specific antiserum nor do they agglutinate horse erythrocytes (cells grown at 37°C do agglutinate horse erythrocytes).[15] The results from ELAC assays of 18°C grown cells showed that extracellular, cell-bound K99 was not produced. A_{400} values obtained were less than 0.05 which is the same as background. Whole cell lysates prepared from cells labeled with ^{35}[S] were prepared from 18°C grown cells and subjected to 2-D gel analysis. An autoradiogram from this experiment is shown in Figure 5. The region where the box is located is where K99 spots would be expected. Clearly no such spots are evident leading to the conclusion that mature K99 subunits were not produced when cells were grown at 18°C. The possibility that a precursor of the K99 subunit was produced at 18°C has not been determined; however, the 2-D gel patterns of K99$^+$ and K99$^-$ cells grown at 18°C did not appear different (not shown).

V. DISCUSSION

Expression of the *E. coli* pilus K99 appears to be a regulated process. Several conditions or metabolic additions can have a striking effect on the amount of K99 detected on the cell surface. It has been previously reported that glucose,[7,8] cyclic AMP,[8] alanine,[6] aeration,[8] and ambient temperature[4,9] either permit or inhibit K99 expression. Experiments presented here demonstrate that, for example, glucose-mediated repression is variable and is dependent upon the basal medium used for cell growth. In a rich medium the glucose-mediated repression is apparent but absent when a minimal medium is used. Since, in a rich medium, the glucose effect can be alleviated by addition of exogenous cyclic AMP[8] it was assumed that K99 expression was subject to classical catabolite repression. The minimal medium does not contain free cyclic AMP while most rich media do. Thus, it is not clear why the catabolite repression of K99 is not more exaggerated than it is in the rich medium (i.e., repression should be very strong in minimal medium).

The results of the experiments presented do lead to the conclusion that K99 pilus outgrowth and K99 subunit synthesis are independently regulated. In minimal medium glucose does induce a demonstratable although incomplete repression of subunit synthesis. Yet the appearance K99 on the cell surface is unaffected. Complete inhibition of subunit synthesis by chloramphenicol also does not affect the appearance of K99 on the cell surface. Thus pilus outgrowth can occur in the absence of pilus subunit synthesis. On the other hand, the converse is also true. Synthesis of subunits in lag phase, early log phase, and stationary phase does not result in the outgrowth of pili. The process of pilus outgrowth as measured by ELAC occurs exclusively in log phase.

The results do not yield a clear-cut answer as to whether expression is indeed controlled by catabolite repression. Pilus outgrowth, however, does not appear to be so regulated.

The interpretation of the results obtained by growing cells at 18°C is that both outgrowth and subunit synthesis are inhibited. Since K99 is a membrane protein it must be transported out of the cell for detection. Transport of proteins is dependent upon the inner membrane and at 18°C the membrane should be in a nonfluid state. Thus subunit export may be prevented. K99 pilus subunits are found primarily in the outer cell membrane and possibly the periplasmic space. Cytoplasmic pools are not detectable and only small amounts are found associated with the inner membrane.[14] Thus synthesis of the subunits may occur on the inner membrane with rapid export. As such, the membrane state dictated by ambient temperature may affect subunit synthesis. The results of 2-D gel analysis of 18°C grown cells show that no K99 is synthesized which is consistent with the above interpretation.

REFERENCES

1. **Brinton, C. C.,** The structure, function, synthesis and genetic control of bacterial pili and a molecular model for DNA and RNA transport in gram negative bacteria, *Trans. N.Y. Acad. Sci.*, 27, 1003, 1965.
2. **Isaacson, R. E.,** Pili of enterotoxigenic *Escherichia coli*, in *3rd Int. Symp. Neonatal Diarrhea*, Acres, S., Ed., Veterinary Infectious Disease Organization, Saskatoon, Saskatchewan, Canada, 1980, 213.
3. **Ørskov, I., Ørskov, F., Smith, H. W., and Sojka, W. J.,** The establishment of K99, a thermolabile transmissible *Escherichia coli* K antigen, previously called "K_{co}", possessed by calf and lamb enterotoxigenic strains, *Pathol. Microbiol. Scand.*, 83, 31, 1975.
4. **Smith, H. W. and Linggood, M. A.,** Further observations on *Escherichia coli* enterotoxins with particular regard to those produced by atypical piglet strains and by calf and lamb strains: the transmissible nature of these enterotoxins and of a K antigen possessed by calf and lamb strains, *J. Med. Microbiol.*, 5, 243, 1971.
5. **Moon, H. W., Nagy, B., Isaacson, R. E., and Ørskov, I.,** Occurrence of K99 antigen on *Escherichia coli* isolated from pigs and colonization of pig ileum by K99$^+$ enterotoxigenic *E. coli* from calves and pigs, *Infect. Immun.*, 15, 614, 1977.

6. **deGraaf, F. K., Boor, P. K., and vanHees, J. E.,** Biosynthesis of the K99 surface antigen is repressed by alanine, *Infect. Immun.*, 30, 125, 1980.
7. **Guinée, P. A. M., Veldkamp, J., and Hansen, W. H.,** Improved Minca medium for the detection of K99 antigen in calf enterotoxigenic strains of *Escherichia coli*, *Infect. Immun.*, 28, 190, 1980.
8. **Isaacson, R. E.,** Factors affecting expression of the *Escherichia coli*, pilus, K99, *Infect. Immun.*, 30, 125, 1980
9. **DeGraaf, F. K., Wientjes, F. B., and Klaasen-Boor, P.,** Production of K99 antigen by enterotoxigenic Escherichia coli: strains of antigen groups 08, 09, 020, and 0101 grown at different conditions, *Infect. Immun.*, 27, 216, 1980.
10. **Isaacson, R. E.,** K99 surface antigen of *Escherichia coli:* purification and partial characterization, *Infect. Immun.*, 15, 272, 1977.
11. **Meyers, J. A., Sanchez, D., Elwell, L. P., and Falkow, S.,** Simple agarose gel electrophoretic method for the identification and characterization of plasmid deoxyribonucleic acid, *J. Bacteriol.*, 127, 1529, 1976.
12. **Levisohn, R., Konisky, J., and Nomura, M.,** Interaction of colicins with bacterial cells. IV. Immunity breakdown studied with colicins Ia and Ib, *J. Bacteriol.*, 96, 811, 1968.
13. **O'Farrell, P. Z., Goodman, H. M., and O'Farrell, P. H.,** High resolution two-dimensional electrophoresis of basic as well as acidic proteins, *Cell*, 12, 1133, 1977.
14. **Isaacson, R. E.,** Regulation of the *Escherichia coli* pilus K99, *Infect. Immun.*, 40, 633, 1983.
15. **Isaacson, R. E.,** unpublished results.

Chapter 14

LECTIN-LIKE BACTERIAL ADHERENCE TO ANIMAL CELLS

Nathan Sharon

TABLE OF CONTENTS

I. INTRODUCTION

Since the 1950s it has been known that the adherence of many strains of *Escherichia coli* and of other enterobacteria to animal cells is inhibited by low concentrations of mannose,* methyl α-mannoside, and mannan, as illustrated in Figure 1 (for a recent review, see Duguid and Old[2]). Such bacteria have been designated as mannose sensitive or "MS". Work carried out during the last 6 years has demonstrated that other sugars inhibit specifically the adherence of various bacteria to animal cells[3] (Table 1). It has recently been shown that the adherence of certain mannose-resistant (MR) strains of *E. coli* is sugar specific, being effectively inhibited by oligosaccharides and glycolipids containing the structure Galα1→4Gal.[4,5]

Ofek et al.[6] studied the interaction of *E. coli* with buccal epithelial cells and were the first to propose that the sugar-specific adherence is mediated by lectins** on the bacterial surface which bind to carbohydrates (e.g., mannose residues) on the animal cells. This proposal was based primarily on the finding that no attachment of the bacteria to the epithelial cells was observed following treatment of the latter with sodium metaperiodate (a reagent known to oxidize sugar residues on cell surfaces) or with concanavalin A (a lectin that binds specifically to mannose or glucose residues on cells). It was also in accord with earlier observations that MS bacteria agglutinate cells of yeast such as *Saccharomyces cerevisiae* and *Candida albicans* that are covered by mannans. Subsequent work by Ofek and Beachey[8] has shown that, as expected, oxidation of yeast cells with periodate abolished their ability to bind MS *E. coli*. It has further been demonstrated that MS *E. coli* bind yeast mannan, and there is a high correlation between the mannan binding capacity of the bacteria and their attachment to animal cells[9] (Figure 2).

II. BACTERIAL PILI AS LECTINS

The MS lectins of *E. coli* may occur on the surface of the bacteria in different forms, most commonly as pili (or fimbriae). The participation of pili in the MS adherence of *E. coli* to cell surfaces was first pointed out by Duguid and Gillies[10] who found a striking correlation between the occurrence of pili and the MS hemagglutination exhibited by the bacteria. The pili responsible for this hemagglutination were designated as "type 1 pili". They consist of arrays of identical protein subunits (molecular weight ~17,000 dalton), known as pilin.[11] Ofek and Beachey[8] fractionated cultures of *E. coli* into nonadherent and adherent populations by adsorption of the latter on buccal epithelial cells followed by elution with methyl α-mannoside (Figure 3). The bacteria thus eluted exhibited a high degree of mannose binding activity and were heavily piliated, whereas the nonadherent bacteria lacked such activity, and were devoid of type 1 pili. Subsequent work has provided direct evidence that the MS lectin of *E. coli* may indeed be in the form of type 1 pili. Isolated and purified type 1 pili were found to agglutinate guinea pig erythrocytes[12,13] and to adhere to monkey kidney cells grown in vitro.[14] All these reactions were specifically inhibited by low concentrations of mannose or methyl α-mannoside. Moreover, purified type 1 pili from *E. coli* or *Salmonella typhimurium* strongly agglutinated yeast cells.[15,16]

III. OTHER FORMS OF BACTERIAL SURFACE LECTINS

Studies in our laboratory have shown that mannose-specific lectins may be present on the bacterial surface in forms other than pili. We have isolated preparations of flagella that

* All sugars are of the D-configuration unless otherwise noted.

** According to a recent definition,[7] a lectin is "a sugar binding protein or glycoprotein of nonimmune origin which agglutinates cells or precipitates glycoconjugates".

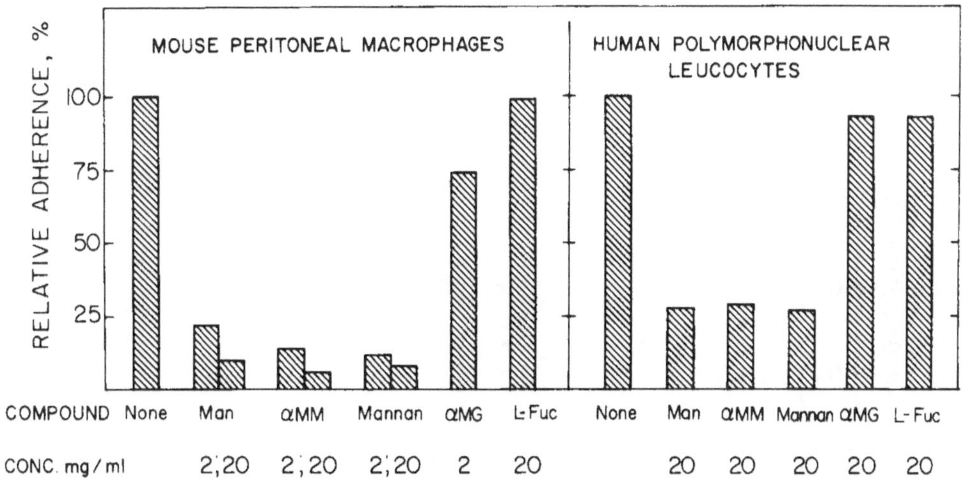

FIGURE 1. Inhibition by carbohydrates of the attachment of *Escherichia coli* 3092 (a K_{12} derivative) to mouse and human phagocytes. αMM, methyl α-mannoside; αMG, methyl α-glucoside. (Data from Bar-Shavit, Z., et al., *Biochem. Biophys. Res. Commun.*, 78, 455, 1977.)

Table 1
SUGAR-BINDING SPECIFICITY OF BACTERIAL ADHERENCE

Sugar	Organism
Mannose	*Escherichia coli* (MS)
	Salmonellae
	Shigellae
	Citrobacter freundii
	Klebsiellae
	Proteus
	Serratia marcescens
	Vibrio cholerae
Galα1→4Galβ1→	*Escherichia coli* (MR)
L-Fucose	*Vibrio cholerae*
Galactose	*Actinomyces viscosus*
	Actinomyces naeslundii
Sialic acid	*Mycoplasma*
N-Acetylglucosamine	*Chlamydia psitacci*
	Chlamydia trachomatis
Galβ1→3GalNAcβ1→3Gal	*Neisseria gonorrhoeae*

Note: For references see Sharon et al.[3] and text.

agglutinate yeast cells from *E. coli* 7343[17,18] and *Serratia marcescens* 8347.[19] The flagellar lectin of *E. coli* 7343 consists of protein subunits with a molecular weight of 36,500 dalton, which is considerably different from that of the type 1 pili (Figure 4); the amino acid composition of the *E. coli* flagellar lectin is also different from that reported for the type 1 pili, for the K99 and K88 antigens, for the colonization factor antigen 1 (CFA1) and the major outer membrane protein Ia of *E. coli*. The flagellar lectin of *S. marcescens* 8347 was found to consist of subunits with a molecular weight of 38,500 dalton; the peptides obtained upon cleavage of this lectin with cyanogen bromide were different from those obtained from the *E. coli* flagellar lectin.[19]

Preliminary experiments carried out in our laboratory by Dr. Thomas J. Pistole (unpublished results) have shown that in certain bacteria, the MS lectin(s) may be tightly attached

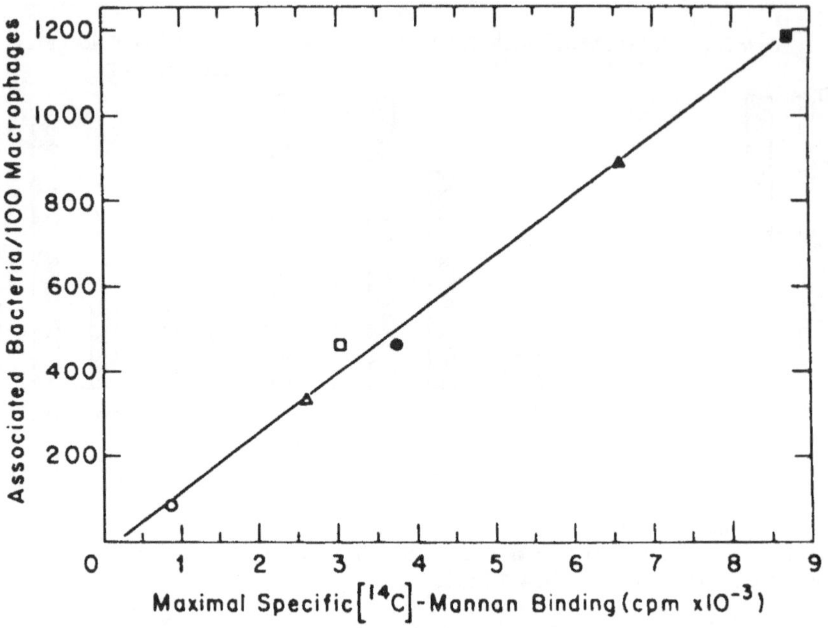

FIGURE 2. Correlation between the [^{14}C]mannan binding capacity and association with macrophages of *E. coli* strains PAT 84 (empty symbols) and 7343 (filled symbols) grown under different conditions: ○ and ●, exponential phase; △ and ▲, cephalexin treated filamentous bacteria; □ and ■, stationary phase bacteria. (From Bar-Shavit, Z., et al., *Infect. Immun.*, 29, 417, 1980. With permission.)

to the outer membrane. For example, homogenization of up to 90 min in an Omnimixer, a procedure known to release type 1 pili and flagella from bacteria, did not release any lectin from a MS strain (1826) of *S. typhimurium*, which is only slightly piliated.

Eshdat et al.[16] have now isolated outer membrane vesicles from *E. coli* 2699, a mannose-specific strain which agglutinated yeast cells whether the bacteria were piliated or not. The outer membrane preparations exhibited stronger agglutinating activity than the pili isolated from the same organism (minimal concentrations that agglutinated yeast cells were 25 μg protein/mℓ and 100 μg protein/mℓ, respectively). Examination by electron microscopy did not reveal any intact or fragmented pili in the outer membrane preparations (Figure 5). These and other findings led Eshdat et al.[16] to suggest that the membrane-bound lectin may be either in the form of pilin monomers inserted into the outer membrane without polymerization, or that it consists of protein subunits different from pilin.

IV. DISSOCIATION AND REASSEMBLY OF *E. COLI* TYPE 1 PILI

Detailed characterization of the structure, organization, and biological properties of type 1 pili is essential for the understanding of the interaction between bacteria and animal cell surfaces and of the role of bacterial lectins in pathogenicity. Such characterization may also be of aid in the development of improved and more specific antipili vaccines. Studies of type 1 pili have, however, been greatly hampered by difficulties encountered in their solubilization and dissociation into biologically functional subunits. For example, type 1 pili are not readily denatured by sodium dodecylsulfate (SDS) even with boiling and the bulk of the protein does not penetrate SDS-polyacrylamide gels.[11] Furthermore, the large size of intact pili (molecular weight $>10^6$) and their filamentous shape have limited the use of gel filtration or of affinity chromatography for their purification and characterization. The dis-

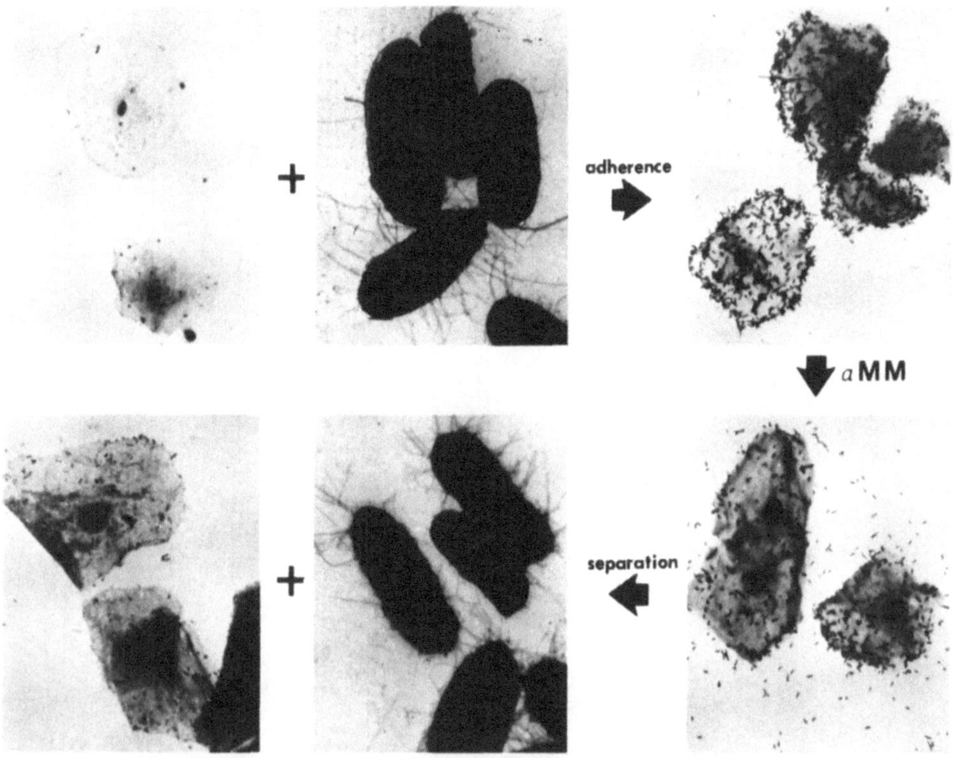

FIGURE 3. Fractionation of *E. coli* by adsorption to epithelial cells and displacement with αMM. Washed epithelial cells (top left) were mixed with a suspension containing few piliated organisms (top center). Nonadherent bacteria were removed by differential centrifugation, leaving only those *E. coli* that were attached to epithelial cells (top right). The adherent organisms were displaced from the surfaces of the epithelial cells by the addition of αMM (bottom right). The displaced bacteria, containing many piliated organisms (bottom center), were separated from the epithelial cells (bottom left) by differential centrifugation. (From Ofek, I. and Beachey, E. H., *Infect. Immun.*, 22, 247, 1978. With permission.)

sociation of the pili into their subunits can be achieved under conditions such as boiling at low pH or chemical modification,[11,13,20] which usually lead to irreversible protein denaturation.

We have found that incubation of pili in saturated guanidine-HCl (>8.3 *M*) at 37°C led to their complete dissociation, as evidenced by nephelometry and electron microscopy.[21] Gel chromatography of the dissociated pili on a Sepharose CL-6B column in the presence of saturated guanidine-HCl yielded a single protein peak with a molecular weight of ~16,000 dalton, corresponding to that of pilin (Figure 6). Dialysis of this peak against 5 m*M* Tris-HCl (pH 8.0) and rechromatography in the same buffer afforded a major protein peak, probably consisting of pilin dimers. The finding that this preparation had no discernible structure in the electron microscope also suggests that it consists of dimers, since trimers might have been expected to resemble rings. The pilin dimer gave a single protein band (molecular weight 16,600) upon polyacrylamide gel electrophoresis in the presence of 0.1% SDS or 10 *M* urea, and penetrated completely into 7% gels in the absence of denaturants.

Exposure of pili to 8.6 *M* guanidine-HCl was sufficiently mild so that the subunits could reassemble into structures resembling short fragments of intact pili, as observed in the electron microscope.[21] No reassembly occurred if the pili were dissociated by acid treatment.[11,20] From the proportion of protein in the incubation mixture that appeared in the void volume of the Sepharose CL-6B column, we estimated that about 30% of the subunits had reassembled when dialyzed against 5 m*M* $MgCl_2$ and 5 m*M* Tris buffer (pH 8.0).[21] Higher yields

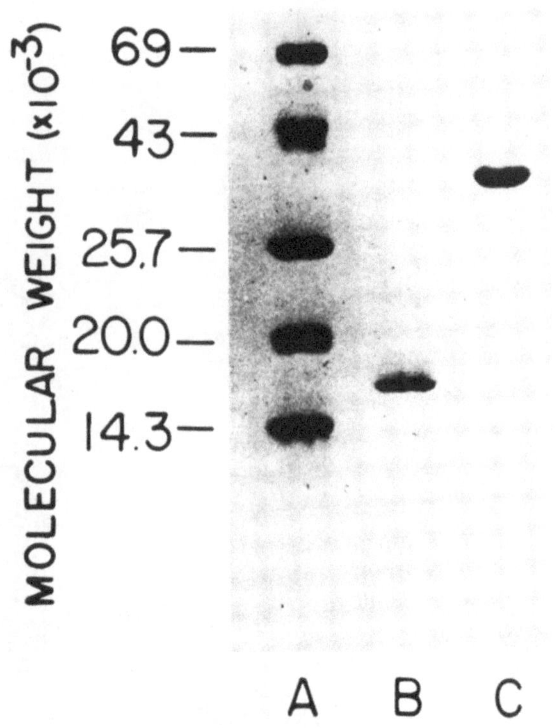

FIGURE 4. Polyacrylamide sodium dodecylsulfate gel electrophoresis of MS lectins isolated in the form of pili (B) and flagella (C). MW markers are in lane A (by courtesy of Dr. Y. Eshdat).

of reassembled pili (approximately 60%) were obtained if dialysis was done against 5 m*M* Tris buffer and 50 m*M* $MgCl_2$, but this was accompanied by protein precipitation. Only trace amounts of reassembled pili (<5%) were detected if the dialysis was carried out in the absence of $MgCl_2$ or in the presence of NaCl for 20 hr at 23°C, and none at 4°C. However, we have not systematically explored the optimal conditions for reassembly and it is possible that a higher yield could be obtained under other conditions. It is also possible that complete reassembly was not obtained because some subunits were irreversibly denatured by the treatment with guanidine-HCl or because the concentrations of subunits fell below some critical level during the reassembly process.

The subunits formed by dissociation with 8.6 *M* guanidine-HCl retained not only their ability to reassemble into intact pili but also their ability to bind to mannose. Thus, about 25% of the subunits obtained after removal of the guanidine-HCl by dialysis were bound to immobilized mannan and could be eluted by methyl α-mannoside (Figure 7). We cannot adequately explain why only part of the subunits bound specifically to the mannan column. Perhaps the majority of the subunits were not renatured after removal of guanidine-HCl to molecules capable of binding mannosyl residues. On the other hand, it is possible that only a portion of the pilin monomers in each intact pilus has mannose-binding activity, implying biological heterogeneity of the individual pili subunits.

Whatever the significance of the low proportion of binding activity, the fact that there was binding at all is additional and perhaps definitive proof that type 1 pili mediate mannose-binding activity of *E. coli*. Earlier attempts at establishing that intact pili bind to mannose could be criticized on the grounds of insufficient purity of the preparations used. In such preparations, the mannose-binding activity might have resided in contaminant proteins. The

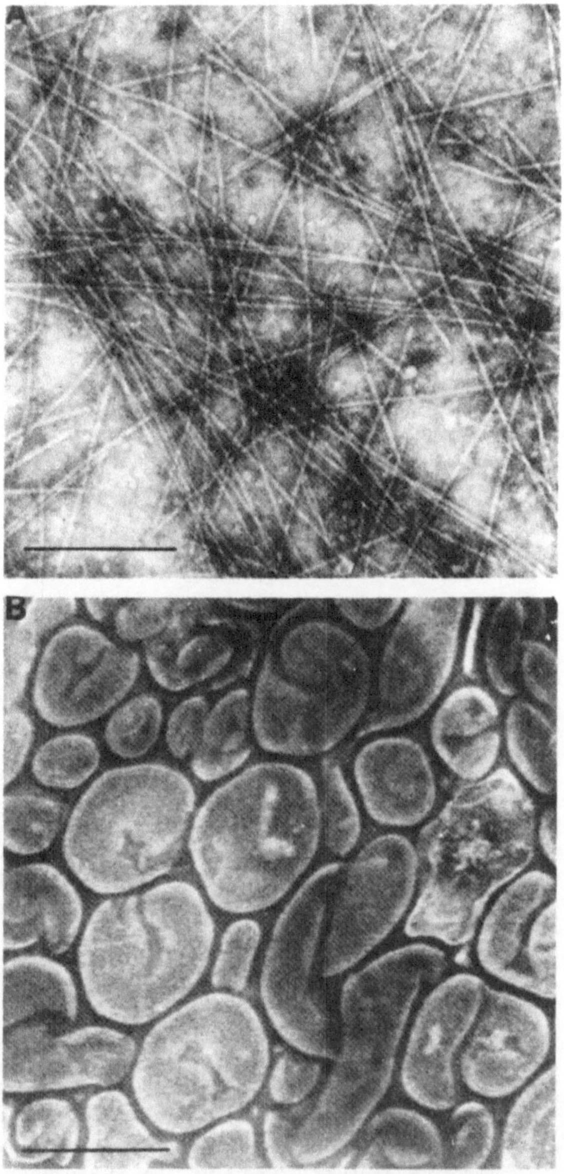

FIGURE 5. Electron micrographs of purified pili (A) and purified outer membranes (B) of *E. coli* 2699. Bar, 0.2 μm. (From Eshdat, Y., et al., *Infect. Immun.*, 34, 980, 1981. With permission.)

fact that other surface constituents such as flagella or outer membranes also have mannose-binding activity makes such an interpretation plausible. The isolation of soluble pili subunits, which behave chromatographically and electrophoretically as a single protein species and possess the ability to bind specifically and reversibly to immobilized mannan, provides strong evidence for the involvement of pili in the mannose-binding activity of *E. coli*.

The pili subunits solubilized by guanidine-HCl may be useful for further study of these appendages. Solubilization and reassembly may provide an efficient method of obtaining highly purified pili. The availability of low molecular aggregates (e.g., dimers) may permit evaluation of the influence of valence on pili-mediated surface phenomena. In addition, such highly purified low molecular weight soluble proteins may be of aid in the development of

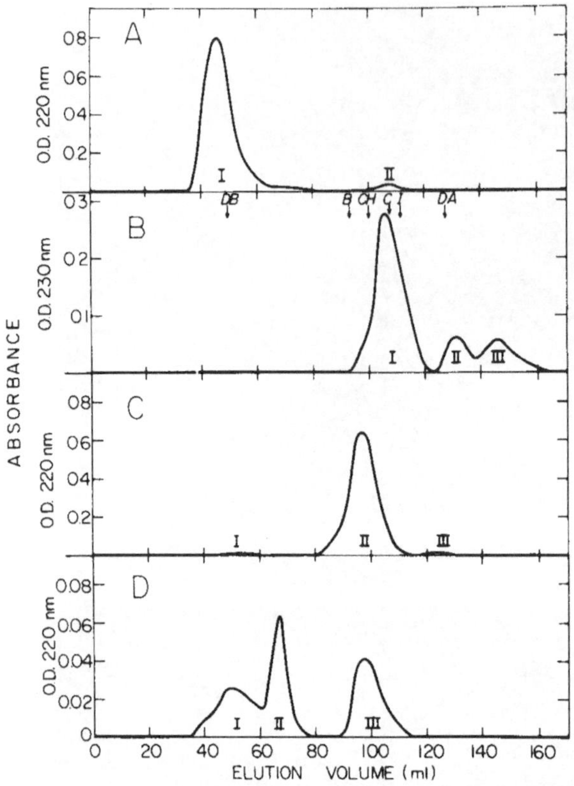

FIGURE 6. Gel filtration on a Sepharose CL-6B column (1.5 × 84 cm). Elution pattern of: (A) purified pili (3.5 mg) in 5 m*M* Tris buffer, pH 8.0; (B) dissociated pili (2.5 mg) in same buffer saturated with guanidine-HCl (protein markers used: DB = dextran blue, BSA = bovine serum albumin, CH = α-chymotrypsinogen, C = cytochrome C, I = insulin and DA = dinitrophenyl–L-alanine); (C) dissociated pili (2.5 mg) in 5 m*M* Tris buffer, pH 8.0, after dialysis of guanidine-HCl against the Tris buffer (the same pattern was obtained when dialysis and elution were done with the Tris buffer in 0.1 *M* NaCl); (D) reassembled pili (0.5 mg) in 5 m*M* Tris buffer, pH 8.0, obtained by dialysis of the dissociated pili against Tris buffer containing 5 m*M* $MgCl_2$ at 23°C. The optical density of the eluate was continuously monitored by a Beckman DB-G spectrophotometer. Fractions of 2 mℓ were collected. (From Eshdat, Y., Silverblatt, F. J., and Sharon, N., *J. Bacteriol.*, 148, 308, 1981. With permission.)

more specific vaccines, and for the isolation of pili receptors from eukaryotic cells. The procedure described here for the complete dissociation of pili can be used as an analytical tool for determination of the purity and homogeneity of pili preparations and for detection of contaminating pili in bacterial outer membrane preparations.[16]

V. CHARACTERIZATION OF THE SUGAR-BINDING SITE OF THE *E. COLI* LECTIN

Until recently, the sugar specificity of MS *E. coli* had been examined only with a limited variety of sugars, mostly monosaccharides. As a result, very little was known about the properties of the combining site of the *E. coli* lectins. The detailed characterization of this

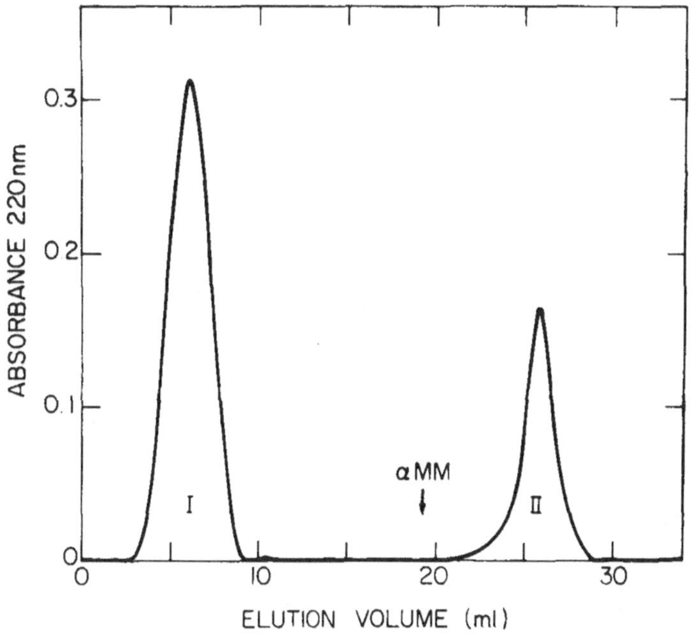

FIGURE 7. Affinity chromatography of pilin dimer (0.4 mg) on a column (1 × 15 cm) of yeast mannan linked to Sepharose 4B. Washing was with 5 m*M* Tris-HCl, pH 8.0, and elution was with 0.5 M methyl α-mannoside in the same buffer. (From Eshdat, Y., Silverblatt, F. J., and Sharon, N., *J. Bacteriol.*, 148, 308, 1981. With permission.)

combining site is important not only for gaining a better insight into the nature of the interaction between bacteria and cell surfaces, but also for the design of more effective inhibitors that will prevent adhesion, colonization of the mucosal surfaces, and infection.

We have examined the inhibitory effect of a large number of mannose-containing oligosaccharides on the agglutination of yeast cells by *E. coli* 346 and the type 1 pili isolated from this organism.[22] The validity of such an approach for the determination of the combining sites of plant and animal lectins is well established.[23] However, no studies of this type had been reported for the bacterial lectins.

Quantitative evaluation of the inhibitory activity of various mannose derivatives was performed by measuring in an aggregometer[8] their effect on the agglutination of yeast cells by the bacteria. Figure 8 shows the results of two typical experiments in which the effect of different concentrations of mannose glycosides on the rate of yeast aggregation by *E. coli* 346 was assayed. The slopes of the lines were used to calculate the rates of aggregation, from which the percent of inhibition was obtained. In all cases, a linear correlation is observed between the percent inhibition and the logarithm of sugar concentration, within the range tested, although not all lines are parallel (Figure 9). For each sugar the concentration causing 50% inhibition was derived from its inhibition line and the relative inhibitory activity in comparison to that of methyl α-mannoside was calculated.

The data summarized in Table 2 show that of the oligosaccharides tested, Manα1→6[Manα1→3]Manα1→6[Manα1→3]ManαOMe and Manα1→6[Manα1→3]-Manα1→6-[Manα1→2Manα1→3]ManαOMe are the best inhibitors of yeast agglutination by *E. coli* 346. The trisaccharide Manα1→3Manβ1→4GlcNAc, as well as the aromatic glycoside p-nitrophenylα-mannoside, are also strong inhibitors. On theother hand, the disaccharides Manα1→2Man, Manα1→3Man and Manα1→6Man are considerably weaker inhibitors, similar to methyl α-mannoside. The combining site of this *E. coli* lectin is therefore

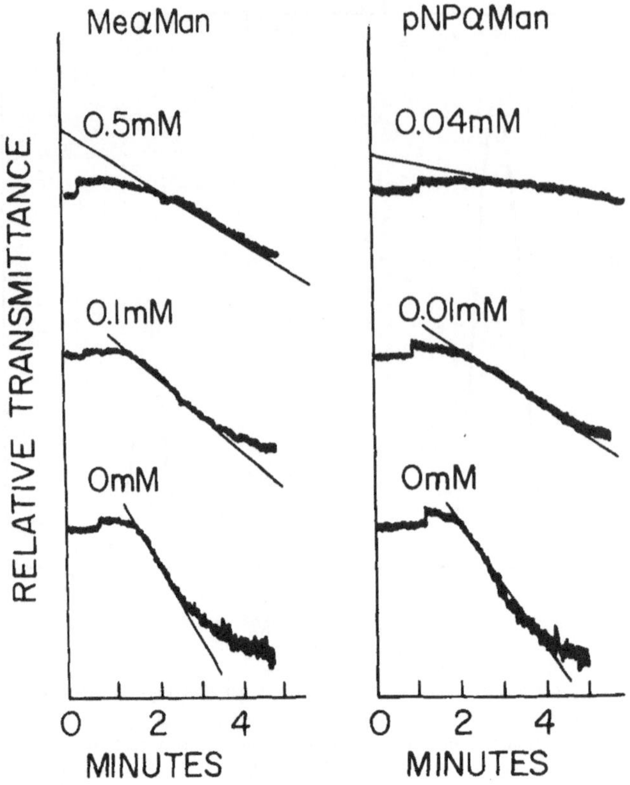

FIGURE 8. Effect of methyl α-mannoside (MeαMan) and *p*-nitrophenyl α-mannoside (pNPαMan) on the aggregation of mannan-containing yeast cells by *E. coli* 346, as measured in a Payton aggregometer. (From Firon, N., Ofek, I., and Sharon, N., *Biochem. Biophys., Res. Commun.*, 105, 1426, 1982. With permission.)

an extended one. The finding that the trisaccharide Manα1→3Manβ1→4GlcNAc is a much better inhibitor than the disaccharide Manα1→3Man, as well as the tetrasaccharide Manα1→2Manα1→3Man-β1→4GlcNAc and the pentasaccharide Manα1→2Manα1→2Man-α1→3Manβ1→4GlcNAc, strongly suggests that the combining site corresponds to the size of a trisaccharide, and that it is in the form of a depression or pocket on the surface of the lectin (Figure 10). Extended carbohydrate-binding sites have been described for enzymes (e.g., lysozyme), several lectins, and antibodies.[23] In the case of the *E. coli* lectin there are probably three subsites, each of which could fit a monosaccharide residue. The fact that there is only little difference between the inhibitory activity of methyl α-mannoside and the disaccharides Manα1→3Man, Manα1→6Man, and Manα1→2Man is an indication that the binding to one of the subsites is weak. The strong inhibition by *p*-nitrophenyl α-mannoside may be due to the existence of a hydrophobic region, adjacent to one of the subsites.

Our data demonstrate that the combining site of the *E. coli* 346 lectin best fits structures found in short oligomannosidic chains of *N*-glycosidically linked glycoproteins.[24] The proposed site is, however, quite different from that of concanavalin A, a well-characterized lectin with closely related sugar specificity.[25] First, while the latter is inhibited by both mannose and glucose, *E. coli* is not inhibited by glucose.[26] Second, Manα1→2Man is a considerably better inhibitor of concanavalin A than is methyl α-mannoside, whereas there is little difference between the inhibitory activity of this disaccharide and methyl α-mannoside on yeast agglutination by *E. coli*. Studies with additional oligosaccharides, as well as detailed examination of molecular models of the various inhibitory compounds, are necessary in

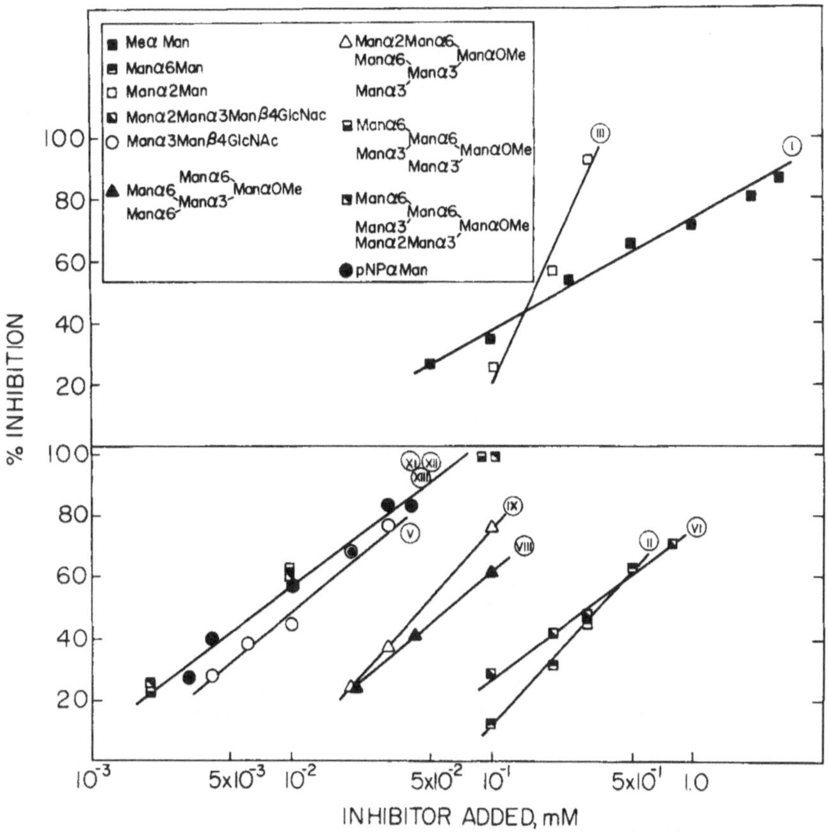

FIGURE 9. Inhibition by mannose derivatives of yeast agglutination by *E. coli* 346. For numbering of sugars tested, see Table 2. (From Firon, N., Ofek, I., and Sharon, N., *Biochem. Biophys. Res. Commun.*, 105, 1426, 1982. With permission.)

order to obtain further insight into the structure of the combining site of the *E. coli* lectin.

The very good correlation found between the relative inhibitory power of the different sugars tested with isolated pili and intact bacteria ($r = 0.975$; $n = 6$; $p < 0.001$) shows that the pili are indeed responsible for the MS adherence of the bacteria to cells.

The variation in the slopes of the different sugars tested (Figure 9) may be accounted for by assuming that the bacterial pili or their subunits exist as a family of lectins (or isolectins) with closely related specificities, as suggested by Kisailus and Kabat[27] for the *Bandeiraea simplicifolia* isolectins.

Work is in progress in our laboratory to establish whether type 1 pili produced by different strains of *E. coli* and other enterobacteria display the same pattern of specificity as the *E. coli* employed in the present study.

VI. PHASE VARIATION IN VIVO

In attempting to clarify the role of the bacterial MS lectins in infection, the well-known ability of *E. coli* and related organisms to undergo reversible transition from a piliated phase to a nonpiliated phase, a phenomenon designated as "phase variation",[11] must be taken into consideration. As shown by studies in vitro, this variation occurs suddenly and at random, with a frequency of about 10^{-3}, which is much higher than mutation. It is affected by various environmental factors, most importantly the number of laboratory passages and the phase and conditions of bacterial growth,[2,28] as also illustrated in Figure 11. Type 1 pili are

Table 2
INHIBITION BY MANNOSE DERIVATIVES OF YEAST AGGLUTINATION BY *E. COLI* 346 AND BY ISOLATED PILI

Me α Man	I	1.00	1.00
Man α 1→ 6Man	II	0.5	n.t
Man α 1→ 2Man	III	1.3	1.8
Man α 1→ 3Man	IV	1.2 *	n.t.
Man α 1→ 3Man β 1→ 4GlcNAc	V	21	30
Man α 1→ 2Man α 1→ 3Man β 1→ 4GlcNAc	VI	0.7	5.5
Man α 1→ 2Man α1→ 2Man α1→ 3Man β 1→ 4GlcNac	VII	0.7 *	n.t.
Man α 1→ 6 \\ Man α 1→ 6 Man α OMe \\ / Man α 1→ 3 / Man α 1→ 3	VIII	3.5	n.t.
Man α 1→ 2Man α 1→ 6 \\ Man α 1→ 6 Man α OMe \\ / Man α 1→ 3 / Man α 1→ 3	IX	4.7	4.8
Man α 1→ 6 \\ Man α OMe / Man α 1→ 3	X	10.5 *	n.t.
Man α 1→ 6 \\ Man α 1→ 6 / \\ Man α 1→ 3 Man α OMe / Man α 1→ 3	XI	30	n.t.
Man α 1→ 6 \\ Man α 1→ 6 / \\ Man α 1→ 3 Man α OMe / Man α 1→ 2Man α 1→ 3	XII	30	36
pNP α Man	XIII	30	48
Gal β1→ 4GlcNAc β1→ 2Man α 1→ 6Man β1→ 4GlcNAc	XIV	0.25 *	n.t.
Gal β1→ 3glcNAc β1→ 3Gal β1→ 4Glc	XV	inactive at 0.8 mM	n.t.
Gal β1→ 4GlcNAc β1→ 2Man	XVI	inactive at 1.6 mM	n.t.
Gal β1→ 4GlcNAc β1→ 6Man	XVII	inactive at 1.6 mM	n.t.
Gal β1→ 4GlcNAc β1→ 2 \\ Man / Gal β1→ 4GlcNAc β1→ 6	XVIII	inactive at 1.6 mM	n.t.

Note: n.t. = not tested.

[a] Because of shortage of material, only two inhibitory concentrations were tested and no line is given in Figure 8.

From Firon, N., Ofek, I., and Sharon, N., *Biochem. Biophys. Res. Commun.*, 105, 1426, 1982.

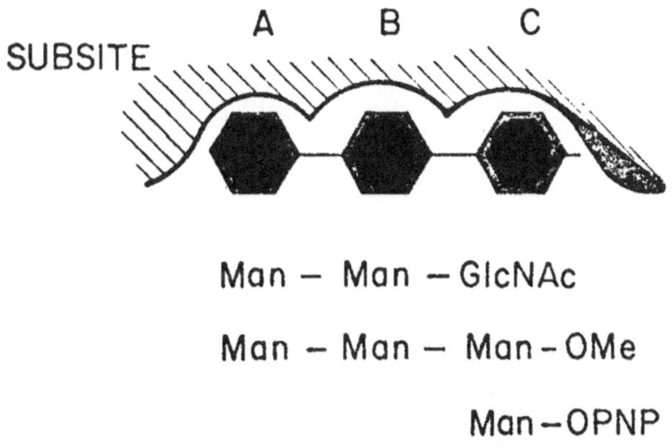

FIGURE 10. Proposed schematic structure of combining site of *E. coli* lectin.

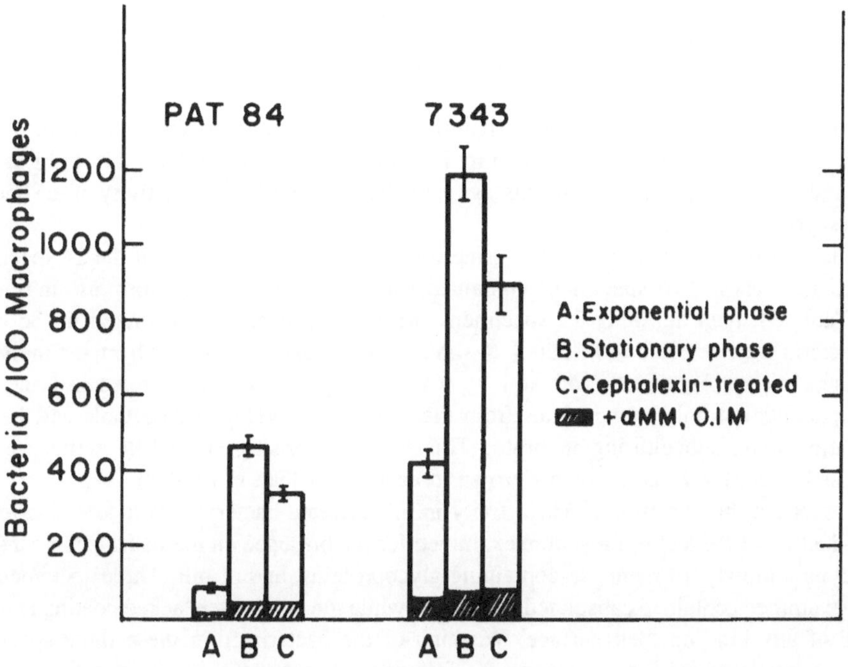

FIGURE 11. Effect of growth conditions on the MS adherence to mouse peritoneal macrophages of *E. coli* strains PAT 84 and 7343. αMM, methyl α-mannoside. (Data from Bar-Shavit, Z., et al., *Infect. Immun.*, 29, 417, 1980.)

best produced by organisms grown for 48 hr in static conditions in broth, but not on agar, while the presence of glucose (1%) in the medium suppresses their formation. Very recently it was demonstrated that phase variation of type 1 pili in *E. coli* is under transcriptional control.[29] It was not known, however, whether phase variation occurs in vivo, and if the MS lectin is formed during the course of natural infection. To obtain information on this problem, we compared the MS lectins of *E. coli* excreted in the urine of patients with urinary tract infections with that of *E. coli* isolated and subcultured in vitro from the same urine.[30,31]

The study group (24 patients) was comprised of elderly individuals — 21 females and 3 males — with a mean age of 75, all of whom were under continuous in-dwelling catheter. The bacteria from the bulk of the urine of the patients were collected by centrifugation,

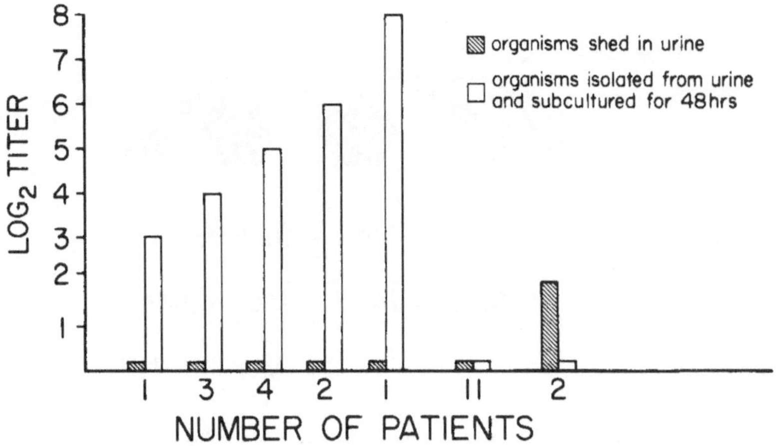

FIGURE 12. Yeast cell aggregation by *E. coli* obtained from patients with urinary tract infections. In all cases yeast agglutination was inhibited by mannose (MS). Hemagglutination of human group A erythrocytes was as a rule MR. (From Ofek, I., Mosek, A., and Sharon, N., *Infect. Immun.*, 34, 708, 1981. With permission.)

washed, and adjusted to contain 10^9 cells/mℓ, of which more than 90% were alive. *E. coli* isolates from each urine were cultivated in 5 mℓ brain heart infusion broth for 48 hr at 37°C under static conditions for only one passage, and the mannose-binding activity of the bacteria was assayed by yeast agglutination.

MS lectins from *E. coli* excreted in urine were detected in only 2 out of the 24 specimens (Figure 12). These two specimens contained somewhat more colony-forming units of *E. coli* (Table 3), than the other 22 specimens. In contrast to bacteria excreted in the urine, MS bacteria were detected in 11 out of 24 strains isolated and subcultured from the individual urine specimens (Figure 12). MR activity was detected in bacteria excreted in urine from seven patients but only four isolates from these patients developed detectable MR hemagglutination upon subculturing in broth. This is not surprising since MR activity is best expressed in certain *E. coli* isolates grown in agar rather than in broth.[31]

To ascertain that the lack of MS activity in the excreted bacteria was indeed due to lack of production of the lectin, they were examined for the presence on their surface of capsules, of coating antibody, of mannose-containing glycoproteins, and of pili. Three specimens out of six examined contained capsulated organisms while the other three lacked coating antibody or pili of any kind on their surface. Washing of the bacteria from these three specimens with a solution of methyl α-mannoside, followed by extensive washing with phosphate buffered saline, failed to reveal any MS activity, suggesting that mannose-containing glycoproteins which might block MS activity of the bacteria were not present on the surface of these organisms.

Our results clearly demonstrated that *E. coli* excreted in the urine of patients lack MS activity. In about half of the cases, the organisms were genotypically capable of producing the MS lectins when subcultured and grown in vitro, suggesting that growth conditions in the bladder do not permit the phenotypic expression of the lectin on the *E. coli* surface.

It is unlikely that the development of MS lectins by strains subcultured in vitro is due to selective outgrowth of spontaneously arising variants of MS bacteria, since for this to occur several successive cultivations are needed.[2,31] Growth in vitro resembles "chemostat conditions" in that it allows continuous proliferation under conditions of constant supply of limited nutrients and removal of excess bacterial population and its deleterious products.[32] It follows that strains of *E. coli* which are genotypically competent to express MS lectins

Table 3
MS ADHERENCE BY *E. COLI* EXCRETED AND GROWING IN VITRO IN URINES OF PATIENTS WITH URINARY TRACT INFECTIONS

Urine specimen	MS adherence and viable counts (CFU/mℓ urine) of bacteria growing in infected urine collected from patients after			
	3 Hr (at 4°C)		24 Hr (at 37°C)	
	CFU/mℓ	MS	CFU/mℓ	MS
I	5×10^5	<1	2×10^8	2
II	2×10^6	<1	5×10^8	3
III	5×10^8	2	n.d.	
IV	2×10^8	2	n.d.	

Note: MS adherence is expressed as the $\log_2$ of highest dilution of bacterial suspension (10^9/mℓ) which agglutinated a standard suspension of yeast cells. Viable counts are expressed as colony forming units (CFU) on agar per milliliter urine sample.

From Ofek, I., Mosek, A., and Sharon, N., *Infect. Immun.*, 34, 708, 1981. With permission.

and were growing in the bladder of the patients examined in this study, were in a phase of growth which represses formation of the lectins, similar to bacteria at the logarithmic phase of growth in vitro. This conclusion is supported by our findings that (1) *E. coli* excreted in urine, which lacked MS lectin, acquired this activity after 24 hr growth in the urine in vitro (Table 3) and (2) the two urine specimens which contained *E. coli* with detectable MS lectin had more bacteria than those with bacteria lacking the lectins, suggesting that in these two specimens the organisms reached stationary phase of growth which permitted the phenotypic expression of the MS lectin.

Since antibodies to type 1 pili of *E. coli* are produced by patients and animals with pyelonephritis,[33,34] our results can be best explained by postulating that infective *E. coli* possessing MS lectins undergo a phase variation at certain stages of the infectious process which does not permit the expression of the lectins. Evidence for the occurrence of phase variation in vivo has also been presented by Stamey,[35] who found that attachment to vaginal epithelial cells of bacteria taken directly from urine was always less than that exhibited by the bacteria after growth in broth. He emphasized that these observations suggest that caution should be exercised in drawing conclusions from in vitro studies to in vivo pathogenesis.

Since the lectins enable the organism to adhere to both epithelial cells and phagocytic cells,[1,9] the bacteria may benefit from the inability to express the MS activity in vivo after infection has been initiated. Thus, the presence of a MS lectin on bacterial surfaces may be crucial in initiating infection, as it enables the organisms to adhere to the epithelial cells (e.g., of the urogenital tract[36,37]) in order not to be swept away by the flow of urine. The absence of the lectins on bacterial surfaces at subsequent stages of the infection in deeper tissue may enable the organisms to escape ingestion and killing by phagocytic cells which are capable of binding the bacteria via the MS lectin. This may be true not only for urinary tract infections, but also for infections of other tissue, such as the gastrointestinal tract.

Urinary tract infections are commonly initiated by the fecal flora. It is of interest, therefore,

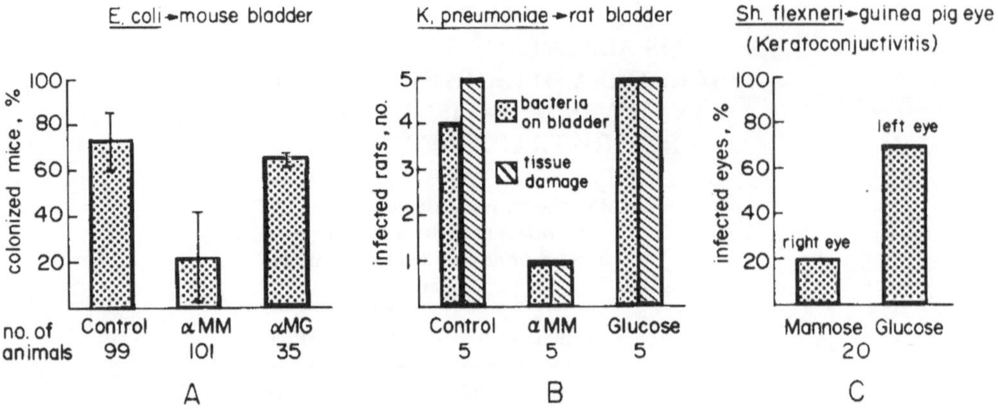

FIGURE 13. Sugar inhibitors of bacterial adherence prevent infection. Data from: (A) Aronson *et al.*[38]; (B) Fader and Davis[40]; (C) Andrade[41]. Sugar concentration used: (A) 10%; (B) 5%; (C) 2%.

to establish whether *E. coli* in this flora possess MS lectins which can potentially bind the organisms to epithelial cells to initiate colonization of the urinary tract surfaces.

VII. EFFECT OF INHIBITORY SUGARS ON EXPERIMENTAL INFECTION

To evaluate directly the role of the MS bacterial lectins in initiating infection, Aronson et al.[38] have tested the possibility of preventing infection by inhibition of adherence. In these experiments, diuretic mice which are highly susceptible to urinary tract infection were used.[39] MS bacteria (5×10^4 *E. coli* 7343 or 1×10^6 *E. coli* G12, both isolated from cases of pyelonephritis in humans) in 0.05 mℓ of saline, or of a sugar solution (10% methyl α-mannoside or methyl α-glucoside) were injected into the bladder of mice. Urine was then collected every 2 or 3 days. Bacteriuria was defined as the presence of 10^5 to 10^6 bacteria/mℓ of urine on days 5 to 16 postinfection; animals were considered nonbacteriuric when three successive urine cultures proved negative.

The results summarized in Figure 13A show that when injection was in methyl α-mannoside, a significant reduction in the extent of bacteriuria occurred, in comparison with the control group. As expected, methyl α-glucoside (which is not an inhibitor of MS adherence) did not affect the number of bacteriuric mice. Inspection under the microscope of stained bladders of mice that were each injected with *E. coli* (10^9 cells) in methyl α-mannoside, revealed a considerably lower number of adherent bacteria than control mice injected with bacteria in saline. It should be noted that exposure of the *E. coli* strains used in this study to a 20% solution of methyl α-mannoside at 37°C for 24 hr did not decrease their viability. When *Proteus mirabilis* was injected into the mice instead of *E. coli*, no effect on the rate of infection was observed by either of the sugars tested, in accordance with the inability of these sugars to inhibit the adherence of *P. mirabilis* to epithelial cells.

In a subsequent study, MS piliated and nonpiliated phase organisms derived from a single strain of *Klebsiella pneumoniae*, were used to examine the effect of pili on experimental infection by the bacteria in rat bladder[40] (Figure 13B). In addition, methyl α-mannoside was tested to determine whether it could protect the bladder surface from adherence of *K. pneumoniae*. In animals infected with piliated bacteria, foci of infection appeared in the majority of bladders examined. Nonpiliated bacteria, in contrast, caused little infection. When bacteria (5×10^7 cells) were injected in the presence of methyl α-mannoside (but not of glucose) a significant decrease in the number of infected rats was observed.

The role of MS adherence in the pathogenicity of infection was also assessed in a different system, by examining the effect of mannose on the ability of a virulent and piliated strain of *Shigella flexneri* (strain 1225/53) to cause experimental keratoconjunctivitis in guinea pigs.[41] For this purpose albino guinea pigs were inoculated in the right eye with the bacteria (9×10^7 cells) suspended in mannose (2%) and in the left eye with the same number of bacteria in glucose (also 2%). Daily observations were then made for 1 week. A remarkable decrease in the infectivity of the bacteria occurred in the presence of mannose, as compared to that of glucose (Figure 13C).

These independent studies clearly demonstrate that MS lectins are a key factor in the ability of various MS bacteria to cause experimental infection, by mediating the attachment of the bacteria to the surfaces of the target tissues. It remains, however, to be seen whether inhibitors of lectin-mediated adherence may also be useful in preventing natural infection.

VIII. CONCLUSIONS

1. Many bacteria possess the ability to adhere to sugars, most commonly to mannose residues, on the surface of epithelial cells.
2. This adherence is mediated by lectins present on the bacterial surface, primarily in the form of pili.
3. Mannose-specific (type 1) pili can be dissociated reversibly into subunits that retain their sugar binding activity.
4. The sugar combining site of *E. coli* lectin corresponds to the size of a trisaccharide, and best fits structures found in short oligomannosidic chains of cell surface glycoproteins.
5. *E. coli* excreted in the urine of patients with urinary tract infection do not possess MS surface lectins, whereas when cultivated in vitro about half of the strains produce the lectin, demonstrating that the bacteria may undergo phase variations in vivo.
6. Methyl α-mannoside or mannose prevent experimental infection by MS strains of *E. coli*, *K. pneumoniae*, or *S. flexneri*.
7. Sugar inhibitors of bacterial adherence may prove to be useful in the prevention of natural infection.

REFERENCES

1. **Bar-Shavit, Z., Ofek, I., Goldman, R., Mirelman, D., and Sharon, N.,** Mannose residues on phagocytes as receptors for the attachment of *Escherichia coli* and *Salmonella typhi*, *Biochem. Biophys. Res. Commun.*, 78, 455, 1977.
2. **Duguid, J. P. and Old, D. C.,** Adhesive properties of enterobacteriacae, in *Bacterial Adherence*, Beachey, E. H., Ed., Chapman and Hall, London, 1980, 186.
3. **Sharon, N., Eshdat, Y., Silverblatt, F. J., and Ofek, I.,** Bacterial adherence to cell surface sugars, in *Adhesion and Micro-Organisms Pathogenicity*, Elliott, K., O'Connor, M., and Whelan, J., Eds., Pitman Press, London, 1981, 119.
4. **Kallenius, G., Mollby, R., Svenson, S. B., Winberg, J., Lundblad, A., Svensson, S., and Cedergren, B.,** The P^K antigen as receptor for the hemagglutinin of pyelonephritic *Escherichia coli*, *FEMS Microbiol. Lett.*, 7, 297, 1980.
5. **Leffler, H. and Svanborg-Eden, C.,** Chemical identification of a glycosphingolipid receptor for *Escherichia coli* attaching to human urinary tract epithelial cells and agglutinating human erythrocytes, *FEMS Microbiol. Lett.*, 8, 127, 1980.
6. **Ofek, I., Mirelman, D., and Sharon, N.,** Adherence of *Escherichia coli* to human mucosal cells mediated by mannose receptors, *Nature (London)*, 265, 623, 1977.
7. **Goldstein, I. J., Hughes, R. C., Monsigny, M., Osawa, T., and Sharon, N.,** What should be called a lectin? *Nature (London)*, 285, 66, 1980.
8. **Ofek, I. and Beachey, E. H.,** Mannose binding and epithelial cell adherence of *Escherichia coli*, *Infect. Immun.*, 22, 247, 1978.

9. **Bar-Shavit, Z., Goldman, R., Ofek, I., Sharon, N., and Mirelman, D.,** Mannose binding activity of *Escherichia coli*: a determinant of attachment and ingestion of the bacteria by macrophages, *Infect. Immun.*, 29, 417, 1980.
10. **Duguid, J. P. and Gillies, R. R.,** Fimbria and adhesive properties in dysentery bacilli, *J. Pathol. Bacteriol.*, 74, 397, 1957.
11. **Brinton, C. C., Jr.,** The structure, function, synthesis and genetic control of bacterial pili and a molecular model for DNA and RNA transport in Gram-negative bacteria, *Trans. N.Y. Acad. Sci.*, 27, 1003, 1965.
12. **Rivier, D. A. and Darekar, M. R.,** Inhibitors of the adhesiveness of enteropathogenic *E. coli*, *Experientia (Basel)*, 31, 662, 1975.
13. **Salit, I. E. and Gotschlich, E. C.,** Hemagglutination by purified type I *Escherichia coli* pili, *J. Exp. Med.*, 146, 1169, 1977.
14. **Salit, I. E. and Gotschlich, E. C.,** Type I *Escherichia coli* pili: characterization of binding to monkey kidney cells, *J. Exp. Med.*, 146, 1182, 1977.
15. **Korhonen, T. K.,** Yeast cell agglutination by purified enterobacterial pili, *FEMS Microbiol. Lett.*, 6, 421, 1979.
16. **Eshdat, Y., Speth, V., and Jann, K.,** Participation of pili and cell wall adhesin in the yeast agglutination activity of *Escherichia coli*, *Infect. Immun.*, 34, 980,1981.
17. **Eshdat, Y., Ofek, I., Yashouv-Gan, Y., Sharon, N., and Mirelman, D.,** Isolation of a mannose-specific lectin from *Escherichia coli* and its role in the adherence of the bacteria to epithelial cells, *Biochem. Biophys. Res. Commun.*, 85, 1551, 1978.
18. **Eshdat, Y., Sharon, N., Ofek, I., and Mirelman, D.,** Structural association of the outer surface mannose-specific lectin of *Escherichia coli* with bacterial flagella, *Isr. J. Med. Sci.*, 16, 479, 1980.
19. **Eshdat, Y., Izhar, M., Sharon, N., and Mirelman, D.,** Flagellar lectin as mediator for adherence of *Serratia marcescens* to eukaryotic cell surfaces, Israel Biochem. Soc. Meeting, Jerusalem, 1981.
20. **McMichael, J. C. and Ou, J. T.,** Structure of common pili from *Escherichia coli*, *J. Bacteriol.*, 138, 969, 1979.
21. **Eshdat, Y., Silverblatt, F. J., and Sharon, N.,** Dissociation and reassembly of *Escherichia coli* type 1 pili (fimbriae), *J. Bacteriol.*, 148, 308, 1981.
22. **Firon, N., Ofek, I., and Sharon, N.,** Interaction of mannose-containing oligosaccharides with the surface fimbrial lectins of *Escherichia coli*, *Biochem. Biophys. Res. Commun.*, 105, 1426, 1982.
23. **Kabat, E. A.,** Dimensions and specificities of recognition sites on lectins and antibodies, *J. Supramol. Struct.*, 8, 79, 1978.
24. **Sharon, N. and Lis, H.,** Glycoproteins, in *The Proteins*, 3rd ed., Vol. V, Neurath, H. and Hill, R. L., Eds., Academic Press, New York, 1982, 1.
25. **Brinton, C. C., Jr.,** The pilation phase syndrome and the uses of purified pili in disease control, in *Proceedings of the 13th Joint U.S.-Japan Conference on Cholera*, Atlanta, GA, Department of Health, Education and Welfare Publication No. 78-1590, National Institutes of Health, Bethesda, 1977, 33.
26. **Duguid, J. P., Clegg, S., and Wilson, M. J.,** The fimbrial and non-fimbrial haemagglutinins of *Escherichia coli*, *J. Med. Microbiol.*, 12, 213, 1979.
27. **Kisailus, E. C. and Kabat, E. A.,** A study of the specificity of *Bandeiraea simplicifolia* lectin I by competitive-binding assay with blood-group substances and with blood-group A and B active and other oligosaccharides, *Carbohyd. Res.*, 67, 243, 1978.
28. **Ottow, J. C. G.,** Ecology, physiology and genetics of fimbriae and pili, *Annu. Rev. Microbiol.*, 29, 79, 1975.
29. **Eisenstein, B. I.,** Phase variation of type 1 fimbriae in *Escherichia coli* is under transcriptional control, *Science*, 214, 337, 1981.
30. **Ofek, I., Mosek, A., and Sharon, N.,** Mannose specific adherence of *Escherichia coli* freshly excreted in the urine of patients with urinary tract infections and of isolates subcultured from the infected urine, *Infect. Immun.*, 34, 708, 1981.
31. **Ofek, I., Goldhar, J., Eshdat, Y., and Sharon, N.,** The importance of MS adhesins (lectins) in infections caused by *Escherichia coli*, *Scand. J. Infect. Dis.*, Suppl. 33, 61, 1982.
32. **Ellwood, D. C. and Tempest, D. W.,** Effects of environment on bacterial wall content and composition, *Adv. Microbiol. Physiol.*, 7, 83, 1972.
33. **Rene, P. and Silverblatt, F. J.,** Serological response to *Escherichia coli* in pyelonephritis, in *Current Chemotherapy and Infectious Disease*, Vol. I, Nelson, J. D. and Grassi, C., Eds., The American Society for Microbiology, Washington, D.C., 1980, 782.
34. **Smith, J. W., Wagner, S., and Swenson, R. M.,** Local immune response to *Escherichia coli* pili in experimental pyelonephritis, *Infect. Immun.*, 31, 17, 1981.
35. **Stamey, T. A.,** *Pathogenesis and Treatment of Urinary Tract Infections*, Williams & Wilkins, Baltimore, 1980, 265.
36. **Schaeffer, A. J., Amundsen, S. K., and Schmidt, L. N.,** Adherence of *Escherichia coli* to human urinary tract epithelial cells, *Infect. Immun.*, 24, 753, 1979.

37. **Van den Bosch, J. F., Verbrom-Sohmer, U., Postma, P., de Graaff, J., and MacLaren, P. D. M.,** Mannose-sensitive and mannose-resistant adherence to human uroepithelial cells and urinary virulence of *Escherichia coli*, *Infect. Immun.*, 29, 226, 1980.
38. **Aronson, M., Medalia, O., Schori, L., Mirelman, D., Sharon, N., and Ofek, I.,** Prevention of colonization of the urinary tract of mice with *Escherichia coli* by blocking of bacterial adherence with methyl α-D-mannopyranoside, *J. Infect. Dis.*, 139, 329, 1979.
39. **Freedman, L. R.,** Experimental pyelonephritis. XIII. On the ability of water diuresis to induce susceptibility to *E. coli* bacteriuria in the normal rat, *Yale J. Med.*, 39, 255, 1967.
40. **Fader, R. C. and Davis, C. P.,** Effect of piliation on *Klebsiella pneumoniae* infection in rat bladders, *Infect. Immun.*, 30, 554, 1980.
41. **Andrade, J. R. C.,** Role of fimbrial adhesiveness in experimental guinea pig keratoconjunctivitis by *Shigella flexneri*, *Rev. Microbiol. (S. Paulo)*, 11, 117, 1980.

Chapter 15

THE ROLE OF PILI AND CAPSULE IN COLONIZATION AND BACTEREMIA WITH *ESCHERICHIA COLI* K1 IN NEONATAL RATS

Nicholas G. Guerina and Donald A. Goldmann

TABLE OF CONTENTS

I. INTRODUCTION

Escherichia coli K1 is a major cause of Gram-negative bacillary bacteremia and meningitis in neonates. The K1 capsular serotype has been recovered from more than 80% of newborns with *E. coli* meningitis and has been associated with a particularly devastating outcome.[1,2] The virulence of *E. coli* K1 has been attributed to the failure of neonatal host defenses to respond adequately to the polysaccharide coat of K1 strains.[3-5]

Rather than studying the fate of *E. coli* K1 after it has reached the neonate's bloodstream, we have chosen to investigate colonization of the neonatal alimentary tract. If colonization of specific epithelial surfaces with *E. coli* K1 is a prerequisite for bacterial invasion, it is possible that an understanding of the colonization process will permit prophylactic intervention before mucoid barriers have been breached.

For our colonization studies, we have modified a previously described neonatal rat model of *E. coli* K1 bacteremia and meningitis.[6] Confirming the work of other investigators, we found that rats fed *E. coli* K1 shortly after birth experienced a high rate of bacteremia. Oral feedings of 6.0×10^6 cfu of *E. coli* K1 delivered atraumatically via an Eppendorf pipette produced bacteremia rates of 20 to 81%, depending on the strain used, whereas other *E. coli* serotypes produced negligible bacteremia. K1 strains promptly, reliably, and stably colonized the oropharynx, duodenum, jejunum, ileum, and colon. Having documented the ability of *E. coli* K1 to colonize the entire neonatal rat alimentary tract, we then attempted to determine whether colonization is the result of adherence of *E. coli* to the oropharyngeal or gastrointestinal epithelium and whether mucosal colonization is mediated by *E. coli* pili.

II. THE PILIATED PHASES OF *E. COLI*

Pili are filamentous protein appendages found on the surface of many bacteria. A given bacterial strain may exist in both a piliated and nonpiliated phase. Bacteria in the piliated phase exhibit many properties which may contribute to their pathogenicity, including adhesion to mammalian surfaces, resistance to phagocytosis, and the capacity to colonize and invade natural hosts.[7]

Among the best characterized *E. coli* pili are the type 1 pili. These pili have been shown to bind receptors on mammalian red blood cells (RBCs) and specific epithelial cells.[8-11] Binding involves the recognition of mannose residues present in the receptor since hemagglutination (HA) by type 1-piliated bacteria or purified type 1 pili is specifically inhibited by free mannose and mannose derivatives.[9-11] Because of this mannose-dependent HA, type 1 pili have been referred to as mannose-sensitive (MS) pili.

E. coli express other types of pili which differ from MS pili in their chemical, physical, and serological properties. These pili hemagglutinate the RBCs of various species in the presence of mannose, a property giving rise to the terms mannose-resistant (MR) or non-mannose-sensitive (NMS) pili. Both of these terms are deficient since neither is descriptive of the actual receptors recognized by these pili. Furthermore, they place in the same group chemically and serologically unrelated pili, even when HA subclassifications based on RBC species specificity are used. However, both terms serve to distinguish these pili from MS pili.* In this chapter, the MS-piliated phase is designated MS+NMS−, the NMS-piliated phase is designated MS−NMS+, and the nonpiliated phase is designated MS−NMS−. Mixed piliated phases can be obtained, but they were not used in the present study.

* The HA classification of *E. coli* pili is further complicated by the existence of non-HA pili. We have not demonstrated these pili on any of the *E. coli* K1 isolates we have studied.

III. CLONING OF THE PILIATED PHASES OF *E. COLI* K1

A given strain of *E. coli* may shift between the MS-piliated and nonpiliated phases. This drifting, known as phase variation, is strongly influenced by growth conditions.[7] This is an important property because it permits the isolation and propagation of cultures containing either MS-piliated or nonpiliated bacteria. For example, when *E. coli* are grown overnight on minimal glucose (MG) agar, the nonpiliated phase is greatly favored, and the identification and selective passage of colonies containing predominantly nonpiliated bacteria leads to a stable nonpiliated culture. However, the identification and selective passage of only colonies containing predominantly MS-piliated organisms results in a culture containing 90 to 100% type 1-piliated bacteria.

The distinguishing colonial features which permit the identification of piliated and nonpiliated colonies on MG agar may be very subtle. This is especially true for *E. coli* K1, perhaps due to the presence of the K1 capsule (Figure 1). Nevertheless, a distinction can usually be made, provided there are piliated phase bacteria in the initial culture. This point is crucial. For example, clinical isolates of *E. coli* are routinely passaged nonselectively on rich laboratory media giving rise to nonpiliated cultures. When these strains are subsequently studied in the laboratory, they may be erroneously reported not to express MS pili unless special growth conditions[7,12] which favor the expression of MS pili are used

E. coli express certain NMS pili on MG agar, but the number produced is very low. However, the expression of NMS pili is greatly enhanced on casamino acid yeast extract (CAYE) agar.[13] If *E. coli* are first cloned into the nonpiliated phase on MG agar, then transferred to CAYE agar, a pure culture of NMS-piliated bacteria is obtained. Thus, it is possible to produce cultures of a given strain of *E. coli* in either the MS+NMS−, MS−NMS+, or MS−NMS− phase.

We have examined the piliated phases of 15 clinical isolates of *E. coli* K1. All of these isolates express MS pili, but only 11 express NMS pili. Two strains, Bla and C94, were selected for extensive study. The piliated phases and serotypes of these and other strains used in this study are listed in Table 1.

IV. ROLE OF PILI AND CAPSULE IN THE VIRULENCE OF *E. COLI* K1

In our preliminary animal studies we found strain B1a to produce higher bacteremia and mortality rates than any of our other strains (81% bacteremia, 45% mortality). We have not detected NMS pili on this strain, suggesting that NMS pili are not virulence factors for *E. coli* K1. To further examine the relationship between piliation and virulence, 1-day-old rats were challenged with 10^6 cfu of C94 and B1a cloned into single-piliated or nonpiliated phases (Table 2). There were no significant differences in colonization and bacteremia among the phases tested for either strain, suggesting that neither MS nor NMS pili are virulence factors for *E. coli* K1. Alternatively, the bacteria could have shifted in vivo to a piliated phase required for mucosal adherence, colonization, and possibly invasion.

To confirm the importance of K1 polysaccharide capsule in the virulence of *E. coli*, an acapsular variant of strain C94 was obtained by serial passage on solid media containing antiserum to the polysaccharide of the group B meningococcus capsule, which is antigenically identical to K1 polysaccharide.[14] Newborn rats were challenged with the piliated phases of both the encapsulated parent and acapsular variant of strain C94 (Table 3). Only the capsular variant produced bacteremia; however, both variants rapidly colonized the oropharynx and intestine in high numbers. Furthermore, three clinical isolates of non-K1 *E. coli* did not produce bacteremia but colonized the oropharynx of neonatal rats (Table 4). Thus, it may be concluded that K1 capsule is a virulence factor necessary for *E. coli* K1 bacteremia, but that it is not required for *E. coli* oropharyngeal colonization.

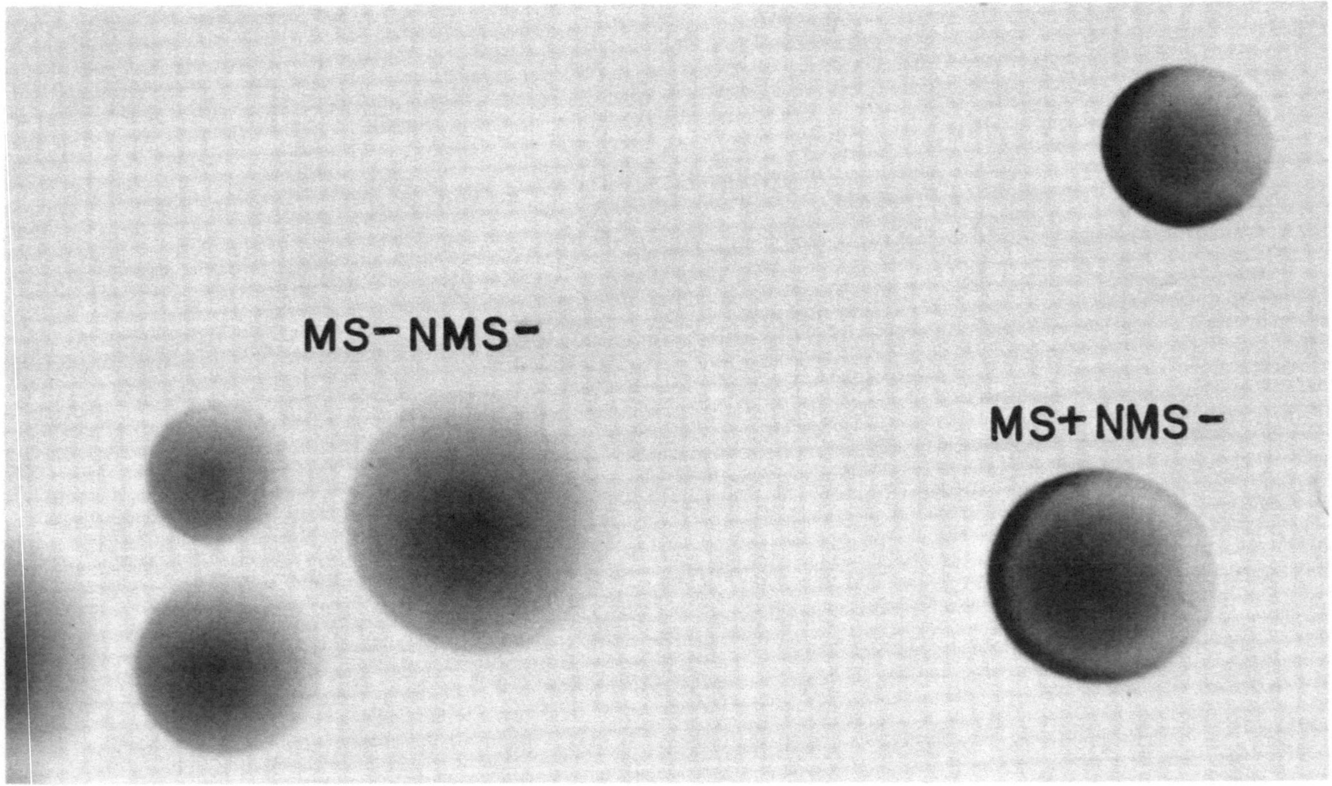

FIGURE 1. The MS-piliated (MS+NMS−) and nonpiliated (MS−NMS−) colonial forms of *E. coli* K1 strain C94. Bacteria from a mixed culture containing piliated and nonpiliated phases were streaked onto a MG agar plate and incubated for 18 to 20 hr at 37°C with room air and humidity. Photographs were taken through a Bausch and Lomb dissecting microscope using transmitted light directed at 20 to 30°C from the vertical. Magnification approximately ×20.

Table 1
CHARACTERISTICS OF *E. COLI* STRAINS USED IN NEONATAL RAT STUDIES

Strain	Serotype[a]	Piliated and nonpiliated phases	Source
B1a	018ac:K1:H7	MS−NMS−, MS+NMS−	CSF (supplied by Dr. John Robbins)
C94	07:K1:H-	MS−NMS−, MS+ NMS−, MS−NMS+	CSF (supplied by Dr. Mary Glode)
346	025:K-:H6	MS−NMS−, MS+NMS−	Urine (supplied by Dr. Frederick Silverblatt)
3781	018ac:K-:H7	MS−NMS−, MS+NMS−, MS−NMS+	Unknown
MT411	023:K92:H4	MS+NMS−	Urine (supplied by Dr. Mary Glode)
F33[b]	Unknown	MS−NMS−	Laboratory (supplied by Dr. Paul Cohen)

[a] All strains were serotyped by Drs. I. and F. Ørskov, Statens Seruminstitut, Copenhagen, Denmark.
[b] Strain F33 is a laboratory K12 strain known to colonize rodents poorly. It was used as a negative control in this study. We have found this strain to be cleared from the oropharynx and small intestine of neonatal rats within 3 days postchallenge.

Table 2
COLONIZATION AND BACTEREMIA IN NEONATAL RATS CHALLENGED WITH PILIATED AND NONPILIATED PHASES OF STRAINS C94 AND B1A

Strain	Phase	N	%Oropharyngeal colonization	Cfu/cm Ileum		%Bacteremia
				+LC	−LC	
C94	MS+NMS−	30	100	7×10^5	$<10^3$	29
	MS−NMS+	30	100	6×10^5	$<10^3$	24
	MS−NMS−	30	100	2×10^6	$<10^3$	26
B1a	MS+NMS−	30	100	2×10^5	$<10^3$	61
	MS−NMS−	30	100	4×10^5	$<10^3$	53

Note: LC = luminal contents. One-day-old rats were fed 10^6 to 10^7 cfu of either the MS+NMS−, MS−NMS+, or MS−NMS− phases of strains C94 or B1a. The animals from each litter were distributed equally among all mothers prior to feeding. Colonization and bacteremia were determined at day five postchallenge. Colonization of the oropharynx was determined for each animal by gentle but thorough swabbing with a sterile cotton-tipped applicator and streaking onto one-half of a MG agar plate. Colonization of the ileum was determined for each animal by plating serial dilutions of a homogenized 1 to 2 cm ileal segment. Each segment selected contained one Peyer's patch. An adjacent segment containing one Peyer's patch was also cultured from each animal after the contents were removed by gentle rinsing with sterile phosphate buffered saline before homogenization. Bacteremia was determined for each animal by plating on MG agar 0.1 to 0.2 mℓ of blood taken by cardiac puncture. All organisms recovered were identified as K1 by the presence of a characteristic halo precipitate formed when grown on meningococcal B antiserum agar.

V. ROLE OF PILIATED PHASES AND CAPSULE IN COLONIZATION WITH *E. COLI*

In order to determine the piliated phase of *E. coli* in colonized neonatal rats, cultures were plated on MG agar. As mentioned previously, the nonpiliated phase of *E. coli* is stable on this medium, and colonies containing MS-piliated organisms in an overnight culture occur

Table 3
COLONIZATION AND BACTEREMIA IN NEONATAL RATS CHALLENGED WITH STRAIN C94 CAPSULAR VARIANTS

Capsule	N	%Oropharyngeal colonization	Cfu/cm Ileum		%Bacteremia
			+LC	−LC	
K1+	30	100	4×10^5	<100	33
K1−	30	100	7×10^4	<100	0

Note: LC = luminal contents. One-day-old rats were fed 10^6 to 10^7 cfu of streptomycin-resistant mutants of either the capsular or acapsular variants of strain C94. Each inoculum consisted of an equal proportion of MS+NMS−, MS−NMS+, and MS−NMS− phases. All cultures were made on MG agar containing 1.0 mg streptomycin per milliliter. The experimental conditions were otherwise identical to those in Table 2.

Table 4
OROPHARYNGEAL COLONIZATION AND BACTEREMIA IN NEONATAL RATS CHALLENGED WITH NON-K1 *E. COLI*

Strain	N	% Oropharyngeal colonization	% Bacteremia
346	30	100	0
3781	30	80	0
MT411	30	100	0
F33	30	0	0

Note: One-day-old rats fed 10^6 to 10^7 cfu of streptomycin-resistant mutants of either strain 346, 3781, MT411, or F33. The inoculum for strains 346 and 3781 consisted of equal numbers of piliated and nonpiliated phase organisms. The inoculum for strain MT411 consisted of only MS-piliated organisms, while that for strain F33 consisted of only nonpiliated organisms. Oropharyngeal colonization and bacteremia were determined as described in Table 2 using MG agar containing 1.0 mg streptomycin per milliliter.

at a very low frequency unless bacteria are in the MS-piliated phase at the time of plating. When neonatal rats fed *E. coli* were cultured on days one and five postchallenge, a rapid shift to the MS-piliated phase was observed for challenge bacteria isolated from the oropharynx, regardless of the feeding phase of the bacteria (Figure 2). This shift occurred for both K1 and non-K1 strains. The MS-piliated phase was maintained in the oropharynx throughout the 5-day colonization period.

It is interesting to note that the majority of challenge bacteria recovered from the ileum were not in the MS-piliated phase at 1-day postchallenge even when bacteria were in this phase at the time of feeding. However, by day five postchallenge the MS-piliated phase predominated (Figure 2). The bacteria recovered from the ileum on day one postchallenge may represent an initial fraction of bacteria which remained in, or shifted to the nonpiliated phase. Since these nonpiliated bacteria may be nonadherent, they would be expected to be

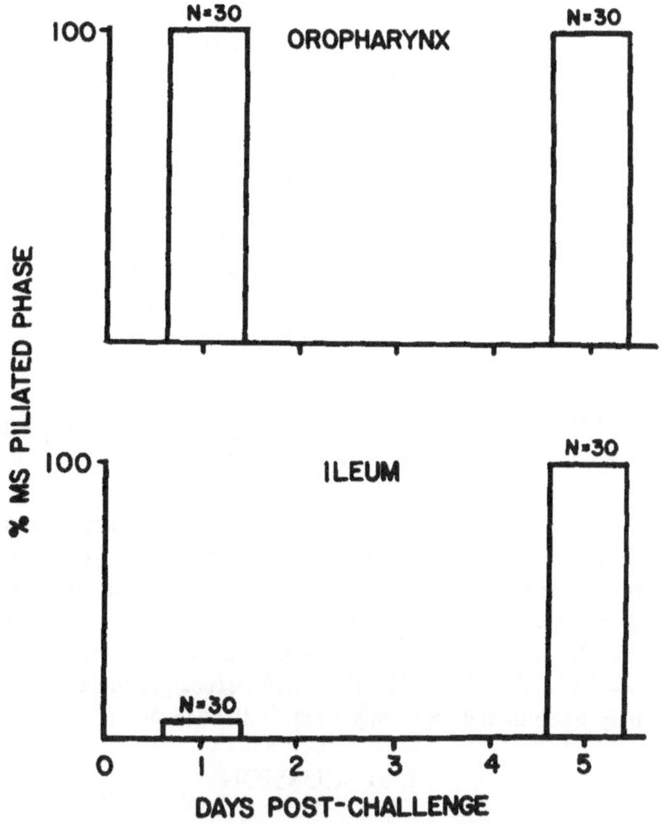

FIGURE 2. In vivo phase variation of *E. coli* K1 and non-K1 *E. coli* in colonized neonatal rats. The experimental conditions were exactly as described for Table 2 except that animals were sacrificed and cultured on days one and five postchallenge, and strains C94, Bla, 3781, and 346 were used. The graph shows the percent of MS-piliated phase bacteria recovered from 30 animals fed MS − NMS − phase bacteria from strain C94. Identical results were obtained for the MS + NMS − and MS − NMS + phases of strain C94, and the piliated and nonpiliated phases of strains Bla, 3781, and 346.

cleared from the small intestine within 3 days postchallenge, the length of time required for clearing of the noncolonizing F 33 control strain (Table 1) from the small intestine. The challenge bacteria recovered from the ileum by day five may represent down stream seeding of MS-piliated organisms from the colonized oropharynx.

VI. SITE OF NEONATAL COLONIZATION BY *E. COLI* K1

Gentle rinsing to remove luminal contents of ileal segments taken from neonatal rats challenged with strains B1a, C94 parent, and C94 acapsular variant resulted in a striking reduction in the number of bacteria recovered (Tables 2 and 3). This indicates that most of the bacteria were probably in the lumen and not adherent to the intestinal mucosa.

Poor mucosal adherence was further demonstrated by light microscopy performed on serial thick sections of ideal segments taken from 5-day-old animals colonized with 10^6 cfu/cm of ileum. No bacteria were observed near the luminal surface of either columnar cells or m-cells on examination of multiple 1-2 segments taken from each of six animals. Furthermore, none of the three phases of *E. coli* strain C94 adhered to ileal mucosa 15 to 30 min after injection into ligated ileal loops of 13-day-old rat pups. Light microscopic examination of ligated ileal loops from $3^1/_2$-day-old rats 30 min after inoculation of bacteria revealed some mucosal damage, sloughing of epithelium, and loss of villi, but there was no evidence of adherence to areas of intact absorptive epithelium.

We did not obtain quantitative oropharyngeal cultures in this study. However, nearly confluent growth was usually observed when oropharyngeal swabs from animals colonized by MS-piliated *E. coli* were streaked directly onto one-half of a culture plate. Furthermore, animals remained colonized by *E. coli* for at least 1 to 2 weeks postchallenge. Although we have not yet attempted to visualize oropharyngeal mucosal colonization with *E. coli* by either light or electron microscopy, the persistence of large numbers of bacteria in the oropharynx suggests that this may be an important site of *E. coli* K1 colonization in neonatal rats.

VII. IN VITRO ADHERENCE OF *E. COLI* K1 TO ISOLATED OROPHARYNGEAL CELLS

Our animal studies suggested that oropharyngeal colonization results from adherence of MS-piliated phase *E. coli* to the oropharyngeal epithelium. To further document this possibility, in vitro adherence assays were performed using isolated oropharyngeal cells with strain C94 cloned into either the MS+NMS−, MS−NMS+, or MS−NMS− phase (Figure 3). Only the MS-piliated phase showed significant adherence, with a mean number of bacteria per epithelial cell of 203 ± 121 (S.D.). As expected, the C94 acapsular variant also adhered strongly to epithelial cells only when in the MS-piliated phase.

VIII. DISCUSSION

Although the small intestine has been shown to be a major site of colonization for various enteropathogenic and enterotoxigenic *E. coli* strains,[13,15-17] the oropharynx appears to be an important site of colonization in *E. coli* K1 infections. Furthermore, our studies indicate that MS pili may play an important role in this colonization.

It is interesting to speculate on the likely site of mucosal invasion by *E. coli* K1. Enteropathogenic *E. coli* have been shown to be taken up by the intestinal columnar cells after forming characteristic adherence complexes between the bacterial outer membrane and the columnar cell plasma membrane.[18] This mechanism of invasion seems unlikely for *E. coli* K1 since we have not been able to demonstrate intestinal mucosal adherence in vivo or in vitro. However, it is possible that *E. coli* K1 directly invades the oropharynx or the adjacent nasopharynx in a manner similar to *Neisseria meningitidis*.[19]

Pilus vaccines have been effective against certain bacterial infections presumably because pilus-specific antibody blocks colonization.[20,21] For example, the colonization of neonatal pig small intestine by enterotoxigenic *E. coli* has been shown to be prevented by pilus-specific antibodies present in the colostrum of immunized sows.[22] In a similar fashion it may be possible to prevent *E. coli* K1 infections by maternal immunization with MS pili. We are currently studying the efficacy of a MS pilus vaccine in the neonatal rat model.

ACKNOWLEDGMENTS

Various portions of these studies were conducted with the following individuals: from the Division of Infectious Diseases, Children's Hospital Medical Center: Herbert Clegg, M.D., Victoria Guerina, T. Woodrow Kessler, Soloman Langermann, and Frank Scannapieco; from the Department of Anatomy, Harvard Medical School: Marian Neutra, Ph.D.

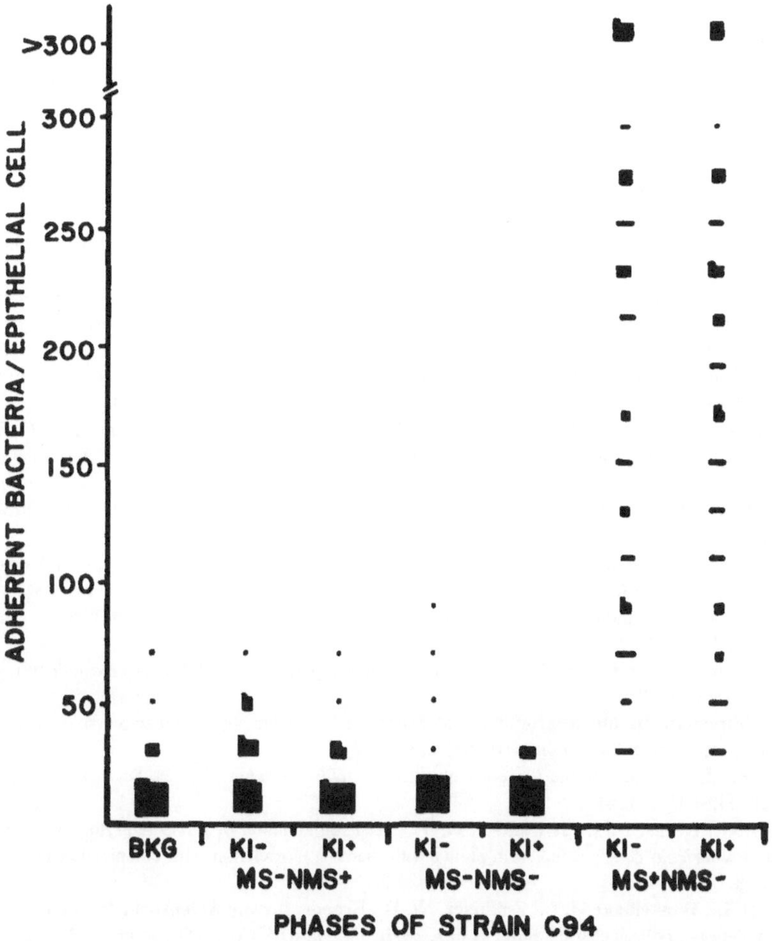

FIGURE 3. In vitro adherence of piliated and nonpiliated phases of strain C94 to isolated oropharyngeal cells. Five-day-old neonatal rats were sacrificed, and the oropharynxes were dissected and swabbed to obtain an epithelial cell suspension free of buccal cells. The oropharyngeal cells were washed three times with sterile phosphate buffered saline pH 7.2 (PBS) and suspended to a final cell concentration of approximately 5×10^4/mℓ. Cloned bacteria were suspended to a final cell concentration of approximately 5×10^8/mℓ and incubated with an equal volume of epithelial cells for 30 min at 37°C. Following incubation, the mixture was filtered through a 5 μMillipore filter to collect epithelial cells and attached bacteria. Free bacteria were further removed by five PBS washes, and each filter was touched to a drop of PBS on a clear microscope slide to transfer the collected epithelial cells. After drying, the cells were fixed with methanol for 30 sec, Gram stained, and examined under the light microscope. The number of bacteria/epithelial cell was determined for the first 30 cells observed using 100X oil immersion. If a cell was folded so as not to reveal more than an estimated 75% of the cell surface, it was rejected from the count. Background (BKG) counts were made on epithelial cells which had been incubated without bacteria but otherwise had been handled in the identical manner as all other samples.

REFERENCES

1. **Robbins, J. B., McCracken, G. H., Gotschlich, E. C., Ørskov, F., Ørskov, I., and Hanson, L. A.,** *Escherichia coli* K1 capsular polysaccharide associated with neonatal meningitis, *N. Engl. J. Med.*, 290, 1216, 1974.
2. **McCracken, G. H., Sarff, L. D., Glode, M. P., Mize, S. G., Schiffer, M. S., Robbins, J. B., Gotschlich, E. C., Ørskov, I., and Ørskov, F.,** Relation between *Escherichia coli* K1 capsular polysaccharide antigen and clinical outcome in neonatal meningitis, *Lancet*, 2, 246, 1974.
3. **Weinstein, R. and Young, L. S.,** Phagocytic resistance of *Escherichia coli* K1 isolates and relationship to virulence, *J. Clin. Microbiol.*, 8, 748, 1978.
4. **Stevens, P., Chu, C. L., and Young, L. S.,** K-1 antigen content and the presence of an additional sialic acid-containing antigen among bacteremic K-1 *Escherichia coli:* correlation with susceptibility to opsononophagocytosis, *Infect. Immun.*, 29, 1055, 1980.
5. **Horwitz, M. A. and Silverstein, S. C.,** Influence of the *Escherichia coli* capsule on complement fixation and on phagocytosis and killing by human phagocytes, *J. Clin. Invest.*, 65, 82, 1980.
6. **Glode, M. P., Sutton, A., Moxon, E. R., and Robbins, J. B.,** Pathogenesis of neonatal *Escherichia coli* meningitis: induction of bacteremia and meningitis in infant rats fed *E. coli* K1, *Infect. Immun.*, 16, 75, 1977.
7. **Brinton, C. C., Jr.,** The piliation phase syndrome and the uses of purified pili in disease control, paper presented at the U.S.-Japan Conference on Cholera, Atlanta, Ga., September, 1977.
8. **Brinton, C. C., Jr.,** The structure, function, synthesis, and genetic control of bacterial pili and a molecular model for DNA and RNA transport in gram negative bacteria, *Trans. N.Y. Acad. Sci.*, 27, 1003, 1965.
9. **Salit, I. E. and Gotschlich, E. C.,** Type I *Escherichia coli* pili: characterization of binding to monkey kidney cells, *J. Exp. Med.*, 146, 1182, 1977.
10. **Salit, I. E. and Gotschlich, E. C.,** Hemagglutination by purified type I *Escherichia coli* pili, *J. Exp. Med.*, 146, 1169, 1977.
11. **Ofek, I., Mirelman, D., and Sharon, N.,** Adherence of *Escherichia coli* to human mucosal cells mediated by mannose receptors, *Nature (London)*, 265, 623, 1977.
12. **Gemski, P., Jr.,** Studies of the Genetics of Piliation in *Escherichia coli*, Ph.D. thesis, University of Pittsburgh, Pittsburgh, 1964.
13. **Evans, D. G., Evans, D. J., Jr., and Tjoa, W.,** Hemagglutination of group A erythrocytes by enterotoxigenic *Escherichia coli* isolated from adults with diarrhea: correlation with colonization factor, *Infect. Immun.*, 18, 330, 1977.
14. **Kasper, D. L., Winkelhake, J. L., Zollinger, W. D., Brandt, B., and Artenstein, M.,** Immunochemical similarity between polysaccharide antigens of *Escherichia coli* 07:K1 (L):NM and group B *Neisseria meningitidis*, *J. Immunol.*, 110, 262, 1973.
15. **Jones, G. W. and Rutter, J. M.,** Role of K88 antigen in the pathogenesis of neonatal diarrhea caused by *Escherichia coli* in piglets, *Infect. Immun.*, 6, 918, 1972.
16. **Bertschinger, H. U., Moon, H. W., and Whipp, S. C.,** Association of *Escherichia coli* with the small intestinal epithelium. I. Comparison of enteropathogenic and nonenteropathogenic strains in pigs, *Infect. Immun.*, 5, 595, 1971.
17. **Cheney, C. P., Schad, P. A., Formal, S. B., and Boedeker, E. C.,** Species specificity of in vitro *Escherichia coli* adherence to host intestinal cell membranes and its correlation with in vivo colonization and infectivity, *Infect. Immun.*, 28, 1019, 1980.
18. **Staley, T. E., Jones, E. W., and Conley, L. D.,** Attachment and penetration of *Escherichia coli* into intestinal epithelium of the ileum in newborn pigs, *Am. J. Pathol.*, 56, 371, 1969.
19. **Stephens, D. S. and McGee, Z. A.,** Interaction of *Neisseria meningitidis* with human nasopharyngeal mucosa: attachment and entry into specialized epithelial cells, paper presented at the 21st Interscience Conf. on Antimicrobial Agents and Chemotherapy, November 4—6, 1981, Chicago, Ill.
20. **Brinton, C. C., Bryan, J., Dillon, J., et al.,** Uses of pili in gonorrhea control: role of bacterial pili in disease, purification and properties of gonococcal pili, and progress in the development of a gonococcal pilus vaccine for gonorrhea, in *Immunobiology of Neisseria gonorrhea*, Brooks, G. F., Gotschlich, E. C., Holmes, K. K., et al., Eds., American Society for Microbiology, Washington, D.C., 1981, 155.
21. **Nagy, B., Moon, H. W., Isaacson, R. E., To, C.-C., and Brinton, C. C.,** Immunization of suckling pigs against enterotoxigenic *Escherichia coli* infection by vaccinating dams with purified pili, *Infect. Immun.*, 21, 269, 1978.
22. **Morgan, R. L., Isaacson, R. E., Moon, H. W., Brinton, C. C., and To, C.-C.,** Immunization of suckling pigs against enterotoxigenic *Escherichia coli*-induced diarrheal disease by vaccinating dams with purified 987 or K99 pili: protection correlates with pilus homology of vaccine and challenge, *Infect. Immun.*, 22, 771, 1978.

Morphology and Bacterial Adherence

Chapter 16

ALTERATIONS IN INTESTINAL STRUCTURE ASSOCIATED WITH MICROORGANISMS

William O. Dobbins, III

TABLE OF CONTENTS

I. INTRODUCTION

The intestinal mucosa, by virtue of its villous structure, presents an extensive area of contact with the intestinal lumen. Intestinal epithelial cells lining villi form an interface between the external environment (the intestinal lumen) and the internal environment (lamina propria). The lamina propria contains an extensive blood vascular and nerve supply and a central lymphatic channel (lacteal). The surface epithelium consists predominately of absorptive cells, with occasional goblet cells interspersed between the absorptive cells.[1] Intestinal crypts contain undifferentiated cells, Paneth cells, enteroendocrine cells, and is the site of intestinal epithelial cell renewal (mitotic activity). Light and electron microscopic observations show that intestinal epithelial cells are tall columnar cells containing a basal nucleus and the usual organelles including modest numbers of mitochondria, approximately equal amounts of smooth and rough endoplasmic reticulum, rather frequent lysosomes, and occasional vesicles and multivesicular bodies.[2] Often seen between intestinal epithelial cells are intraepithelial lymphocytes, cells which may have an important role in the immune function of the intestine.[3] Intraepithelial lymphocytes appear to be almost exclusively T-cells and the majority are the suppressor/cytotic ($T5/8^+$) variety as determined by immunofluorescent staining with monoclonal antibodies.[4] Other cells that may be seen between intestinal epithelial cells include macrophages, PMN leukocytes, and eosinophils.[5] These latter cells appear to be important constituents of the epithelium only in pathological conditions.

II. THE APICAL SURFACE AND ITS SURFACE COAT

The apical surface of intestinal absorptive cells is a highly specialized area which has distinct and unique properties of structure and function (Figure 1). The basal portion of the apical surface is called the terminal web and it consists of numerous interlocking filaments which gives structural support to the microvilli.[6] The only organelles found within the terminal web are occasional secretory vesicles, vesicles necessary for transport of secretory IgA.[7] Numerous microvilli project from the terminal web into the intestinal lumen, and increase the intestinal epithelial cell surface area by a factor of 15 to 25. Filaments contained within the core of microvilli have been shown to contain an actin-like material which may promote contraction of microvilli. All microvilli are included within a continuation of the plasma membrane which has the standard trilaminar structure found on all plasma membranes. A surface coat (glycocalyx) consisting of numerous fine filaments projects from the outer lamella of the plasma membrane.[1] This surface coat is exceedingly prominent over microvilli of intestinal epithelial cells (Figure 2), and is somewhat less prominent over microvilli of goblet and undifferentiated crypt cells. The surface coat is also less well developed over microvilli of surface mucus cells of the stomach and of epithelial cells in the colon. The surface coat is most prominent at microvillus tips and is sparsely present on the lateral surfaces of microvilli. The surface coat appears to be an effective barrier to large particles, such as bacteria and fat droplets. It does not appear to restrict movement of colloidal materials, large molecules such as thorium dioxide, or small molecules. Histochemical reactions of the surface coat indicate that it is composed of a sulfated, weakly acidic mucopolysaccharide. It is intensely periodic acid-Schiff (PAS) positive and reacts with Alcian blue and colloidal iron stains for acid mucopolysaccharides. Some 25 or more enzymes have been localized to the surface coat and its plasma membrane, including alkaline phosphatase, adenosine triphosphatase, thiamine triphosphatase, numerous disaccharidases, amylase, and aminotransferase. The microvillus surface coat is continuously synthesized in the cell, transported to the cell surface, and lost into the lumen. It is distinctively different from other components of the polysaccharide coat found on the lateral and basal plasma membranes. So-called "surface forming" vesicles, many of which can be seen within the terminal web,

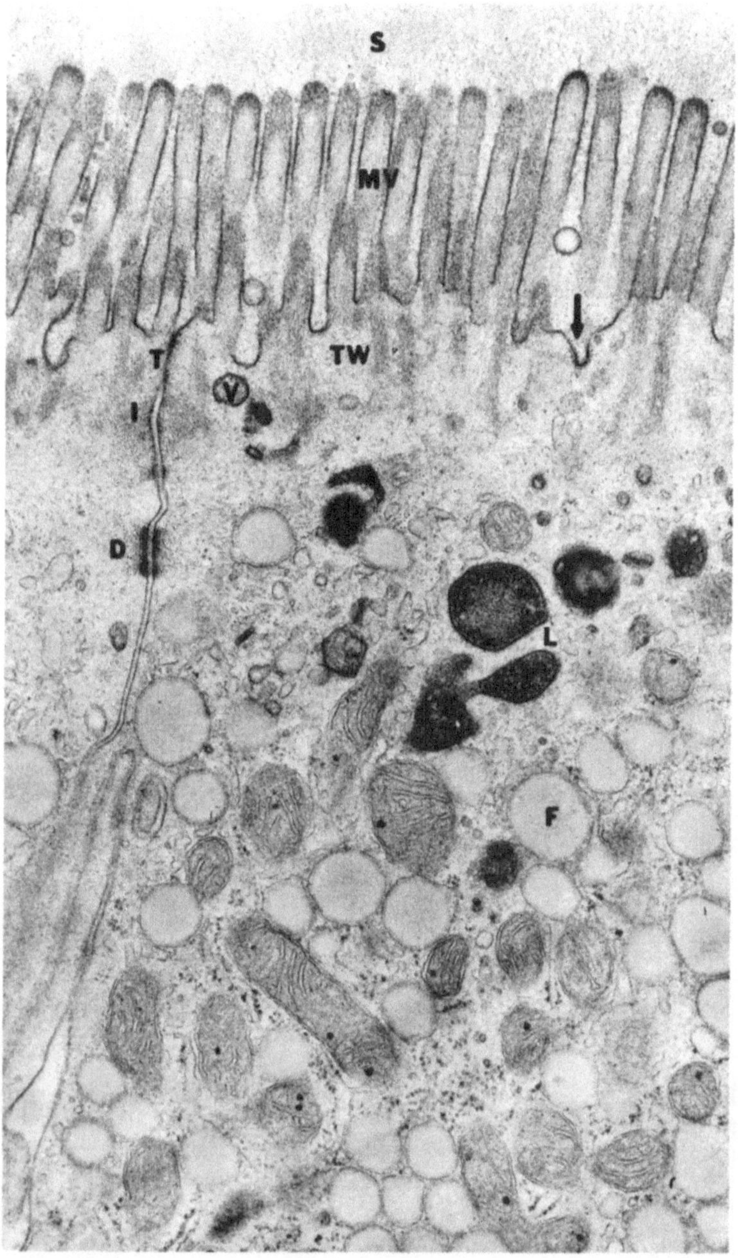

FIGURE 1. Electron micrograph of apical portion of two intestinal absorptive cells in normal man during fat absorption. Microvilli (MV) are closely and regularly spaced and possess a very prominent surface coat (S). Microvillous core tubules extend into the terminal web (TW). There is a characteristic junctional complex consisting of tight junction (T), intermediate junction (I), and desmosome (D). Membrane invaginations at the base of microvilli, called apical pits (arrow), have been shown to be involved in ferritin uptake but not in fat uptake. Apical pits eventually contribute their contents to apical lysosomes (L). Note, the "secretory" vesicle (v) to the right of the junctional complex, an organelle important for transport of secretory IgA. Droplets of resynthesized triglycerides (F) can be seen within profiles of the smooth endoplasmic reticulum. Bicarbonate buffered osmium tetroxide fixation. Double stained. Magnification × 20,000.

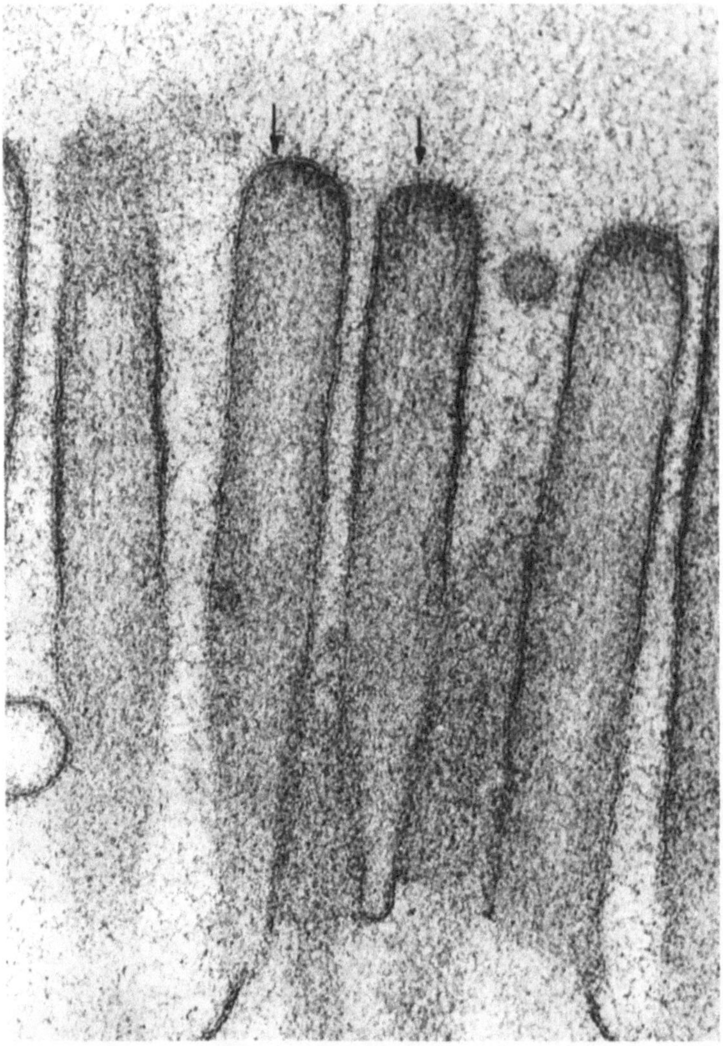

FIGURE 2. Higher magnification of portion of microvilli illustrated in Figure 1. The microvilli are enclosed by a trilaminar membrane consisting of dense inner and outer leaflets and a less dense inner stratum. The outer leaflet is slightly less dense and less thick than the inner leaflet. The surface coat is conspicuous and consists of fine filaments attached to the outer dense leaflet of the plasma membrane (arrows). Magnification × 80,000.

provide membranes for the expansion and elongation of microvilli. These vesicles appear to originate from dilatations of the Golgi cisternae at the mature face of this organelle. The membrane of these vesicles closely resembles the plasma membrane, and these vesicles contain polysaccharide-rich material that stains in a manner identical to that of the surface coat. These data clearly indicate that the microvillous surface coat is a product of the intestinal epithelial cell, it is not a layer of absorbed mucus or other material, and it is an intricate part of the dynamic plasma membrane.

Adjacent intestinal epithelial cells are attached by a characteristic junctional complex (Figure 1). This complex consists of a tight junction, intermediate junction, desmosomes, and, uncommonly, of gap junctions. A detailed description of each component of these junctions is not important to a conference relating to bacterial adherence and thus will not be given here. Descriptions of the junction can be readily obtained.[1,6]

Finally, note in such a conference as this must be made of a unique cell, the M-cell, an epithelial cell found over lymphoid tissue in Peyer's patches, cecal patches, and the appendix. M-cells (microfold cells) are specialized epithelial cells found between intestinal absorptive cells and are actively involved in transport of antigen from the intestinal lumen to underlying lymphoid tissue.[8]

III. INTERACTIONS WITH MICROORGANISMS

Morphological alterations in the microvillus border of intestinal epithelial cells may be produced by a variety of microorganisms. Some of these alterations are clearly the morphological representation of adherence. However, these alterations may not represent clinically important pathological alterations. A number of organisms colonize the intestine. These organisms are particularly likely to adhere to blind invaginations of the epithelium, or to enter the crypts of Lieberkuhn, only then establishing a form of contact relationship with the epithelial surface. Such organisms include protozoans, spiral-shaped microorganisms, yeast, fusiform bacteria, and lactobacilli.

A. Filamentous Bacteria

The best example of an organism that attaches directly to epithelial cells is that of a segmented filamentous bacterium commonly found in the rat ileum.[9] These bacterial filaments may cover the surface of ileal villi in a dense hair-like mat. The individual bacterial filament attaches to epithelial cells by a specialized attachment segment or "hold fast." At the attachment site the hold fast displaces microvilli and invaginates, but does not penetrate, the epithelial cell membrane. The invagination only takes place to the level of the terminal web. The cell wall of the bacterium is closely applied to the outer leaflet of the trilaminar plasma membrane of the epithelial cell. The attachment site is further modified by the presence of two dense layers in the apical cytoplasm. The outer layer is represented by a dense homogeneous plaque, 30 to 50 nm thick, which is found just below the area of contact between bacterial cell walls and epithelial cell membranes. The dense plaque itself is surrounded by a filamentous layer, the filaments apparently being derived from those of the terminal web. The morphological changes in intestinal epithelial cells appear to be limited to the terminal web and microvillus area and do not seem, at least as determined morphologically, to adversely affect the epithelial cells. When there is very dense colonization, the alterations at the microvillus border are very severe and it seems likely that the function of these cells may then be altered. This attachment of bacteria to intestinal epithelium has been described, appropriately, as a prokaryotic-eukaryotic cell junction.[10]

B. Spirochetes

An association of bacteria and spirochetes to the luminal plasma membrane of colonic epithelial cells in man has been noted[2] (Figure 3). Figure 4 shows the presence of rod-shaped bacteria parallel to and between microvilli of colonic epithelial cells found in normal man. Actual attachment of bacteria to the plasma membrane of human epithelial cells, similar to that seen in rodents, has not been observed. Even though bacteria and spirochetes are aligned parallel to and between microvilli of epithelial cells, there is no evident attachment of these organisms to the plasma membrane, nor is there an increased density of the cytoplasm just below the points of contact. This close association of bacteria and spirochetes with microvilli of colonic epithelial cells in man has not been related to any apparent pathologic condition. Further, it can be observed at light microscopy and has been reported to occur in as many as 5 to 10% of colonic appendiceal specimens.[2] Similar observations have also been noted in the colon of monkey and have been described as "spirochetosis of the colon."

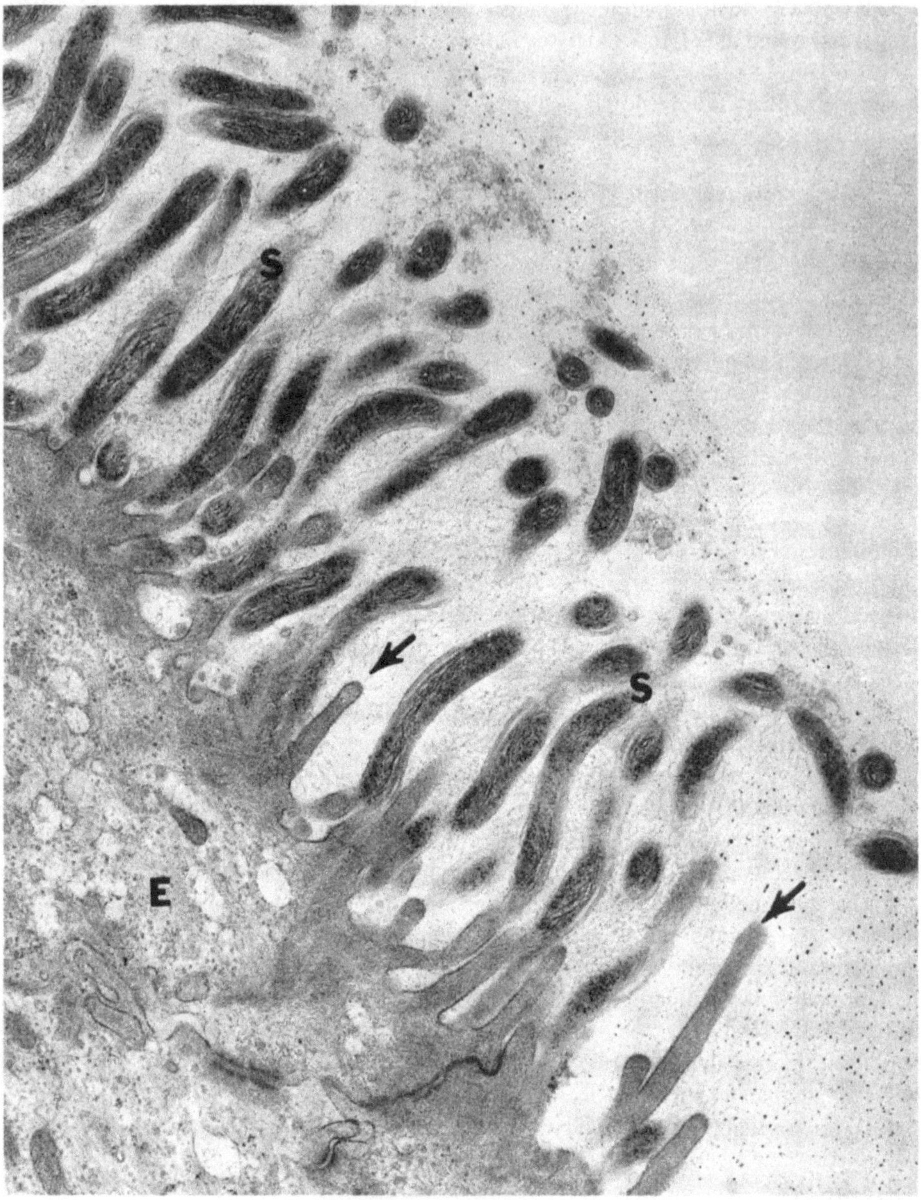

FIGURE 3. Electron micrograph of normal colonic epithelial (E) cells showing a striking parallel alignment of spirochetes (S) to microvilli. The spirochetes are so profuse that it is difficult to identify the microvilli (arrows). Fixed in osmium, double stained. Magnification × 16,925.

C. Protozoa

The human intestine is often parasitized by the protozoan, *Giardia lamblia*. These organisms may occasionally result in extensive injury to the intestinal mucosa, the crypt cells being particularly susceptible to injury. Figure 5 shows several trophozoites of *Giardia lamblia* adjacent to epithelial cells in a human intestinal biopsy. Note that the sucker plate of one of the organisms appears to be penetrating microvilli. Following penetration, it has been demonstrated that these organisms may invade the epithelial cell and even be found within the lamina propria. Owen has shown recently that intraepithelial lymphocytes and

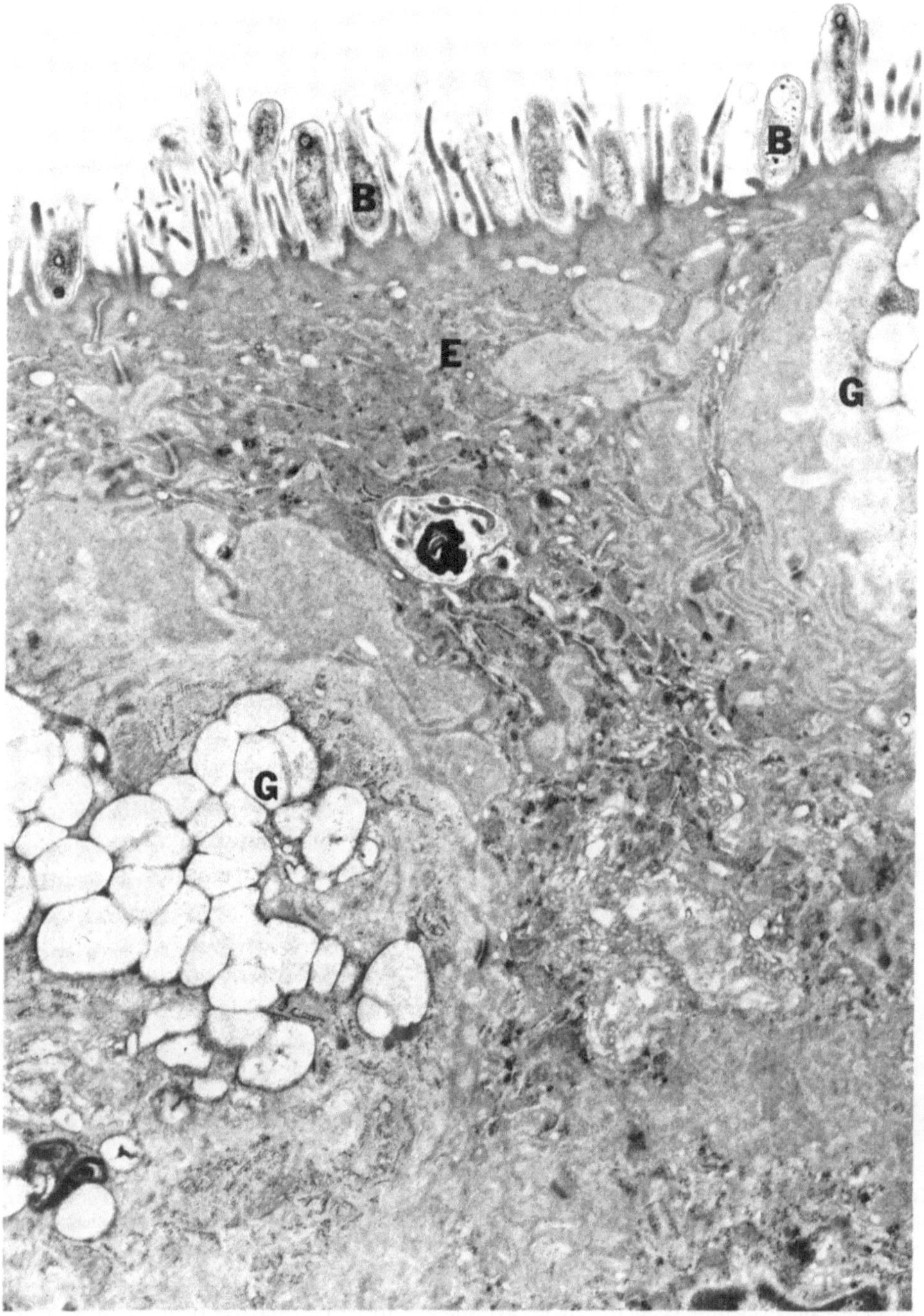

FIGURE 4. Electron micrograph of normal colonic epithelial (E) and goblet (G) cells showing noninvasive bacteria (B) aligned parallel to epithelial cell microvilli. Fixed in osmium, double stained. Magnification × 8970.

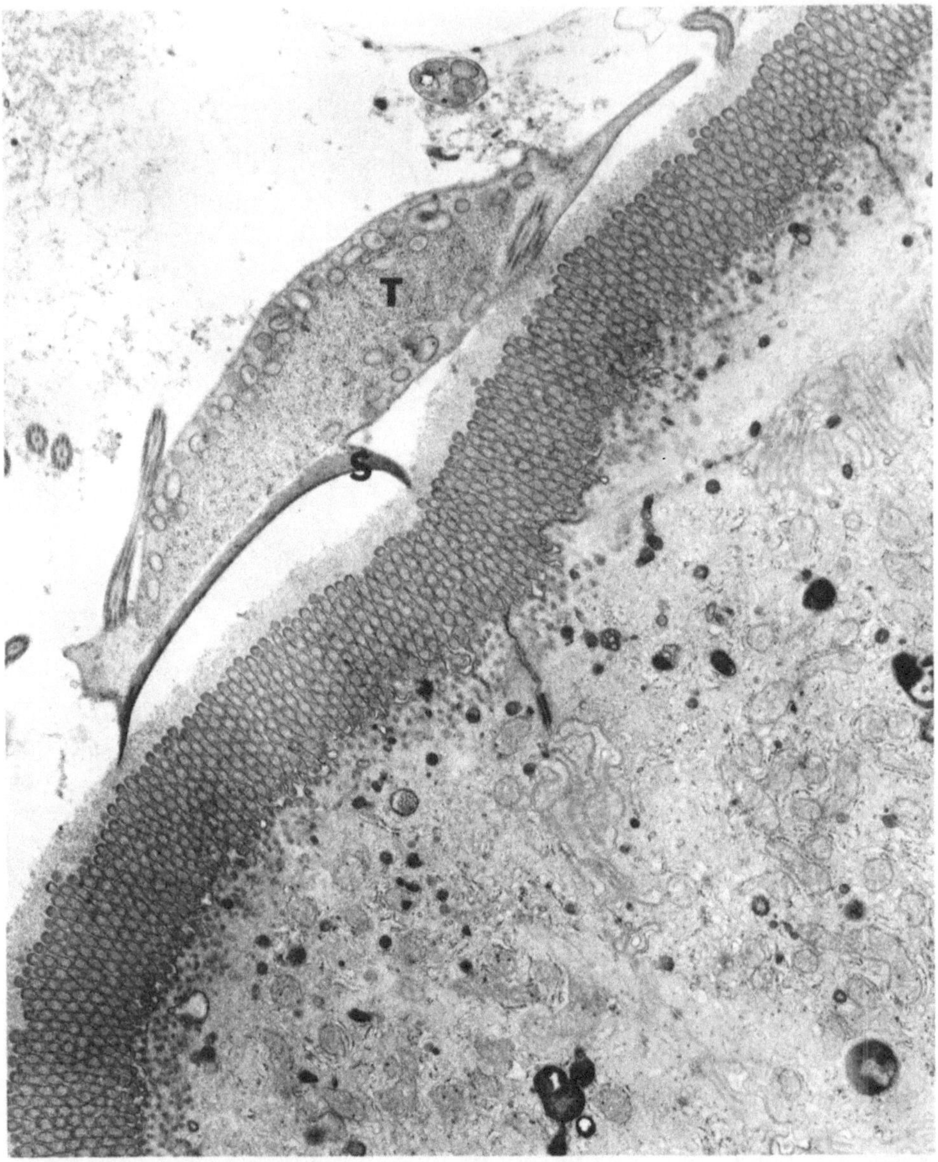

FIGURE 5. Electron micrograph showing trophozoite (T) of *Giardia lamblia* in intestinal lumen of a patient with chronic diarrhea. Sucker plate (S) appears to be in process of "attaching" to the surface coat and microvilli. Note the prominence of the surface coat. Fixed in osmium, double stained. Magnification × 12,285.

macrophages play an important role in protecting the intestinal mucosa of the mouse from *Giardia muris*.[11]

Cryptosporida are protozoan parasites that may produce mucosal abnormalities of the small intestine in a variety of species including man.[12] *Cryptosporidium* is located extracellularly, as illustrated at light microscopy in Figure 6, and appears to be intimately attached to the microvillus plasma membrane of intestinal epithelial cells as illustrated at electron microscopy (Figure 7). Cryptosporidiosis is an uncommon human disease and has been reported in five patients only, four of whom had "immune compromise." Its clinical manifestation is profuse, watery diarrhea. The diarrhea may result from the unique attachment of cryptosporidia and alteration of the sodium pump (cyclic AMP?) of the epithelial cell.

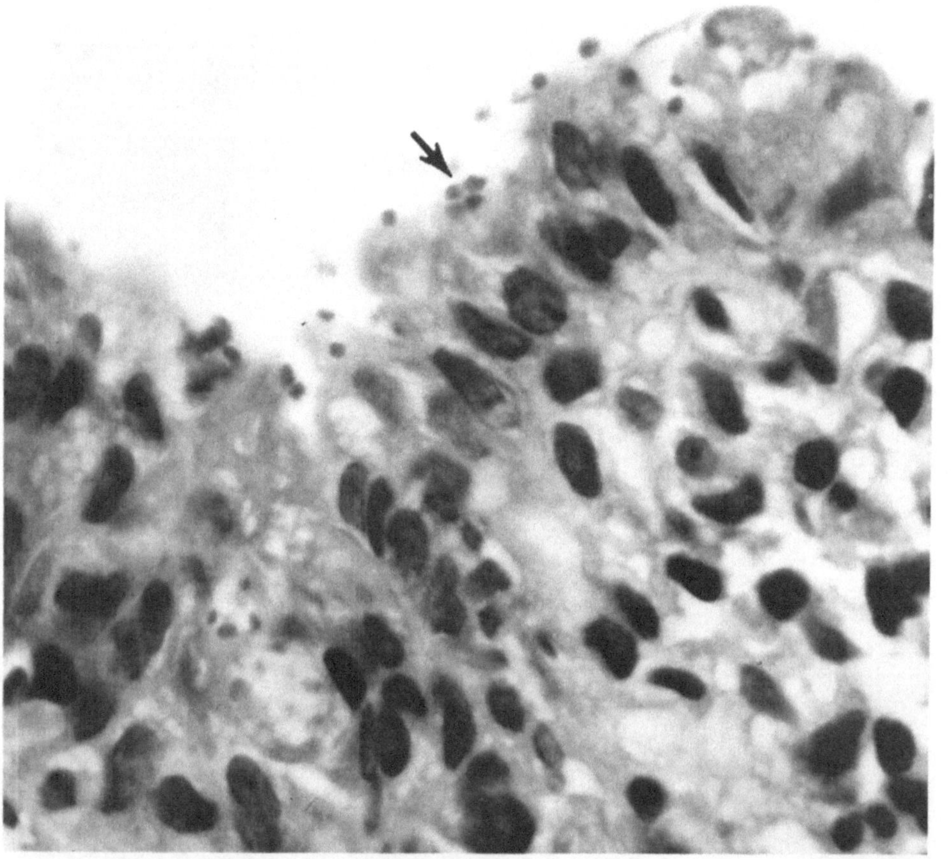

FIGURE 6. Light microscopic photomicrograph which illustrates numerous cryptosporidial trophozoites (arrow) at and presumably attached to the apical surface of intestinal absorptive cells of an intestinal biopsy obtained from a man with severe watery diarrhea. H & E stain. Magnification × 1000.

In contrast, *Isospora belli*, another member of the *coccidia* family which includes the cryptosporidium, invades the intestinal epithelial cell and is found in an intracytoplasmic position.

D. Whipple's Disease

Whipple's disease is the only bacterial disease of man in which invasion of intestinal epithelial cells has been documented by electron microscopic studies.[2] This is a systemic disease in which the intestine is characteristically severely involved, and in which invasion of intestinal epithelial cells by Whipple bacilli results in diarrhea and steatorrhea. The organisms under discussion up to this point all appear to adhere to the intestinal mucosa prior to producing pathological changes, if any. In contrast, there is no evidence that Whipple bacilli attach to intestinal epithelial cells, or that these bacilli even invade the intestinal epithelial cell from the lumen. There is a striking predisposition for Whipple bacilli to involve the intestinal lamina propria and to be taken up by macrophages. The intestinal mucosa becomes filled with PAS-positive macrophages, the characteristic light microscopic, and usually diagnostic finding in this condition. Further, it would appear that Whipple bacilli invade the intestinal epithelial cell through the basal surface adjacent to the lamina propria, rather than through the apical surface.

The characteristic bacterial structure of the organisms as observed in the untreated patient,

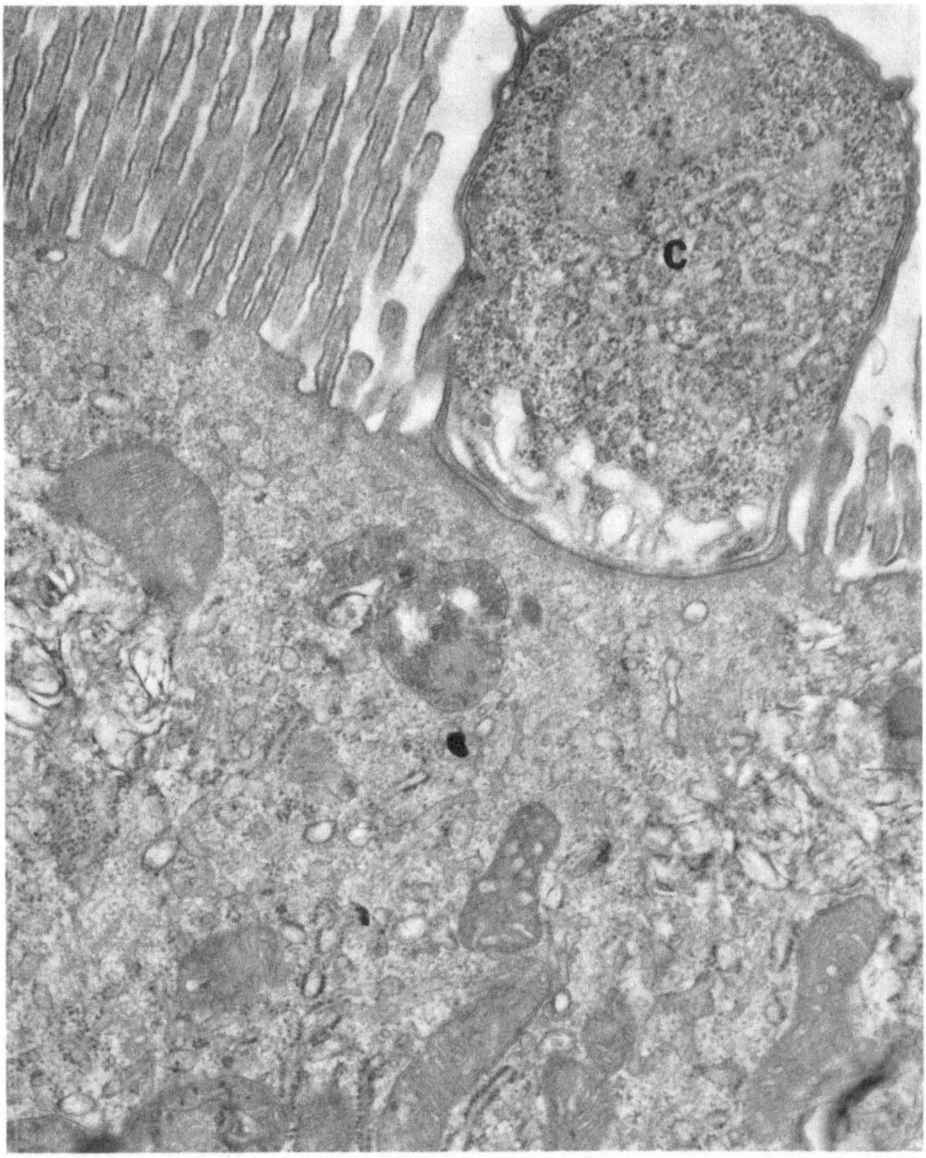

FIGURE 7. Electron microscopic photomicrograph illustrating a cryptosporidial (C) species found in mouse. It is intimately attached to the plasma membrane of an intestinal absorptive cell. Note that the microvilli have been displaced by the organism and that there appears to be little evidence of injury to the epithelial cell. Fixed in osmium, double stained. Magnification × 25,000.

the absence of the organisms after antibiotic therapy, and the uniformly good clinical results of antibiotic treatment leave little doubt that Whipple's disease is of bacterial etiology. However, the Whipple bacillus has not been cultured in vitro, nor has the disease been reproduced in animals. The failure to culture such an organism which is the source of modest human suffering and occasional death, should be a challenge to members of this audience. Given the state of technology in biomedical sciences, it is a bit of an indictment not to be able to culture an organism that clearly is the cause of a human disease. Culture of this organism is required in order to better understand the disease, and hopefully to develop simpler diagnostic techniques than those currently used, i.e., intestinal biopsy, lymph node biopsy, or brain biopsy.

REFERENCES

1. **Dobbins, W. O., III,** Human intestinal epithelium as a biological membrane, in *Pathology of Cell Membranes,* Trump, B. F., Ed., Academic Press, New York, 1975, 429.
2. **Dobbins, W. O., III,** Diagnostic pathology of the intestine and colon, in *Diagnostic Electron Microscopy,* Vol. I, Trump, B. F. and Jones, R. T., Eds., John Wiley & Sons, 1978, 253.
3. **Selby, W. S., Janossy, G., and Jewell, D. P.,** Immunohistological characterization of intraepithelial lymphocytes of the human gastrointestinal tract, *Gut,* 22, 169, 1981.
4. **Janossy, G., Tidman, W., Selby, W. S., Thomas, J. A., Granger, S., Kung, P. C., and Goldstein, G.,** Human T lymphocytes of inducer and suppressor type occupy different microenvironments, *Nature (London),* 288, 81, 1980.
5. **Austin, L. L. and Dobbins, W. O., III,** Intraepithelial leucocytes of the intestinal mucosa in normal man and in Whipple's disease: a light and electron microscopic study, *Digest Dis. Sci.,* 27, 311, 1982.
6. **Hull, B. E. and Staehelin, L. A.,** The terminal web. A reevaluation of its structure and function, *J. Cell. Biol.,* 81, 67, 1979.
7. **Wagura, H., Nakane, P. K., and Brown, W. R.,** Translocation of dimeric IgA through neoplastic colon cells in vitro, *J. Immunol.,* 123, 2359, 1979.
8. **Owen, R. L. and Jones, A. L.,** Epithelial cell specialization within human Peyer's patches: an ultrastructural study of intestinal lymphoid follicles, *Gastroenterology,* 66, 189, 1974.
9. **Erlandsen, S. L. and Chase, D. G.,** Morphological alterations in the microvillous border of villous epithelial cells produced by intestinal microorganisms, *Am. J. Clin. Nutr.,* 27, 1277, 1974.
10. **Wagner, R. C. and Barrnett, R. J.,** The fine structure of prokaryotic-eukaryotic cell junctions, *J. Ultrastruct. Res.,* 48, 404, 1974.
11. **Owen, R. L., Allen, C. L., and Stevens, D. P.,** Phagocytosis of giardia muris by macrophages in Peyer's patch epithelium in mice, *Infect. Immun.,* 33, 591, 1981.
12. **Weinstein, L., Edelstein, S. M., Madara, J. L., Falchuk, K. R., McManus, B. M., and Trier, J. S.,** Intestinal cryptosporidiosis complicated by disseminated cytomegalovirus infection, *Gastroenterology,* 81, 584, 1981.

Chapter 17

ULTRASTRUCTURAL STUDIES OF THE INTERACTION OF BACTERIA WITH INTESTINAL CELL SURFACES

Marian R. Neutra

TABLE OF CONTENTS

I. THE INTESTINAL CELL SURFACE

The luminal plasma membrane of intestinal absorptive cells displays a rich and heterogeneous array of glycoconjugates.[1] By electron microscopy, this array, termed the "glycocalyx," is visible as a filamentous mat extending 0.2 μm or more from the tips of microvilli[2] (Figure 1). Although the exact arrangement of specific integral and peripheral membrane glycoproteins in the intestinal cell "glycocalyx" is not known,[3] it is well established that these cell surface oligosaccharides provide a variety of potential binding sites for bacterial toxins,[4,5] microorganisms,[6-9] and experimental ligands such as plant lectins.[10] The application of fluorescein-conjugated lectins to intestinal tissue sections has demonstrated reproducible regional and cellular variations in the saccharide composition of the glycocalyx.[9-13] The exact distribution and availability of lectin binding sites in the glycocalyx of living intestinal cells in the intact animal has not been systematically defined. The striking species, regional and cellular specificity of bacterial attachment in the living intestine, reviewed in this volume and elsewhere, underscores the fact that available binding sites on luminal cell surfaces in vivo must indeed vary dramatically.[7-9,14] Application of ferritin-conjugated lectins to the surface of the living mucosa permits the identification of available binding sites at the ultrastructural level (Figure 2). With this approach, one can determine whether specific sites are located near the membrane surface or in more peripheral portions of the glycocalyx.

II. INTERACTIONS WITH MICROORGANISMS

Polymeric surface macromolecules on bacterial cells, such as the pili that display lectin-like saccharide specificity,[16-20] may also extend a considerable distance from the bacterial cell membrane so that interacting bacterial and intestinal cell surfaces could be separated by variable distances.[21] At one extreme, superficial associations could involve only the most peripheral bacterial proteins and glycocalyx components at a distance of tens of nanometers from the intestinal cell membrane, and such "loose" associations may be sufficient for bacterial immobilization and perhaps for colonization. On the other hand, interactions of cell surface components that are closely associated with, or integrated in bacterial and intestinal cell membranes may be required for the formation of the close, semipermanent attachment sites in which prokaryote and eukaryote are separated only by a few nanometers. Complex structural modifications of the intestinal cell seem to be required for the maintenance of semipermanent attachment sites, but not for more superficial associations. To distinguish between superficial and intimate attachment, and to define the cellular specializations involved, ultrastructural methods are essential.

When attachment or adherence leads to phagocytosis of the bacterium and to destruction of the intestinal cells (as observed in *Salmonella typhimurium* and *Shigella flexneri* infections,[22-25]) the prokaryote-eukaryote interaction is relatively transient and thus is difficult to define ultrastructurally. The associations of bacterium and host that are maintained over time without lethal damage to either participant may thus provide the most useful models for studying the process of bacterial attachment to intestinal cells. Certain nonpathogenic bacteria attach to intestinal cells with strict species and cellular specificity, but without compromising host cell integrity. The ultrastructural characteristics of such attachments have been described in relatively few cases.

A. Gram-Negative Bacteria in the Rat Colon

In the rat colon, an elongated, Gram-negative microbe is attached by one end in a uniform indentation of the apical surface of crypt epithelial cells.[26] Electron microscopy of thin sections revealed that both the bacterium and the host show structural specializations that

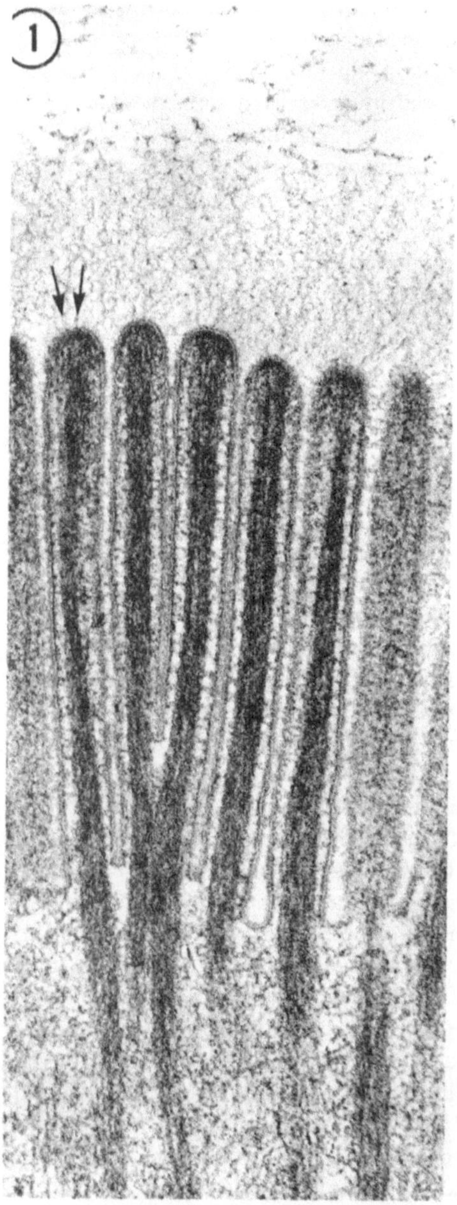

FIGURE 1. Microvilli on the luminal surface of an ileal absorptive cell from a cat. Filamentous cell surface macromolecules extend from the plasma membrane. At the tips of microvilli, these filaments (arrows) blend into the thick layer of membrane-associated glycoconjugates called the "glycocalyx" (Quick-frozen, freeze-substituted preparation courtesy of Dr. Susumu Ito). Magnification × 50,000.

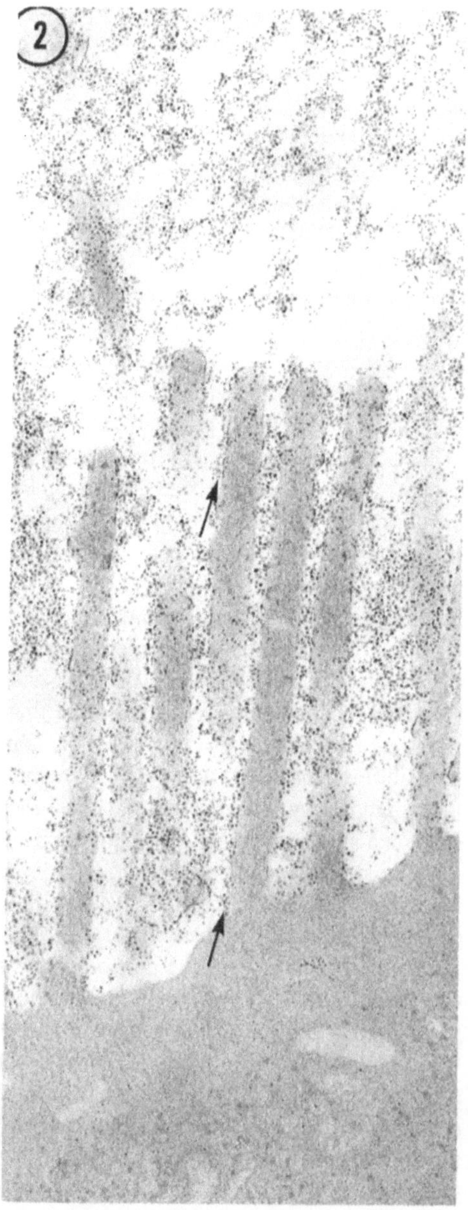

FIGURE 2. Microvilli on the luminal surface of a duodenal absorptive cell from a rabbit, exposed to affinity-purified, ferritin-conjugated ricin for 15 min prior to fixation. The lectin has bound to saccharide sites throughout the glycocalyx, including sites located close to the plasma membrane (arrows). (See also Neutra, M. R., Hall, T. S., Guerina, N. G., and Nicolson, G. L., *Gastroenterology*, 82, 1137, 1982). Magnification × 49,650.

serve to maintain the attachment. The bacterial cell wall is rigid and thickened at the blunt attached end, and the host cell cytoplasm forms a stable, cup-shaped depression lined by the plasma membrane that accommodates the bacterial tip; a uniform 30 to 40 nm gap between bacterial and host cell membranes is crossed by an ordered array of filamentous material apparently arising from both membranes. Another striking example, described in the distal ileum of rodents and chickens, is the segmented, filamentous Gram-positive organism that attaches via a conical "hold fast" to villus cell apical surfaces.[27-31] At the indented attachment site, the bacterial capsular surface and the host cell plasma membrane appear to be in direct contact. Immediately under the host cell membrane, thin sections revealed three layers of aggregated cytoplasmic components: first, a narrow dense layer presumed to be submembrane proteins, then an aggregate of thin filaments, and finally a layer of tonofilaments.[29] Freeze-fracture replicas of this portion of the intestinal cell membrane revealed a dramatic increase in the number of E-face intramembrane particles,[29-31] suggesting that both the submembrane proteins and specific macromolecules in the host cell membrane were recruited to the site in response to the binding of bacterial surface components. In addition, intramembrane particle aggregates were observed in the bacterial membrane at the attachment site.[31] The chemical nature of any of these intramembrane or submembrane cytoplasmic components that participate in attachment has not been explored.

B. Spirochetes and Flagellated Bacteria in Monkey Colon

Only one other intestinal-bacterial attachment site has been studied by both conventional EM and freeze-fracture methods. Ultrastructural studies of the large intestines of monkeys, humans, and other mammals revealed that two types of organisms — a spirochete and a flagellated bacterium — coexist on the apical surfaces of mature absorptive cells, and that they both indent the host cell apical membrane.[32-36] Further EM and freeze-fracture studies in this laboratory revealed that the attachment sites of the two organisms differ dramatically, even on a single intestinal cell[37] (Figure 3). The attached end of the spirochete lies embedded in the host cell terminal web, held in a deep, sleeve-like invagination of the host cell plasma membrane and adhering to adjacent microvilli (Figure 4a). A uniform 6 to 8 nm gap containing fibrillar cell coat material separates the spirochete outer membrane from apical, microvillar membranes. The portion of the bacterial outer membrane that participates in the attachment site is marked in freeze-fracture replicas by a series of circumferential P-face particle arrays (Figure 4b). At the embedded tip of the organism, the freeze-fractured outer membrane displays a small, distinctive set of closely-packed ridges or grooves. In thin sections, a condensation of terminal web microfilaments is seen in the host cell cytoplasm surrounding the tip of the spirochete (Figure 4a). The flagellate, in contrast, lies in a shallow indentation of the intestinal cell membrane and does not adhere to microvilli (Figure 5a). Only a narrow cuff of bacterial membrane around the terminal flagellum participates in the "junction" with the host cell; in this region, bacterial and intestinal cell membranes are again separated by a 6 to 8 nm gap that contains cell surface material, but that does not exclude exogenous macromolecular tracers (Figure 5c). The narrow attached cuff of bacterial outer membrane is marked in freeze-fracture replicas by a loose band of heterogeneous intramembrane particles (Figure 5b). Thus, two different bacteria evoke two dramatically different kinds of attachment sites on a single host cell, and the shape of the sites corresponds to specialized areas on the bacterial outer membrane (Figure 6). This suggests that distinctive bacterial surface proteins may provide the "signals" that induce host cell modifications.

C. Bacterial/Protozoal Associations in the Termite

Attachment of two types of bacteria to a single eukaryotic cell is known to occur among the bacterial-protozoal associations found elsewhere in nature. In the hindgut of the termite, for example, several elaborate prokaryotic-eukaryotic relationships exist, but only one of

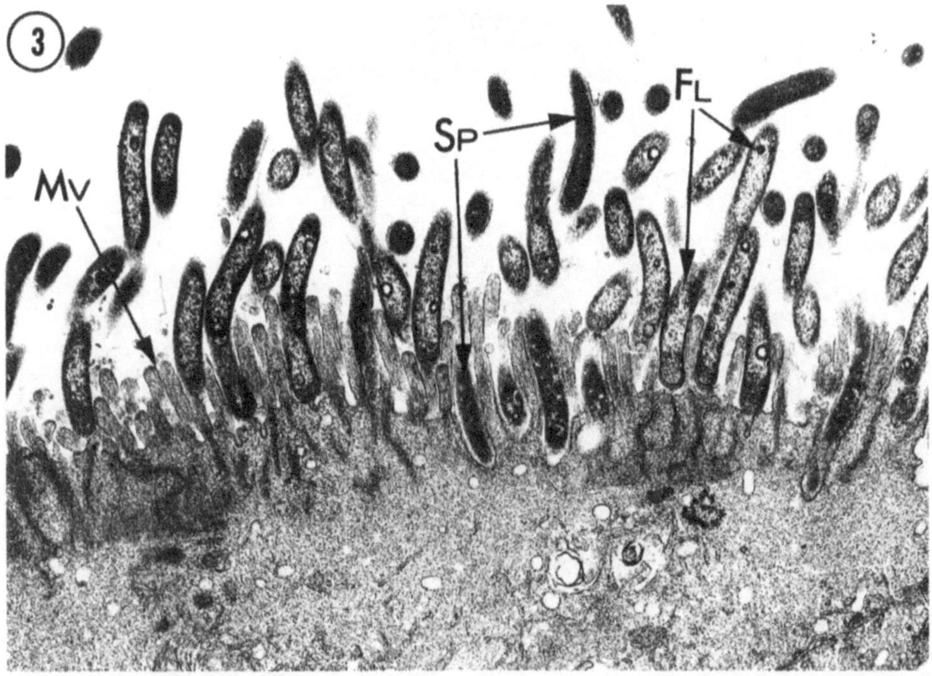

FIGURE 3. Two types of Gram-negative bacteria adhere to the luminal surface of a single colonic cell from a healthy monkey. Both are attached by one end to the cell surface between microvilli (Mv) and do not appear to damage the host cell. Spirochetes (Sp) deeply indent the host cell membrane, whereas the flagellated bacteria (Fl) are more superficially attached. Magnification × 16,614. (From Neutra, M. R., *J. Ultrastruct. Res.*, 70, 186, 1980. With permission.)

these has been studied by both thin section and freeze-fracture electron microscopy.[40] In this case, rod-shaped and filamentous Gram-negative bacteria are arranged in orderly, alternating rows on the surface of a devescovinid protozoan host. The two prokaryotes occupy morphologically distinct attachment sites: one lies along a ridge on the host cell surface whereas the other is partially embedded in a trough-like invagination (Figure 7). In both cases, condensations of microfilamentous material appear in the host cell cytoplasm immediately subjacent to the bacterium, and these elements, presumably cytoskeletal in nature, appear to maintain the modified architecture of the host cell surface. Freeze-fracture replicas of the host cell membrane show special intramembrane particle arrays along the edges of both attachment sites. Whether bacterial surface specializations are also present is not known, but the patterns of bacterial replication in this system suggested to Tamm[40] that cell surface "signals" on the bacterial outer membrane are responsible for induction of the host cell specializations. In this and other protozoal systems, it seems that both bacterium and host benefit from the relationship, and that the morphological adaptations of both indicates the mutual consent of the symbiotic partners.[38-40]

III. RECOGNITION, ADHESION, AND STABILIZATION OF STRUCTURAL ALTERATIONS

The intestinal cell, however, derives no known benefit from the attached, nonpathogenic bacteria for which it obligingly modifies its luminal surface. Indeed, host cell accommodation is observed even when the association leads to disease. For example, the epithelial cells of

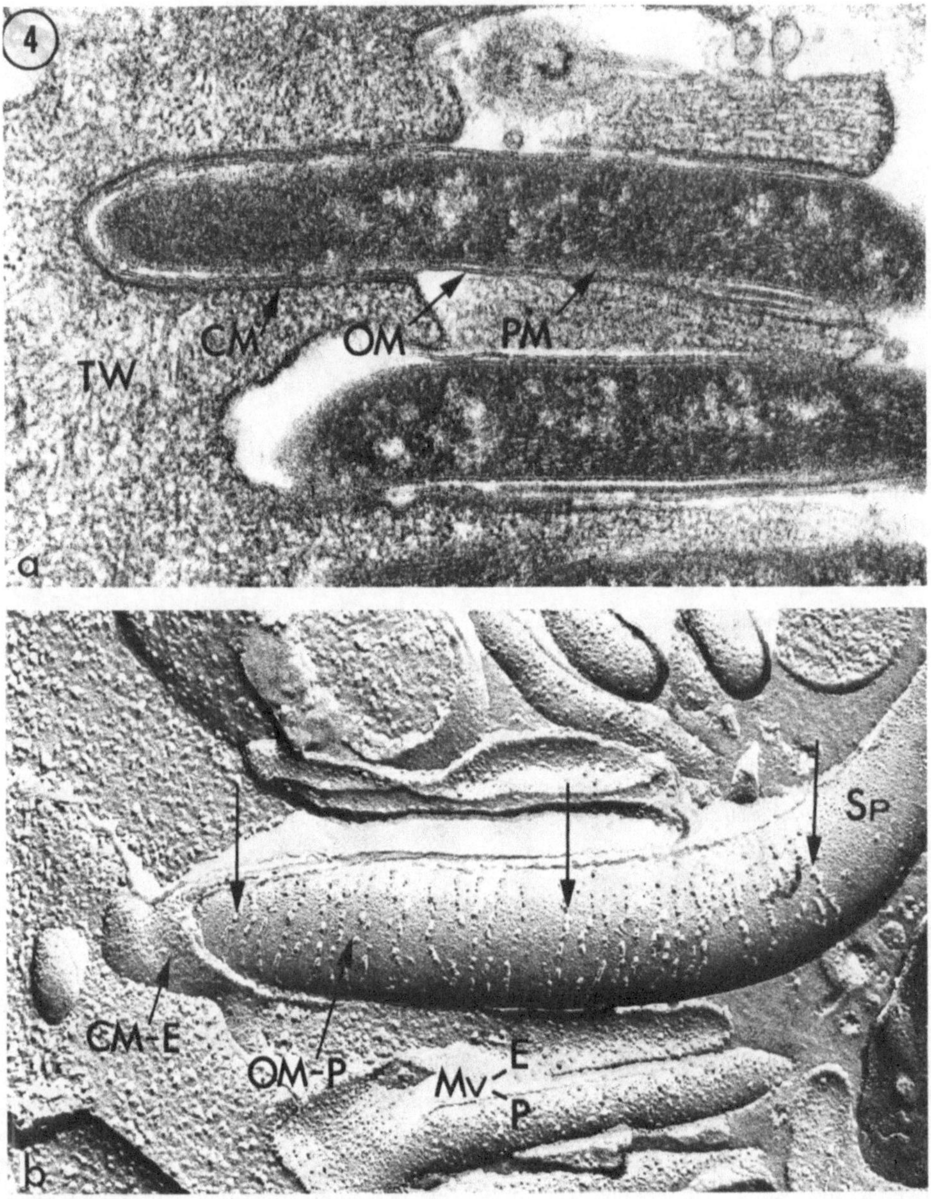

FIGURE 4. Attachment site of the spirochete in monkey colon. (a) The spirochete outer membrane (OM) adheres to the invaginated host cell membrane (CM) and to adjacent microvilli (Mv), separated by a 6 to 8 nm gap. The spirochete plasma membrane (PM) is also visible. Thin filaments and intermediate filaments of the host cell terminal web (TW) are seen subjacent to the invaginated cell membrane. Magnification × 83,340. (b) The tip of a freeze-fractured spirochete (Sp) is hidden in a sleeve of host cell plasma membrane (CM-E face). E and P faces of adjacent microvilli (Mv) are seen. The fracture plane has passed through the bacterial outer membrane, exposing its P face (OM-P). The attached portion of the membrane displays a series of circumferential rows of intramembrane particles (arrows). Magnification ×75,660. (From Neutra, M. R., *J. Ultrastruct. Res.*, 70, 186, 1980. With permission.)

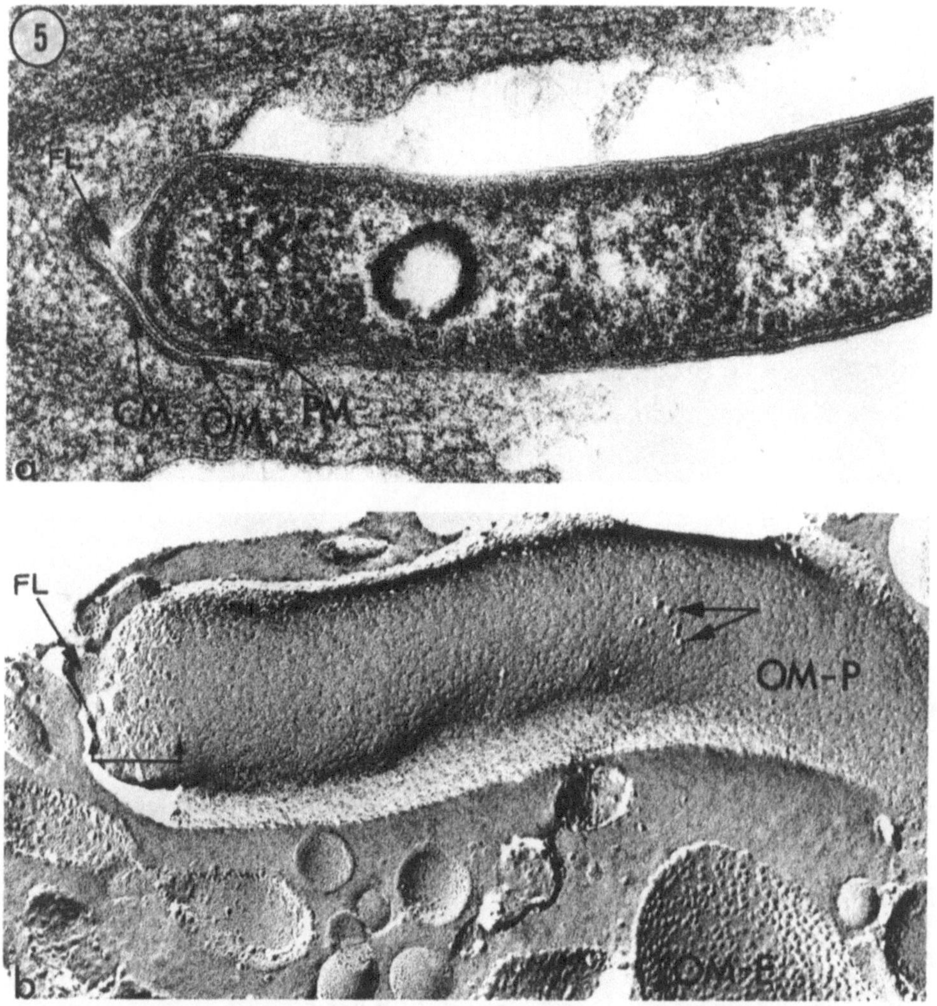

FIGURE 5. Attachment site of the flagellated bacterium in monkey colon. (a) The rounded end of the bacterium adheres to the host cell membrane (CM) between microvilli. The terminal flagellum (FL) indents the apical cell surface. Bacterial outer membrane (OM) and intestinal cell membrane (CM) are separated by 6 to 8 nm in the attachment zone. Magnification × 113,635. (b) The freeze-fractured outer membrane P face (OM-P) of the bacterium is marked at the attachment zone (bracket) by heterogeneous particles that encircle the terminal flagellum (Fl). Magnification ×66,766. (c) Horseradish peroxidase passes through the narrow gap between the host cell membrane (CM) and the bacterial outer membrane (OM. An intermediate layer separates the outer membrane and the plasma membrane (PM) of the flagellate. (Magnification ×79,544. (From Neutra, M. R., *J. Ultrastruct. Res.* 70, 186, 1980. With permission.)

rabbit ileum infected with *E. coli* strain RDEC-1, a pathogenic organism, form pedestal-like structures supported by a cytoskeletal scaffold in response to bacterial attachment.[41] In the neonatal pig, an initial stage of *E. coli* infection involves the formation of tight bacterial-intestinal cell attachment sites supported by submembrane host cell filament aggregates.[42] In this case, attachment is followed by internalization into membrane-limited vacuoles and lysosomes and then by degeneration of the host cell. The active participation of the epithelial cell both in attachment and in phagocytosis, then, clearly does not indicate a symbiotic or commensal relationship. Rather, the bacteria seem to have tricked the host cell into rear ranging membrane and cytoskeletal components that were intended for a more benign purpose.

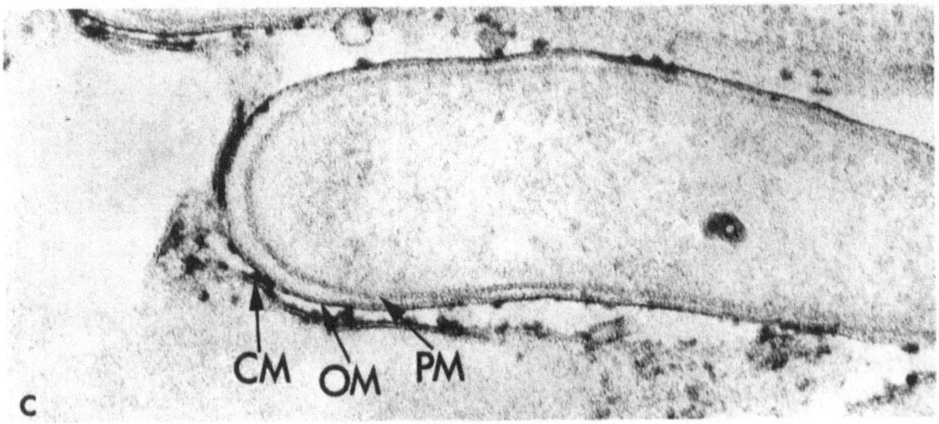

FIGURE 5C

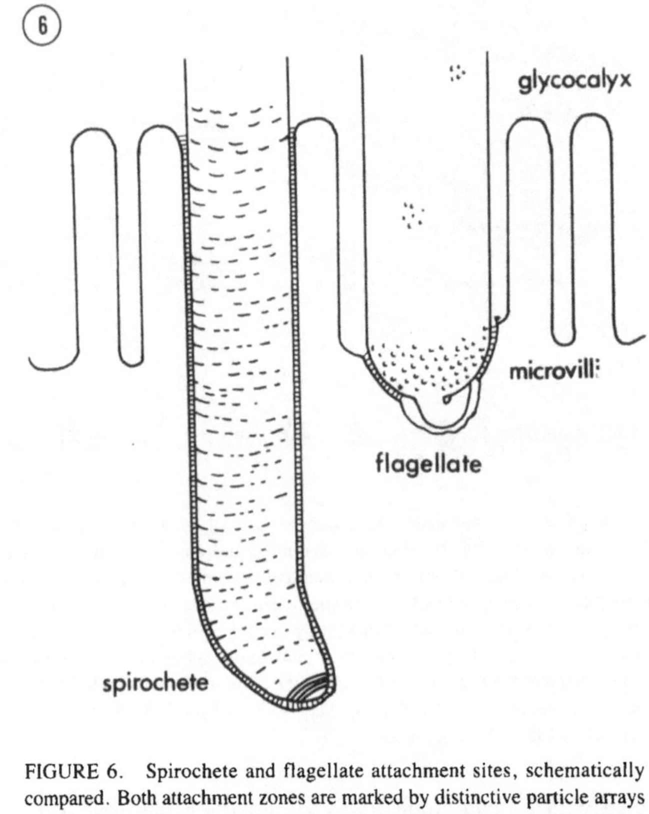

FIGURE 6. Spirochete and flagellate attachment sites, schematically compared. Both attachment zones are marked by distinctive particle arrays seen in freeze-fracture replicas of the bacterial outer membrane. (From Neutra, M. R., *J. Ultrastruct. Res.*, 70, 186, 1980. With permission.)

It is clear that mutual bacterial and intestinal cell "recognition" and adhesion is a prerequisite both for invasion and for the formation of stable attachment sites. After initial adhesion, some unknown span of time seems to be required for the formation of the elaborate, intimate associations of nonpathogenic bacteria with intestinal cells described in freeze-fracture studies, but progressive stages in the development of such sites have not been observed. Mutual recognition of cell-specific surface macromolecules is also the initial event

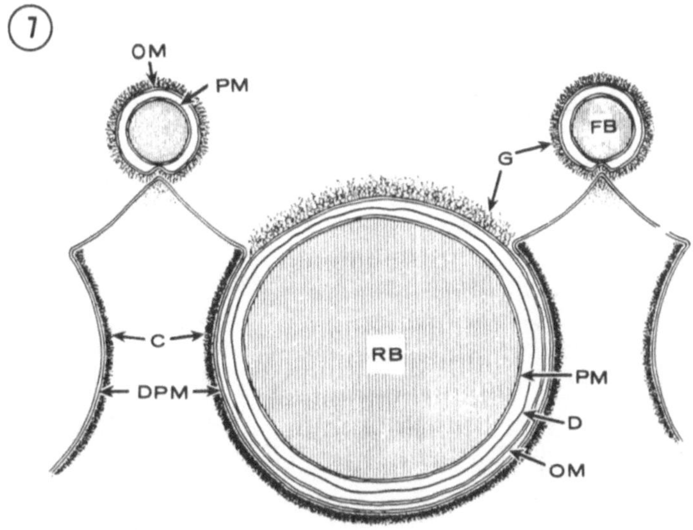

FIGURE 7. Schematic drawing showing the major features of the attachment sites of two types of Gram-negative bacteria on the surface of a devescovinid protozoan in the termite hindgut. Rod-shaped bacteria (RB) lie in a deep trough-like invagination of the host cell surface whereas thin, fusiform bacteria (FB) are held on longitudinal ridges, here shown in cross-section. The surfaces of both organisms are coated by a dense glycocalyx (G) that is visible in the attachment zone of the fusiform bacterium, but not in that of the rod bacterium. Cytoplasmic proteins form distinctive coats (C) under the host cell membrane at both attachment sites. OM =

in the association of embryonic cells during tissue assembly. Cell binding assays in vitro demonstrated that such associations occur in two steps: first, a reversible recognition step dependent on the specificity of cell surface macromolecules and second, a stabilization step leading to closer, more permanent intercellular bonds.[43] Stabilization is time and temperature-dependent in a manner similar to intramembrane diffusional phenomena, suggesting that tight bond formation may depend on the mobility of cell surface macromolecules.[43] The stabilization of the semipermanent bacterial-intestinal cell associations described by freeze-fracture also appears to require rearrangements of intramembrane molecules of both partners. Although multiple cell surface components may be involved in tight adhesion, such rearrangements could be triggered by a recognition-dependent, initial attachment event. Unfortunately, intestinal spirochetes, flagellated bacteria, and other attached species have not been cultured in a free form before freeze-fracture examination, so we cannot rule out the possibility that specialized domains on the bacterial membrane existed prior to attachment.

A. Arrested Phagocytosis

Ultrastructurally, semipermanent bacterial-host cell attachment sites resemble a stage of phagocytosis that has been arrested. Elegant experimental evidence from macrophage systems suggests that completion of phagocytosis requires that the relevant binding sites be distributed over the entire surface of the particle to be engulfed.[44,45] The engulfment of large particles, microorganisms or opsonized cells by peritoneal macrophages appears to involve a rapid, zipper-like sequential binding of macrophage membrane receptors to ligands on the surface of the particle. Binding usually progresses until the particle surface is completely covered by cytoplasmic extentions of the macrophage, and these then fuse with each other to close the phagosome. Partial phagocytosis was experimentally produced using lymphocytes coated with immunoglobulin that had been allowed to "cap" on one area of the lymphocyte surface.

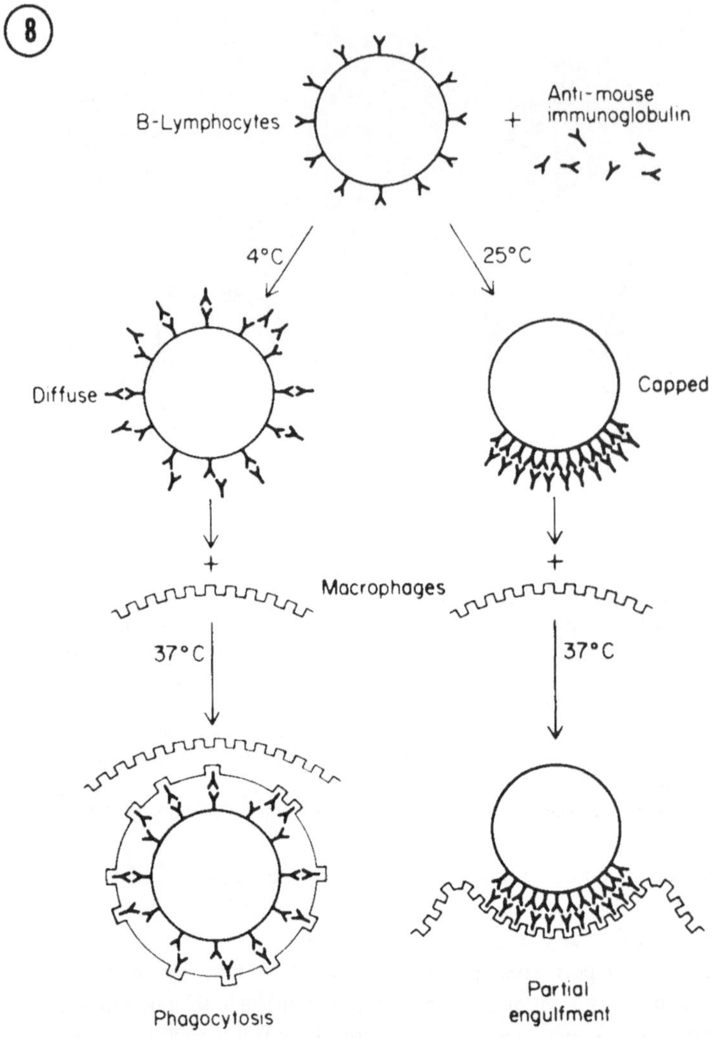

FIGURE 8. Diagram summarizing the "zipper" model of phagocytosis, based on experiments in which IgG-coated lymphocytes were presented to peritoneal macrophages. Diffusely-coated lymphocytes were phagocytosed by macrophages (left). Capped lymphocytes were only partially engulfed and remained bound to the macrophage surface (right). (From Silverstein, S. C., Michl, J., and Sung, S. S. J., *Transport of Macromolecules in Cellular Systems*, Dahlem Konferenzen, Berlin, 1978, 250. With permission.)

When presented to macrophages, capped lymphocytes were only partially engulfed. Binding was limited to the portion of the lymphocyte membrane bearing the immunoglobulin ligand, and the lymphocyte remained attached to cup-like invaginations in the macrophage surface[44,45] (Figure 8). These images suggest that the ability of certain bacterial species to remain attached on eukaryotic cells without being phagocytosed may also depend on a limited, localized array of ligands on the bacterial cell surface.

B. Cytoskeletal Alterations

The relationship between membrane-integrated glycoproteins, membrane-associated proteins, and underlying cytoskeletal elements has been subject to intense investigation.[46,47]

Again, the function of this integrated system in phagocytic cells may be relevant to understanding the fates of intestinal bacteria. Professional phagocytes, like intestinal cells, exhibit a striking recruitment of cytoplasmic filaments under the invaginated membrane involved in partial or complete engulfment of a cell or microorganism.[45] In the case of the macrophage, some of these filaments are known to consist of actin;[48-50] their local polymerization may depend in part upon the local release of an actin-binding protein from the plasma membrane in response to a transmembrane signal, as such release was shown to occur in response to the occupation of overlying Fc receptors.[50] The dynamics and modulation of actin-based networks are complex, involving a variety of actin-associated proteins, local intracellular pH changes, calcium ion activity, and other factors.[51] The biochemical and ultrastructural architecture of the normal intestinal microvillous border and terminal web region has been described in considerable detail,[52-56] and is reviewed in this volume.[57] The rearrangement of cytoskeletal components that accompanies formation of bacterial attachment sites on intestinal cells, however, has not yet been explored.

C. Mucin Interactions

It is generally accepted that most microorganisms in the intestinal lumen are prevented from contacting absorptive cell surfaces by their interaction with mucins and other glycoconjugates that are shed from absorptive cells or secreted by goblet cells into the lumen.[7,21,58] Although both microorganisms and macromolecules may be effectively entrapped in the mucus gel, the intestinal mucus layer is discontinuous and a certain percentage of organisms and ligands undoubtedly have access to the intestinal cell glycocalyx. For example, cationized ferritin, a large, polyvalent protein (mw 500,000) binds to most cells on the mucosal surface within a few minutes after being introduced into the lumen[59] (Figure 9). In this light, it is remarkable that so few microorganisms actually succeed in attaching to intestinal cell surfaces.

IV. DISSOCIATION AND SHEDDING

An initial association of bacteria with brush border macromolecules may occur relatively frequently, but may be followed by dissociation and release rather than by more intimate attachment or phagocytosis.[7,21] Under normal in vivo conditions, intestinal cell surface macromolecules are continually shed from the brush border membrane into the lumen at rates that vary according to their size and their position in or on the membrane.[60,61] Continual replacement of membrane constituents and glycocalyx macromolecules occurs through intracellular synthesis,[62,63] intracellular transport of membrane vesicles,[63,64] and insertion, by exocytosis, into the apical cell surface.[64] Retrieval of membrane by endocytosis and recycling of luminal surface components has not been demonstrated in absorptive cells of adult intestine, although membrane proteins are conserved and renewed in this way by many other types of cells.[65,66] The unique position of the intestinal epithelial cell would make the routine endocytosis of luminal membrane a risky activity, with the inherent danger of incorporating opportunistic, foreign luminal ligands, and organisms along with the internalized membrane.

Shedding of intestinal cell surface macromolecules is dependent in part upon their cleavage by adsorbed pancreatic proteases,[67] so that loss of glycoconjugates (and hence of bacterial binding sites) would proceed in normal small intestine, at higher rates than in cultured explants,[68] whether or not bacteria are associated with the glycocalyx. Shedding also occurs in the colon.[69] Accordingly, macromolecular tracers that bind to the glycocalyx and are visible ultrastructurally are generally not endocytosed but are gradually sloughed from the microvillus surface. Cationized ferritin, for example, binds electrostatically[70] to negatively charged sugars on the surfaces of the microvillus membrane and glycocalyx of adult intestinal absorptive cells, but it is not endocytosed.[59] Instead, the polyvalent ligand aggregates in multiple clusters along the microvillus and at the microvillus tip, and the clusters appear to be eventually sloughed off into the lumen.[59] Conceivably, intestinal cells could use a similar strategy to deal with the polyvalent ligands of bacterial surfaces.

An alternative means of shedding intestinal cell surface binding sites involves the "budding" of small membranous vesicles, containing integral membrane enzymes and other glycoconjugates, from the microvilli.[71] Such small vesicles are commonly observed in the immediate vicinity of microvilli on normal absorptive cells. Their frequency increases dramatically under in vitro conditions,[71] and also increases in a dose-dependent fashion when the luminal surface is exposed to lectins specific for the saccharides present on the brush border.[72] The "budding" process allows loss of apical membrane without compromise of plasma membrane integrity, and could thus serve to prevent further attachment, colonization, or penetration of the epithelial barrier.

V. INTERACTIONS IN NEONATAL INTESTINE

The shedding process is not as effective in neonatal intestine as in the adult, and this difference, along with chemical differences in cell surface glycoconjugates, may partly explain the vulnerability of the neonate to bacterial colonization and invasion. Neonatal microvillus membranes, like those of the adult, display multiple binding sites for nonspecific, "opportunistic" ligands such as cationized ferritin,[73] but the pattern of mobility of these binding sites in the plane of the membrane is unlike that of adult cells. In the neonatal rat jejunum, binding sites on the lower half of the microvillus membrane move, without clustering, into the apical tubulovesicular system.[73,74] In this system, bound macromolecules are separated from fluid-phase proteins,[74,75] and cationized ferritin is transported in small vesicles across the cell to the basolateral membrane.[73,74] The nonspecific membrane-bound tracer thus takes advantage of the transcellular membrane shuttle intended for the transport of maternal IgG. Extensive ultrastructural tracer studies have established that maternal IgG binds specifically to Fc receptors on the proximal half of the microvillus membrane and is incorporated and transported along the unidirectional route that was visualized by cationized ferritin.[74,75] The distal half of the jejunal microvillus membrane, in contrast, binds IgA;[76] cationized ferritin on this portion of the membrane is not endocytosed. In the neonatal rat ileum, in contrast to the jejunum, IgG receptors are not present.[75] We have found that binding sites for cationized ferritin on ileal absorptive cell microvilli show a bidirectional pattern of mobility: cationized ferritin binding sites move and cluster both at the tip and at the base of each microvillus (Figures 9 and 10). Although some tracer is shed from microvillus tips, some aggregates of tracer at the microvillus base may be eventually incorporated into the endocytic complex of the cell and into the extensive lysosomal system. Enteropathogenic *E. coli* follows a similar route in the neonatal ileum.[42]

Little is known of the multiple, sequential events that allow certain strains of bacteria to evade the shedding of macromolecules and membrane from the intestinal cell surface and then to proceed to form tight attachment sites. If "successful" organisms such as *E. coli* RDEC-1 in the ileum and nonpathogenic spirochetes in the colon bear a fixed array of surface binding factors in an "attachment domain" of their own outer membrane, these could serve to immobilize the complementary host cell membrane receptors and thus to prevent receptor mobility, clustering, sloughing, and loss. Ultrastructural tracer studies may eventually be useful in identifying ligands capable of inducing the unusual rearrangements of host cell receptors and cytoskeletal elements that lead to the formation of elaborate bacterial attachment sites.

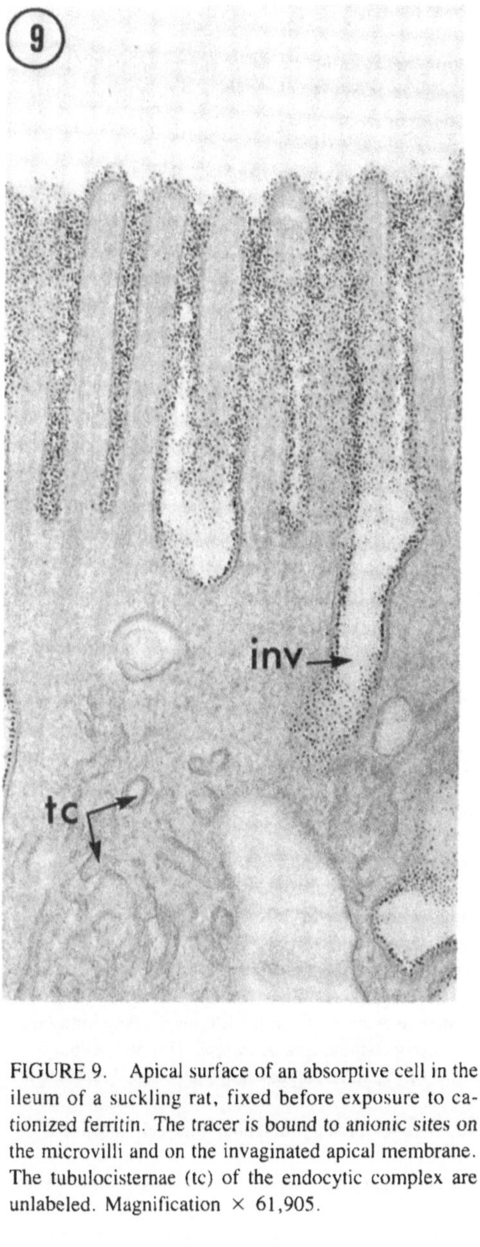

FIGURE 9. Apical surface of an absorptive cell in the ileum of a suckling rat, fixed before exposure to cationized ferritin. The tracer is bound to anionic sites on the microvilli and on the invaginated apical membrane. The tubulocisternae (tc) of the endocytic complex are unlabeled. Magnification × 61,905.

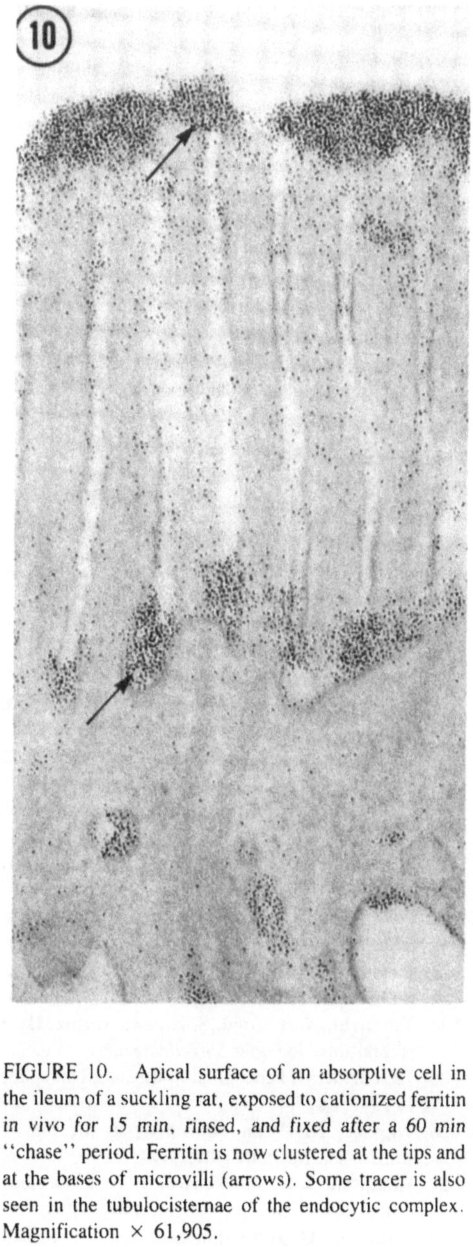

FIGURE 10. Apical surface of an absorptive cell in the ileum of a suckling rat, exposed to cationized ferritin *in vivo* for 15 min, rinsed, and fixed after a 60 min "chase" period. Ferritin is now clustered at the tips and at the bases of microvilli (arrows). Some tracer is also seen in the tubulocisternae of the endocytic complex. Magnification × 61,905.

REFERENCES

1. **Kim, Y. S.,** Intestinal glycoprotein structure, in *Attachment of Organisms to the Gut Mucosa*, Boedecker, E., Ed., CRC Press, Boca Raton, Fla., 1984.
2. **Ito, S.,** Structure and function of the glycocalyx, *Fed. Proc. Fed. Am. Soc. Exp. Biol.*, 28, 12, 1969.
3. **Ugolev, A. M. and DeLaey, P.,** Membrane digestion. A concept of enzymic hydrolysis on cell membranes, *Biochim. Biophys. Acta*, 300, 105, 1973.
4. **Hollenberg, M. D., Fishman, P. H., Bennett, V., and Cuatrecasas, P.,** Cholera toxin and cell growth: role of membrane gangliosides, *Proc. Natl. Acad. Sci. U.S.A.*, 71, 4224, 1974.

5. **Draper, R. K., Chin, D., and Simon, M. I.,** Diphtheria toxin has the properties of a lectin, *Proc. Natl. Acad. Sci. U.S.A.*, 75, 261, 1978.
6. **Gottschalk, A., Belyavin, G., and Biddle, F.,** Glycoproteins as influenza virus hemagglutinin inhibitors and as cellular virus receptors, in *Glycoproteins, Their Composition, Structure and Function*, Gottschalk, A., Ed., Elsevier, Amsterdam, 1972, 1082.
7. **Gibbons, R. J.,** Adherence of bacteria to host tissue, *Microbiology*, 1977, 395, 1977.
8. **Smith, H.,** Microbial surfaces in relation to pathogenicity, *Bacteriol. Rev.*, 41, 475, 1977.
9. **Sharon, N., Eshdat, Y., Silverblatt, J., and Ofek, I.,** Bacterial adherence to cell surface sugars, in *Adhesion and Microorganism Pathogenicity*, Ciba Foundation Symposium 80, Pitman Medical Ltd., London, 1981, 119.
10. **Etzler, M. E. and Branstrator, M. L.,** Differential localization of cell surface and secretory components in rat intestinal epithelium by use of lectins, *J. Cell Biol.*, 62, 329, 1974.
11. **Essner, E., Schreiber, J., and Griewski, R. A.,** Localization of carbohydrate components in rat colon with fluoresceinated lectins, *J. Histochem. Cytochem.*, 26, 452, 1978.
12. **Freeman, H. J., Etzler, M. E., Garrido, A. B., and Kim, Y. S.,** Alterations in cell surface membrane components of adapting rat small intestinal epithelium: studies with lectins after massive proximal jejuno-ileal resection and jejuno-ileal transposition, *Gastroenterology*, 75, 1066, 1978.
13. **Etzler, M. E. and Branstrator, M. L.,** Cell surface components of intestinal epithelial cells and their relationship to cellular differentiation, in *Development of Mammalian Absorptive Processes*, Ciba Foundation Series 70, Excerpta Medica, Elsevier/North-Holland, Amsterdam, 1979, 51.
14. **Gibbons, R. J., Spinell, D. M., and Skoke, Z.,** Selective adherence as a determinant of the host tropisms of certain indigenous and pathogenic bacteria, *Infect. Immun.*, 11, 238, 1976.
15. **Dubos, R., Schaedler, R. W., Costello, R., and Hoet, P.,** Indigenous, normal and autochthonous flora of the gastrointestinal tract, *J. Exp. Med.*, 122, 67, 1965.
16. **Jones, G. W. and Freter, R.,** Adhesive properties of *Vibrio cholerae*, *Infect. Immun.*, 14, 240, 1976.
17. **Hirschberger, D., Mirelman, D., and Thaler, M.,** The effects of mannose and its derivatives on interactions of pathogenic *E. coli* with newborn intestinal mucosa, *Gastroenterology*, 72, 1069A, 1977.
18. **Ofek, I., Mirelman, D., and Sharon, N.,** Adherence of *Escherichia coli* to human mucosal cells mediated by mannose receptors, *Nature (London)*, 265, 623, 1977.
19. **Costerton, J. W., Geesey, G. G., and Cheng, K. J.,** How bacteria stick, *Sci. Am.*, 238, 86, 1978.
20. **Salit, I. and Gotschlich, E.,** Hemagglutination by purified type I Escherichia coli pili, *J. Exp. Med.*, 146, 1169, 1977.
21. **Freter, R.,** Mechanisms of association of bacteria with mucosal surfaces, in *Adhesion and Microorganism Pathogenicity*, Ciba Foundation Symposium 80, Pitman Medical Ltd., London, 1981, 36.
22. **Takeuchi, A.,** Electron microscope studies of experimental *Salmonella* infection. I. Penetration into the intestinal epithelium by *Salmonella typhimurium*, *Am. J. Pathol.*, 50, 109, 1967.
23. **Takeuchi, A.,** Electron microscope observations on penetration of the gut epithelial barrier by Salmonella typhimurium, *Microbiology*, 1975, 174, 1975.
24. **Takeuchi, A., Formal, S. B., and Sprinz, H.,** Experimental acute colitis in the Rhesus monkey following peroral infection with *Shigella flexneri*: an electron microscope study, *Am. J. Pathol.*, 52, 503, 1968.
25. **Takeuchi, A.,** Penetration of the intestinal epithelium by various microorganisms, *Curr. Top. Pathol.*, 54, 1, 1971.
26. **Wagner, R. C. and Barrnett, R. J.,** The fine structure of prokaryotic-eukaryotic cell junctions, *J. Ultrastruct. Res.*, 48, 404, 1974.
27. **Hampton, J. C. and Rosario, B.,** The attachment of microorganisms to epithelial cells in the distal ileum of the mouse, *Lab. Invest.*, 14, 1464, 1965.
28. **Chase, D. G. and Erlandsen, S. L.,** Evidence for a complex life cycle and endospore formation in the attached, filamentous, segmented bacterium from murine ileum, *J. Bacteriol.*, 127, 572, 1976.
29. **Chase, D. G. and Erlandsen, S. L.,** The morphology of cell-microbe junctions at the apical surface of rat ileal enterocytes, *J. Cell Biol.*, 79, 262a, 1978.
30. **Snellen, J. E. and Savage, D. C.,** Freeze-fracture study of the filamentous, segmented microorganism attached to the murine small bowel, *J. Bacteriol.*, 134, 1099, 1978.
31. **Madara, J. L.,** Structural specializations of absorptive cell membranes at filamentous organism-absorptive cell attachment sites in mammalian small intestine, in *Attachment of Organisms to the Gut Mucosa*, Boedecker, E. C., Ed., CRC Press, Boca Raton, Fla., 1984.
32. **Harland, W. A. and Lee, F. D.,** Intestinal spirochetosis, *Br. Med. J.*, 3, 718, 1967.
33. **Gear, E. V. and Dobbins, W. O.,** Rectal biopsy, *Gastroenterology*, 55, 522, 1968.
34. **Lee, F. D., Kraszewski, A., Gordon, J., Howie, J. G. R., McSenney, D., and Harland, W. A.,** Intestinal spirochetosis, *Gut*, 12, 126, 1971.
35. **Takeuchi, A. and Zeller, J. A.,** Ultrastructural identification of spirochetes and flagellated microbes at the brush border of the large intestinal epithelium of the rhesus monkey, *Infect. Immun.*, 6, 1008, 1972.

36. **Takeuchi, A., Jervis, H. R., Nakazawa, H., and Robinson, D. M.,** Spiral shaped organisms on the surface colonic epithelium of the monkey and man, *Am. J. Clin. Nutr.*, 27, 1287, 1974.
37. **Neutra, M. R.,** Prokaryotic-eukaryotic cell junctions: attachment of spirochetes and flagellated bacteria to primate large intestinal cells, *J. Ultrastruct. Res.*, 70, 186, 1980.
38. **Cleveland, L. R. and Grimstone, A. V.,** The fine structure of the flagellate Mixotricha paradoxa and its associated microorganisms, *Proc. R. Soc.*, B159, 668, 1964.
39. **Bloodgood, R. A. and Fitzharris, T. P.,** Specific associations of prokaryotes with symbiotic flagellate Protozoa from the hindgut of the termite Reticulitermes and the wood-eating roach Cryptocercus, *Cytobios*, 17, 103, 1976.
40. **Tamm, S. L.,** The ultrastructure of prokaryotic-eukaryotic cell junctions, *J. Cell Sci.*, 44, 335, 1980.
41. **Takeuchi, A., Inman, L. R., O'Hanley, P. D., Cantey, J. R., and Lushbaugh, W. B.,** Scanning and transmission electron microscopic study of *Escherichia coli* 015 (RDEC-1) enteric infection in rabbits, *Infect. Immun.*, 19, 686, 1978.
42. **Staley, T. E., Jones, W., and Corley, L. D.,** Attachment and penetration of Escherichia coli into intestinal epithelium of the ileum in newborn pigs, *Am. J. Pathol.*, 56, 371, 1969.
43. **McClay, D. R., Wessel, G. M., and Marchase, R. B.,** Intercellular recognition: quantitation of initial binding events, *Proc. Natl. Acad. Sci. U.S.A.*, 78, 4975, 1981.
44. **Griffin, F. J. Jr., Griffin, J. A., and Silverstein, S. C.,** Studies on the mechanism of phagocytosis. II. The interaction of macrophages with anti-immunoglobulin IgG coated bone marrow derived lymphocytes, *J. Exp. Med.*, 144, 788, 1976.
45. **Silverstein, S. C., Michl, J., and Sung, S.-S. J.,** Phagocytosis, in *Transport of Macromolecules in Cellular Systems*, Silverstein, S. C., Ed., Dahlem Konferenzen, Berlin, 1978, 245.
46. **Nicolson, G. L.,** Transmembrane control of the receptors on normal and tumor cells. I. Cytoplasmic influences over cell surface components, *Biochim. Biophys. Acta*, 457, 57, 1976.
47. **Pollard, T. D.,** Contractile proteins and cell surface dynamics, in *Transport of Macromolecules in Cellular Systems*, Silverstein S. C., Ed., Dahlem Konferenzen, Berlin, 1978, 489.
48. **Reaven, E. P. and Axline, S. G.,** Subplasmalemmal microfilaments and microtubules in resting and phagocytosing cultivating macrophages, *J. Cell Biol.*, 59, 12, 1973.
49. **Wang, E., Eddson, P. J., Goldberg, A. R., and Bianco, C.,** Endocytic redistribution of contractile proteins in mouse peritoneal macrophages, *J. Cell Biol.*, 75, 255a, 1977.
50. **Hartwig, J. H., Davies, W. A., and Stossel, T. P.,** Evidence for contractile protein translocation in macrophage spreading, phagocytosis, and phagolysosome formation, *J. Cell Biol.*, 75, 936, 1977.
51. **Schliwa, M.,** Proteins associated with cytoplasmic actin, *Cell*, 25, 587, 1981.
52. **Bretscher, A. and Weber, K.,** Localizations of actin and microfilament-associated proteins in the microvilli and terminal web of the intestinal brush border by immunofluorescence microscopy, *J. Cell Biol.*, 79, 839, 1978.
53. **Begg, D. A., Rodewald, R., and Rebhun, L. I.,** The visualization of actin filament polarity in thin sections. Evidence for the uniform polarity of membrane-associated filaments, *J. Cell Biol.*, 79, 846, 1978.
54. **Matsudaira, P. T. and Burgess, D. R.,** Identification and organization of the components in the isolated microvillus cytoskeleton, *J. Cell Biol.*, 83, 667, 1979.
55. **Mooseker, M. S. and Stephens, R. E.,** Brush border actinin? Comparison of two proteins of the microvillus core with actinin by two-dimensional peptide mapping, *J. Cell Biol.*, 86, 466, 1980.
56. **Yin, H. L., Albrecht, H., and Fattorum, A.,** Identification of gelsolin, a Ca-dependent regulatory protein of actin gel-sol transformation, and its intracellular distribution in a variety of cells and tissues, *J. Cell Biol.*, 91, 901, 1981.
57. **Burgess, D. R.,** in *Attachment of Organisms to the Gut Mucosa*, Boedecker, E. C., Ed., CRC Press, Boca Raton, Fla., 1984.
58. **Forstner, G., Wesley, A., and Forstner, J.,** Clinical Aspects of Gastrointestinal Mucus, in *Mucus in Health and Disease*, in press.
59. **Jersild, R. A. and Crawford, R. W.,** The distribution and mobility of anionic sites on the brush border of intestinal absorptive cells, in press.
60. **Alpers, D. H.,** The relation of size to the relative rates of degradation of intestinal brush border proteins, *J. Clin. Invest.*, 51, 2621, 1972.
61. **Alpers, D. H.,** Protein turnover in intestinal mucosal villus and crypt brush border membranes, *Biochem. Biophys. Res. Commun.*, 75, 130, 1977.
62. **Forstner, G. G.,** (1-^{14}C) glucosamine incorporation by subcellular fractions of small intestinal mucosa, *J. Biol. Chem.*, 245, 3584, 1970.
63. **Bennett, G., Leblond, C. P., and Haddad, A.,** Migration of glycoprotein from the Golgi apparatus to the surface of various cell types as shown by radioautography after labeled fucose injection into rats, *J. Cell Biol.*, 60, 258, 1974.

64. **Michaels, J. E. and Leblond, C. P.,** Transport of glycoproteins from Golgi apparatus to cell surface by means of "carrier" vesicles, as shown by radioautography of mouse colonic epithelium after injection of ^{3}H-fucose, *J. Microscopie Biol. Cell,* 25, 243, 1976.
65. **Schneider, Y.-L., Tulkens, P., DeDuve, C., and Trouet, A.,** Fate of plasma membrane during endocytosis. II. Evidence for recycling (shuttle) of plasma membrane constituents, *J. Cell Biol.,* 82, 466, 1979.
66. **Farquhar, M. G. and Palade, G. E.,** The Golgi apparatus (complex)-(1954-1981)-from artifact to center stage, *J. Cell Biol.,* 91, part 2, 77s, 1981.
67. **Alpers, D. H. and Tedesco, F. I.,** The possible role of pancreatic proteases in the turnover of intestinal brush border proteins, *Biochim. Biophys. Acta,* 401, 28, 1975.
68. **Berteloot, A., Chabot, J.-G., and Hugon, J. S.,** Turnover of mouse intestinal brush border membrane proteins and enzymes in organ culture, *Biochim. Biophys. Acta,* 678, 423, 1981.
69. **Lamont, J. T. and Ventola, A.,** Synthesis and secretion of colonic glycoproteins. Evidence for shedding in vivo of low molecular weight membrane components, *Biochim. Biophys. Acta,* 629, 553, 1980.
70. **Danon, D., Goldstein, L., Marikovsky, Y., and Skutelsky, E.,** Use of cationized ferritin as a label of negative charges on cell surfaces, *J. Ultrastruct. Res.,* 38, 500, 1972.
71. **Berteloot, A., Chabot, J. G., and Hugon, J. G.,** Organ culture of the adult mouse intestine. V. Vesiculation of the brush border membrane during the culture, *Biol. Cell,* 42, 1981.
72. **Lorenzsonn, V. and Olsen, W. A.,** In vivo responses of rat intestinal epithelium to intraluminal dietary lectins, *Gastroenterology,* 82, 838, 1982.
73. **Jersild, R. A.,** Restricted mobility and endocytosis of anionic sites on newborn rat jejunal brush border membranes, *Anat. Rec.,* 202, 61, 1982.
74. **Rodewald, R.,** Distribution of immunoglobulin G receptors in the small intestine of the young rat, *J. Cell Biol.,* 85, 18, 1980.
75. **Abrahamson, D. R. and Rodewald, R.,** Evidence for the sorting of endocytic vesicle contents during the receptor-mediated transport of IgG across the newborn rat intestine, *J. Cell Biol.,* 91, 270, 1981.
76. **Brown, W.,** Routes of IgA entry into the intestine, in *Attachment of Organisms to the Gut Mucosa,* Boedeker, E., Ed., CRC Press, Boca Raton, Fla., 1984.

Chapter 18

THE ASSOCIATION OF MICROORGANISMS WITH THE TISSUES AND THE MUCOUS "BLANKET" OF THE GASTROINTESTINAL SYSTEM

J. W. Costerton, K.-J. Cheng, and K. R. Rozee

TABLE OF CONTENTS

I. INTRODUCTION

Microbiology has been profoundly altered, during the last 5 years, by the concept that the predominant mode of bacterial growth is actually in a biofilm adherent to surfaces within the environment. This mode of growth has been documented and found to be predominant in soil,[1] in fresh water streams,[2] in marine systems,[3] in industrial aquatic systems,[4] in the normal human urethra,[5] in the diseased lung[6] and bladder,[7] and on inert prostheses in the human body.[8] Moreover, bacteria living in these diverse environments have been found universally to be surrounded by an extracellular polysaccharide glycocalyx[9,10] that both protects them from antibacterial factors (bacteriophage, phagocytic cells, etc.) and mediates their formation of glycocalyx-enclosed microcolonies within the adherent biofilm. This glycocalyx is lost by most bacteria, on serial subculture "in vitro".[10] It has been demonstrated that these bacteria deep within biofilms are partially protected from chemical biocides,[11] by the ion-exchange function of their glycocalyces,[10] and that the physiology of adherent bacteria may actually differ[12] from cells of the same strain growing in the free-floating "planktonic" mode. Further, bacteria living in biofilms often form "consortia" in which they cooperate physiologically with other bacteria[13] and even with the colonized tissue.[14] Thus, firm evidence has accumulated to show that bacteria in nature and disease grow in a protected mode in which they live in glycocalyx-enclosed microcolonial consortia within adherent biofilms. It is sobering to realize that the microbiological approach to the gastrointestinal tract has been to obtain pure cultures of its autochthonous[15] and pathogenic microbiota and to extrapolate from data obtained in studies of these planktonic organisms to processes involving bacteria in biofilms on the digestive tissue and the digesta. We have instituted a morphological program in which bacteria are examined "*in situ*" on gut surfaces to study their location and interrelationships, and in which pure cultures are obtained and used only to identify the organisms and to establish their basic physiological capabilities (e.g., urease production).

II. THE MICROBIOTA OF NONSECRETORY GUT EPITHELIA

The term "glycocalyx" was first used to describe the polysaccharide coat at the surface of eukaryotic cells[16] and subsequently borrowed[17] to describe the similar fibrous polysaccharide coat on bacterial cells. Thus, bacteria actually colonize the superficial glycocalyces of the nonsecretory epithelial tissues of the gut (Figure 1), and we can see a thick biofilm containing many different bacterial morphotypes on the surface of the bovine reticulorumen.[14,15,18] Scanning electron microscopy (SEM) shows the distribution of these adherent organisms (Figure 2), and allows their quantitation on different regions of the gastrointestinal epithelium,[19] but the dehydration inherent in preparation for electron microscopy causes a dramatic collapse and disappearance of their glycocalyces, which are 99% water of hydration.[20] On the other hand, transmission electron microscopy (TEM) shows the condensed fibrils of the bacterial glycocalyx very clearly, when these polyanionic structures are stained with ruthenium red[21] (Figure 1) and the extensive exopolysaccharides of both the tissue and the adherent bacterial population are discernible in these preparations. A prolific and varied adherent bacterial flora has been described in the nonsecretory regions of the cockroach gut,[22] in which distinct bacterial morphotypes have been seen to be very specifically associated with different tissues of the intestine. Similar morphological studies have associated various specific morphotypes of bacteria and yeasts with nonsecretory gut tissues of the mouse[23] and the chicken,[24] and clear morphological differences have been noted[14] between the adherent and the planktonic bacterial populations of the bovine rumen. These specific associations suggest a possible role of adherent bacteria in the physiological functioning of certain digestive organs and the adherent population of the bovine rumen has been shown to digest and "recycle" cells of the stratified squamous epithelium, to scavenge oxygen by

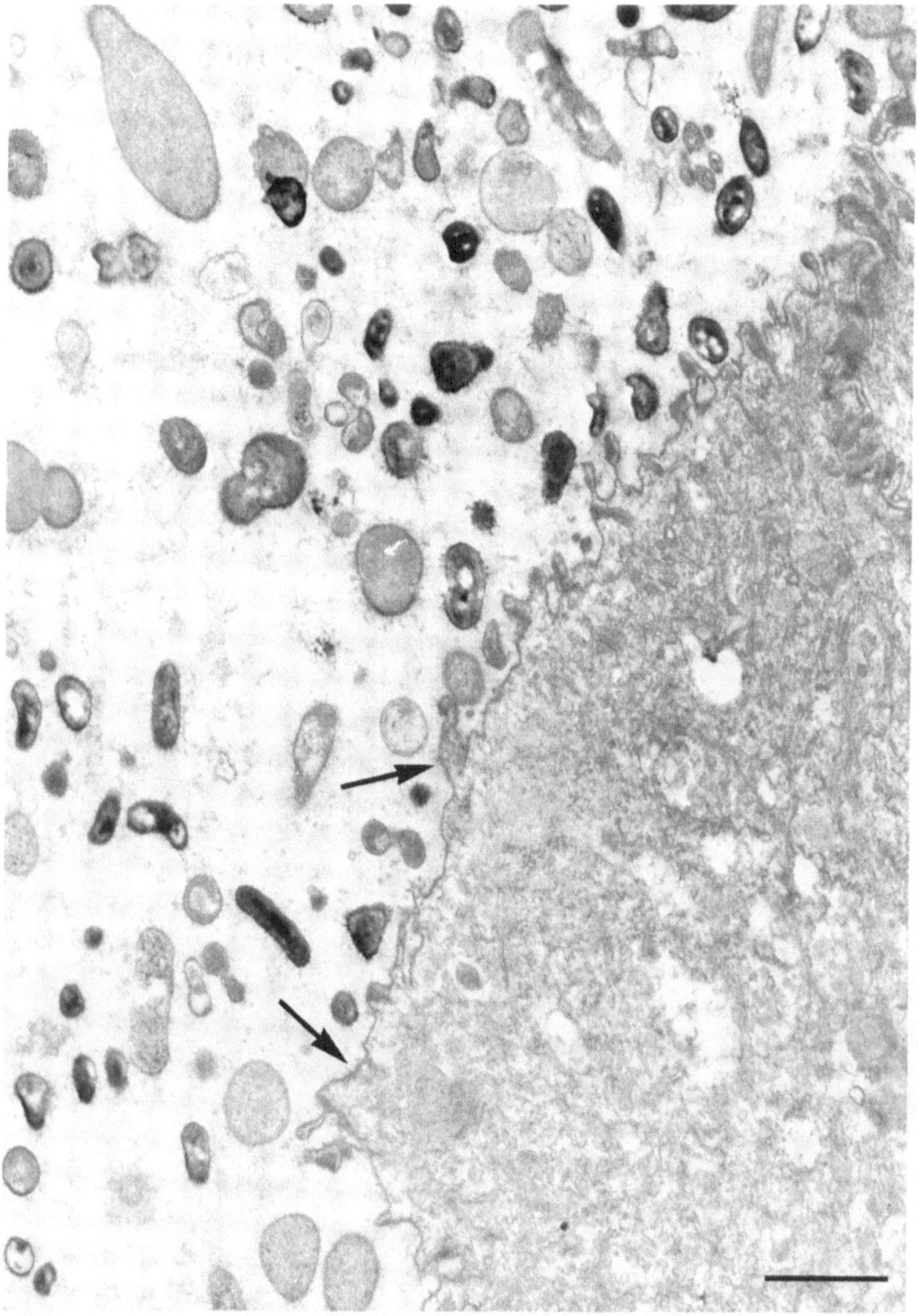

FIGURE 1. Transmission electron micrograph (TEM) of a section of a ruthenium red-stained preparation of the nonsecretory epithelium of the reticulorumen of a yearling steer. Note the very extensive bacterial colonization of this gastrointestinal tissue and the presence of a condensed electron-dense fibrillar matrix in the spaces between these bacteria. The condensed fibrillar glycocalyx of the epithelial cells (arrows) is also evident in this micrograph. The bar represents 1.0 μm.

FIGURE 2. Scanning electron micrograph (SEM) of the surface of a thiocarbohydrazide-metallized critical-point-dried preparation of the nonsecretory epithelium of the reticulorumen of a yearling steer. Note the extensive colonization of the surface of these epithelial cells by bacteria and the condensation of their enveloping glycocalyces to form condensed amorphous masses (arrows). The bar indicates 5.0 μm.

growing as facultative anaerobes, and to produce urease[14] for the conversion of urea to ammonia at the tissue surface. This physiologically cooperative adherent bacterial population is established early in the life of a calf and persists throughout the dietary changes of the usual bovine development sequence.[14] The introduction of selected adherent species into newborn lambs (4 hr) has been shown significantly to enhance early weight gain[49] in these animals.

III. THE MICROBIOLOGY OF SECRETORY GUT EPITHELIA

Direct morphological examination of the adherent bacteria becomes much more difficult when the colonized tissue is covered with the "mucous blanket" that is characteristic of secretory gut epithelia. This mucous blanket probably corresponds to the "unstirred layer" described by gut physiologists,[25] and estimated by them to be approximately 400 μm thick in its fully hydrated state, and it is radically condensed during dehydration and largely removed during conventional processing for both SEM and TEM. Consequently, morphological studies have concentrated on organisms whose avid adherence to, or even insertion into, the tissue commits them to remain with the tissue when the mucous blanket is lost during processing. The fascinating filamentous bacteria seen in the mouse ileum[26,27] (Figure 3), and the spiral organisms[28] that may completely replace the microvilli[29] in monkeys and humans, are examples of especially adherent bacteria, and many protozoa have also developed very effective holdfast mechanisms (Figure 4) that mediate their adhesion to the tissue even if the mucous blanket is stripped away.

The technical solution to the problem of dehydration condensation of hydrated polysaccharides during processing for electron microscopy has been partially solved by reacting these fibrous matrices with bivalent "stabilizers", before drying, so that they retain their original configuration and dimensions. The first successful stabilization of the bacterial glycocalyx was accomplished by Birdsell et al.[30] using lectins to stabilize the teichoic acid "capsule" of *Bacillus subtilis,* and specific antibodies have since been used to stabilize the glycocalyces of a wide variety of bacteria "in vitro"[31,32] and "in vivo"[33] on the tissue surface. The K30-type exopolysaccharide of enteropathogenic *Escherichia coli* has been demonstrated using this technique (Figure 5), as these pathogenic bacteria colonized the gut of a newborn calf, and the remarkable size of this glycocalyx is clearly seen, as is its integration into the mucous blanket. We have modified this stabilization technique by producing antibodies to the glycoprotein component of the mucous of the mouse ileum,[34] and we have used this antiserum to minimize the condensation of the mucous blanket during dehydration for electron microscopy. When combined with very gentle handling of gut specimens, this stabilization technique allows the retention of the mucous blanket on secretory gut tissue (Figure 6) and thus allows us to examine the large autochthonous[15] microbial population that lives in this specialized habitat. It is very germane to note that several fusiform bacteria have a distinct ecological preference for the mucous blanket habitat,[35,36] that many intestinal organisms show positive chemotaxis to mucous components,[37] and that many bacterial organisms are only motile in the viscous medium of the gut mucous.[38] The distinction between the truly adherent bacterial population of the mouse ileum and the much more numerous autochthonous population of the mucous blanket is clearly seen in Figures 4, 6, and 7 in which this heavily colonized structure is partially detached from a microvillar surface colonized only by specially adherent bacteria (Figure 7) and protozoa (Figure 4).

When the mucous blanket is fully retained, by antibody stabilization and/or especially careful handling, the villi are not seen directly but they deform the mucous blanket to form gently rounded contours (Figure 8). The surface of the mucous blanket is somewhat compacted, by dehydration, and bacteria and protozoa can be seen in this consolidated surface (Figure 9). However, the most useful micrographs are those showing rare and fortuitous

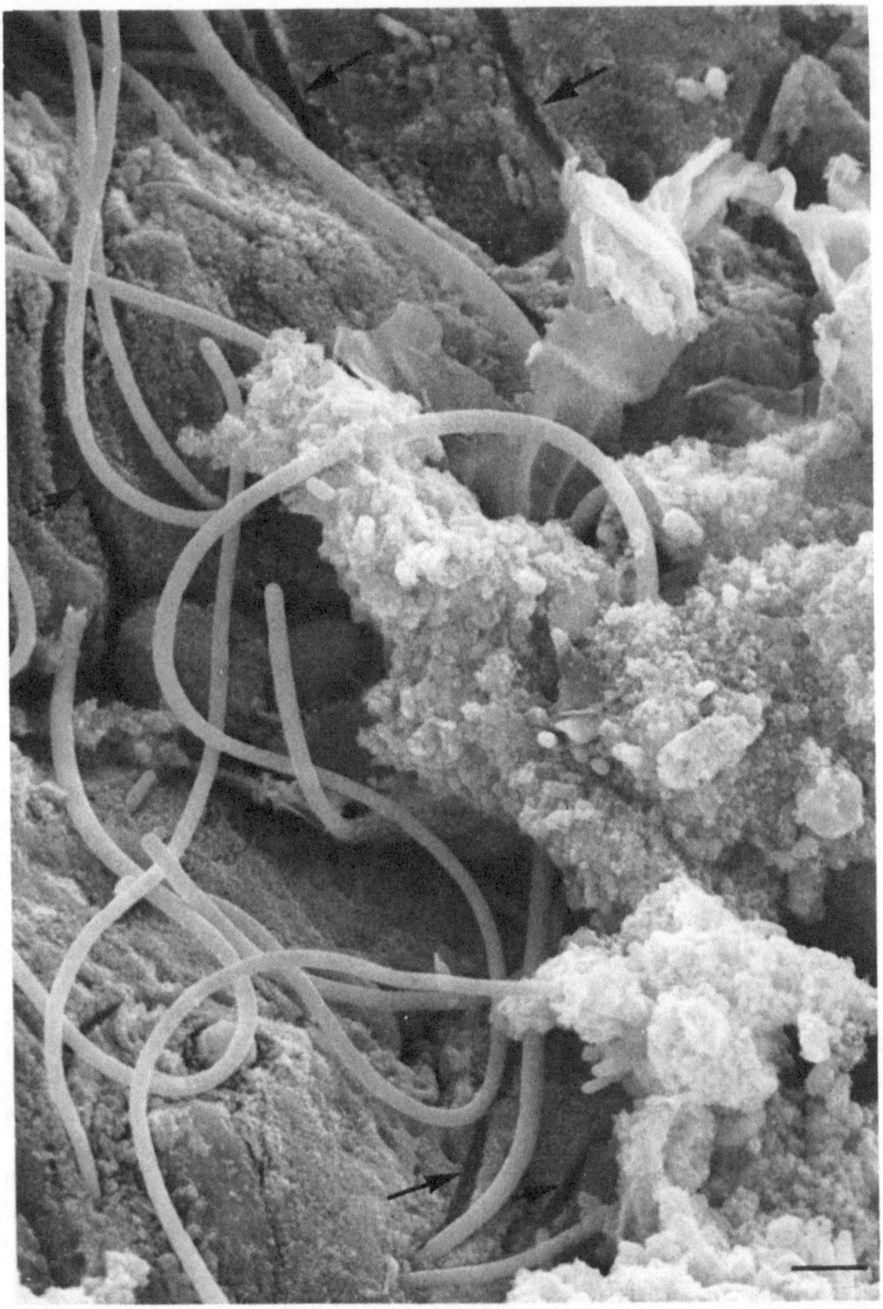

FIGURE 3. SEM of a thiocarbohydrazide-metallized critical-point-dried preparation of the ileum of a 24-day-old mouse showing the insertion of filamentous bacteria into the cells of this epithelial tissue. Note that the microvillar surfaces of these epithelial cells are indented (arrows) to form "trenches" that indicate that these very adherent bacteria may sometimes be pressed against the tissue cell surface. The bar indicates 5.0 μm.

FIGURE 4. SEM of a thiocarbohydrazide-metallized critical-point-dried preparation of the ileum of a 24-day-old mouse showing the avid adhesion of a Giardia cell to the epithelium (arrow), and a varied population of protozoa and bacteria in the apposed mucous blanket population. The highly specialized adhesion apparatus of a detached Giardia cell is visible at G. The bar indicates 5.0 μm.

cross-sections of the mucous blanket (Figure 10) where dehydration cracks have formed (Figures 8 and 9) and the full approximately 50 μm thickness of this structure is seen with the myriad microbial morphotypes that inhabit this viscous layer. Because the mucopeptides of the mucous blanket and the exopolysaccharide of the bacterial glycocalyces are radically condensed (from 400 μm to 50 μm), even in the presence of stabilizing antimucous antibodies, the whole mucous blanket is compacted and its polysaccharide matrix components

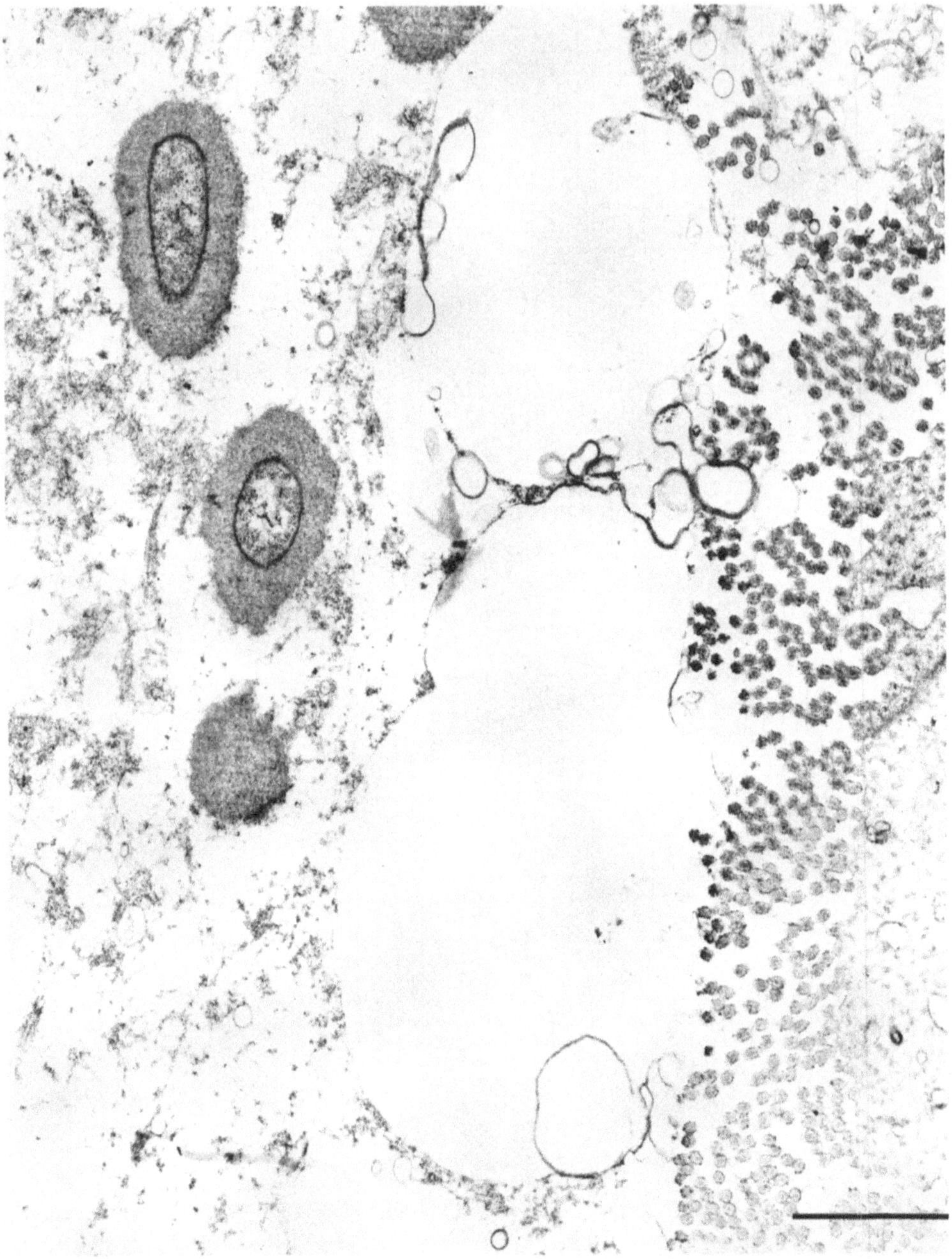

FIGURE 5. TEM of a ruthenium red-stained preparation of the ileum of a newborn calf infected with a $K99^+$ $K30^+$ strain of enterotoxigenic *Escherichia coli*. The specimen has been treated with specific antisera prepared against the K30 capsular polymer and the glycocalyces of these cells are stabilized in their real dimensions and real interrelationships within the fibrous mucous blanket. The bar represents 1.0 μm.

are usually seen as amorphous or fibrous masses (Figure 3, 4, and 6 to 12). Two predominant protozoan morphotypes were identified and the stabilized mucous blanket of the mouse ileum (Figures 4, 6, and 10), and many dozens of bacterial morphotypes could be discerned (Figures 4, 6, and 10 to 12). Individual bacterial morphotypes were seen to be dispersed throughout the mucous blanket, and no obvious aggregations of a single morphotype were noted, but some bacteria appeared to associate with certain protozoa or with cells of the very adherent

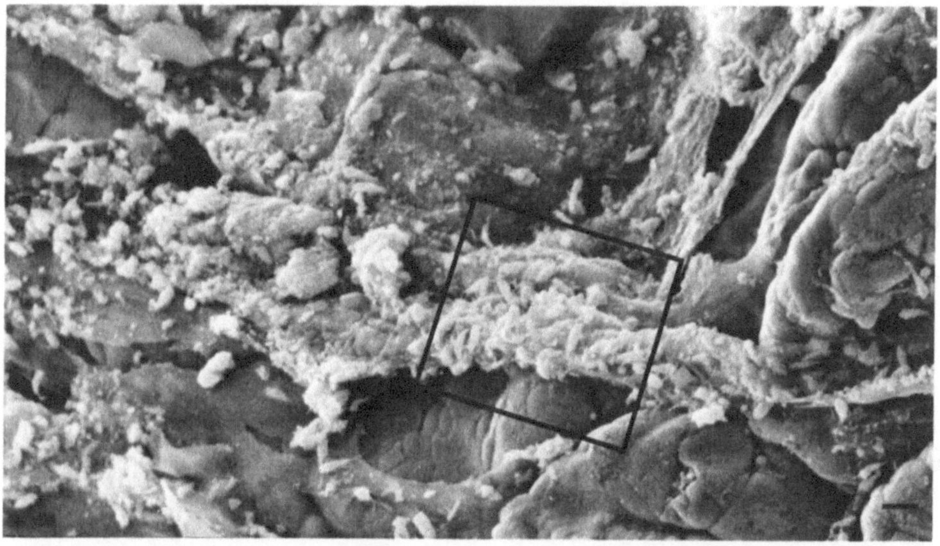

A

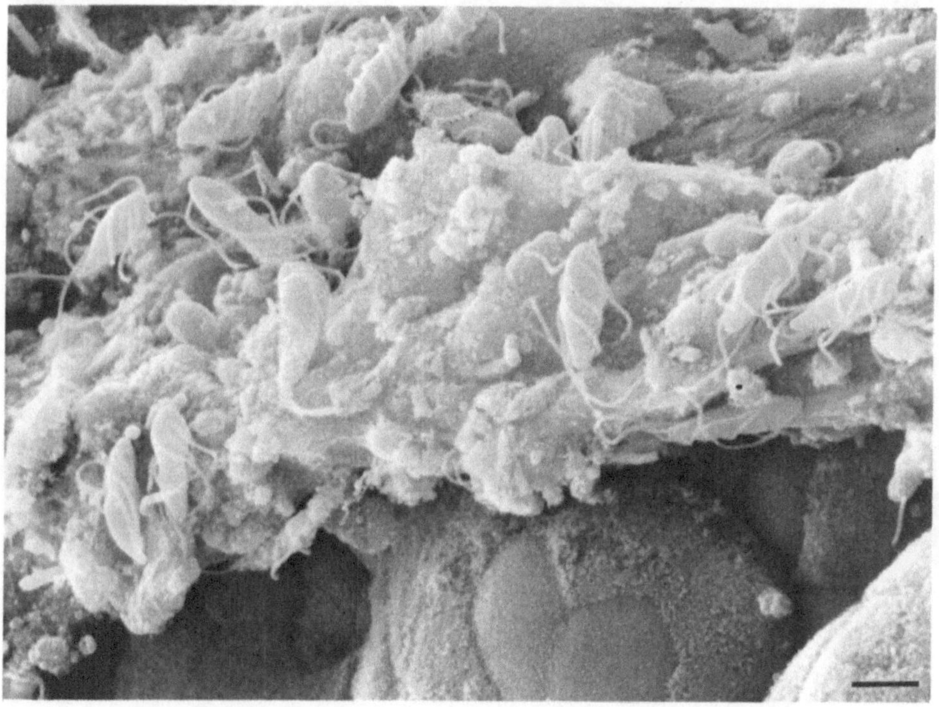

B

FIGURE 6. SEMs of the thiocarbohydrazide-metallized critical-point-dried antibody-stabilized mucous blanket on the ileal epithelium of a 24-day-old mouse. Figure 6A shows the partial retention of the mucous blanket on this epithelial tissue and Figure 6B shows the mucous blanket separating from the tissue surface. Note that virtually all of the protozoa and bacteria colonizing this tissue remain associated with the mucous blanket when this coherent structure is detached from the tissue surface. The bars represent 5.0 μm. The box indicates the enlarged area.

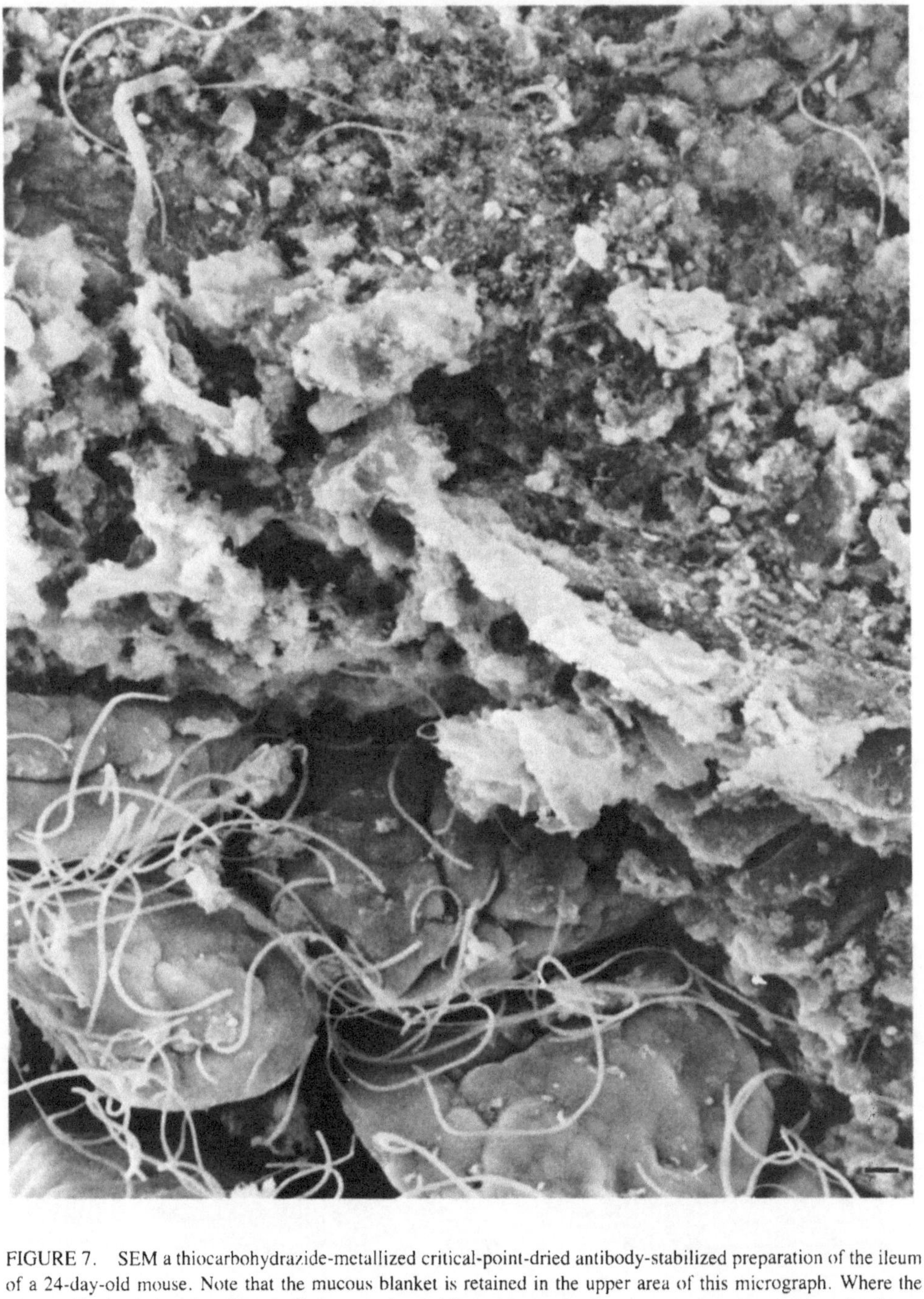

FIGURE 7. SEM a thiocarbohydrazide-metallized critical-point-dried antibody-stabilized preparation of the ileum of a 24-day-old mouse. Note that the mucous blanket is retained in the upper area of this micrograph. Where the mucous blanket is lost the adherent filamentous bacterial species are clearly seen to remain inserted into the tissue, and these long filaments are also seen to traverse the mucous blanket and protrude at its lumenal surface, even when this structure is retained. The bar represents 10 μm.

filamentous species seen in Figure 11. However, the general distribution of protozoa and bacteria appeared to be essentially random (Figure 12) within the fibrous biofilm constituted by the colonized mucous blanket. The mucous blanket is not completely retained on the glycocalyx surface of the mouse ileum, in specimens prepared for TEM, but the fibrous nature of this matrix and the variety and distribution of bacteria are clearly seen in Figure 13. Taken together, these micrographs present an unequivocal body of morphological data

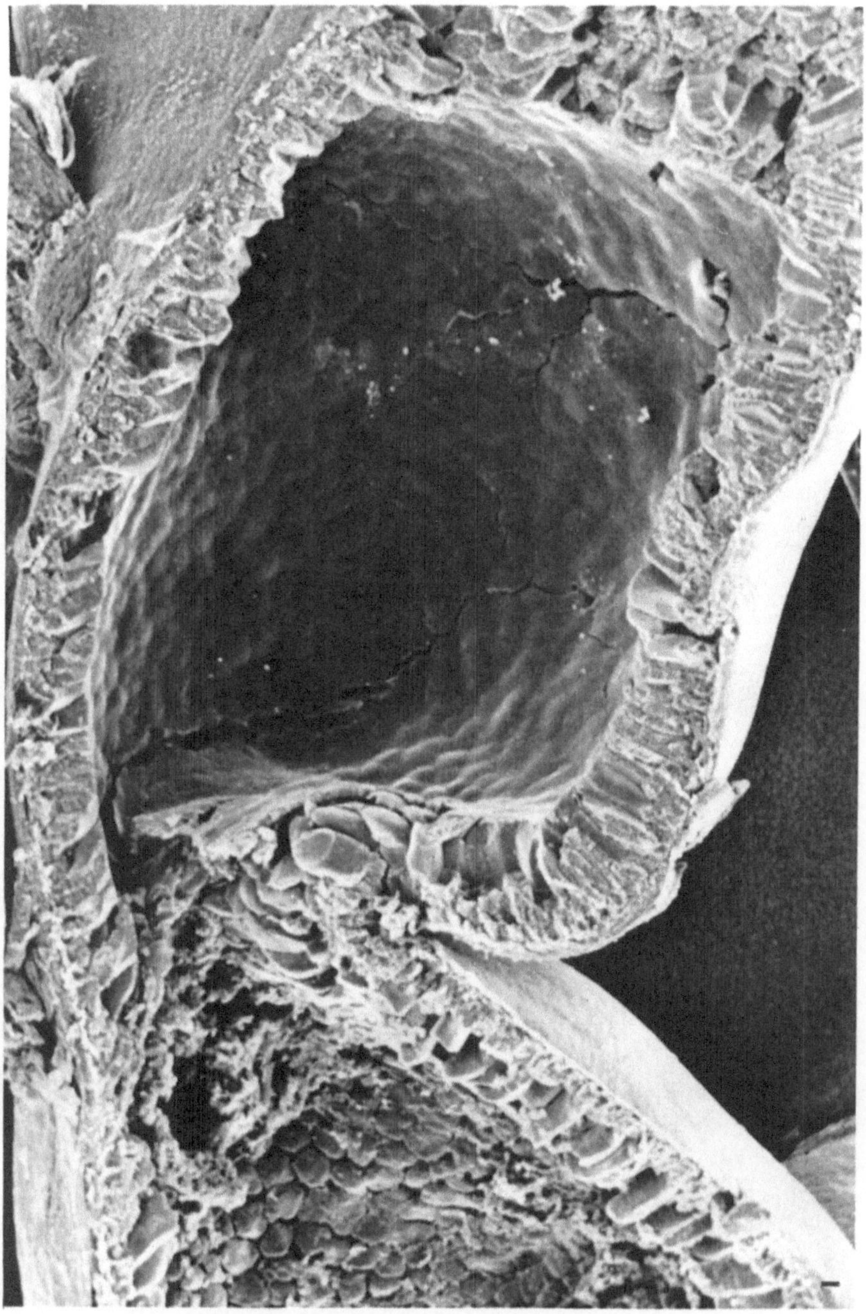

FIGURE 8. SEM of a thiocarbohydrazide-metallized critical-point-dried specimen from a 24-day-old mouse in which especially careful handling had retained the mucous blanket intact in the part of the ileum seen in the upper part of this micrograph. The mucous blanket is cracked by dehydration but it is retained and the villi are only seen as deformations of this coherent mucous layer, in contrast to the area seen in the lower part of the micrograph in which the mucous blanket is removed and the villi are clearly visible. The bar indicates 50 μm.

FIGURE 9. The surface of the mucous blanket seen in the upper portion of Figure 8 showing the degree of compaction produced by dehydration during processing, the formation of a dehydration crack, and the presence of bacteria in the surface of this coherent and continuous mucous layer. The bar indicates 5.0 μm.

that states that the mucous blanket is continuous over the epithelial surface of the mouse ileum (Figure 8). This thick hydrated layer (approximately 400 μm) is condensed during dehydration for electron microscopy to form a thinner structure (approximately 50 μm) within which rare dehydration cracks (Figures 8 and 9) allow us to see within this continuous biofilm of bacteria and protozoa embedded in mucous and glycocalyx polymers (Figures 3, 4, 6, 7, 10 to 13).

While we must presume that the mucous blanket and the majority of its autochthonous

FIGURE 10. SEM of a thiocarbohydrazide-metallized critical-point-dried antibody-stabilized preparation of the ileum of a 24-day-old mouse showing the full height of one wall of a dehydration crack in the mucous blanket, from the blanket surface (s) to the surface of the villus (v). Note that this virtual "section" through the mucous blanket shows a rich morphological variety of bacteria and protozoa embedded in an amorphous matrix of condensed polysaccharides. The bar represents 5 μm.

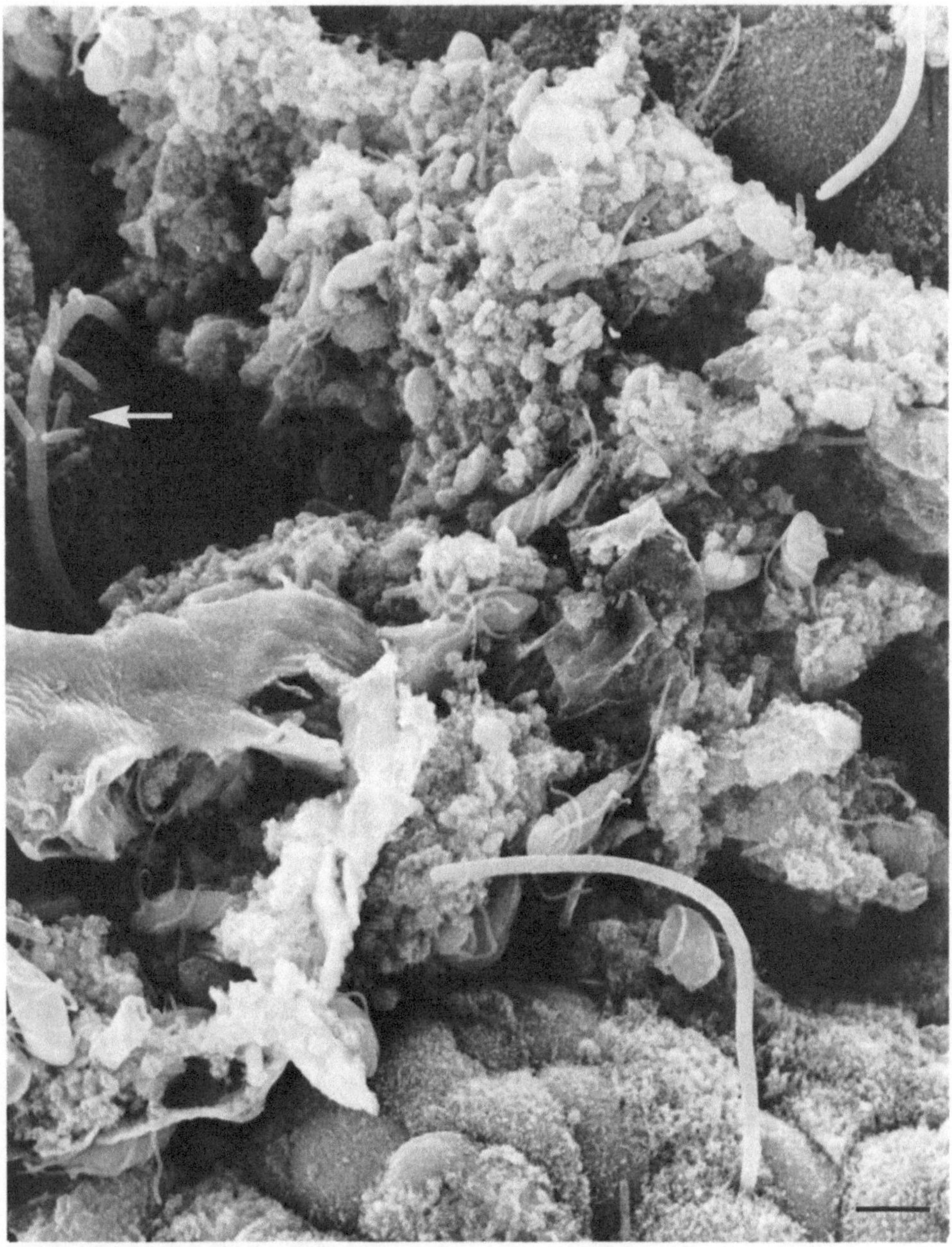

FIGURE 11. SEM of a thiocarbohydrazide-metallized critical-point-dried antibody-stabilized preparation of the ileum of a 24-day-old mouse showing a fragment of the mucous blanket retained in the spaces between villi. A variety of bacteria and protozoa are seen within the amorphous matrix of this fragment and some rod-shaped bacteria are seen to be associated (arrow) with one of the bacterial filaments protruding from the denuded microvillar surface of the villi. The bar represents 5 μm.

inhabitants are mobile within the gut, some of the very adherent protozoa (Figure 4) and filamentous bacteria (Figures 3, 7, 11) are clearly immobile. That these very adherent bacteria may be pressed into the microvillar surface of the epithelial cells by the passing mucous is indicated by impressions in the microvillar tissue surface clearly seen in Figure 3, and the protruding filaments seen where the mucous blanket is intact in Figure 7 indicate that these

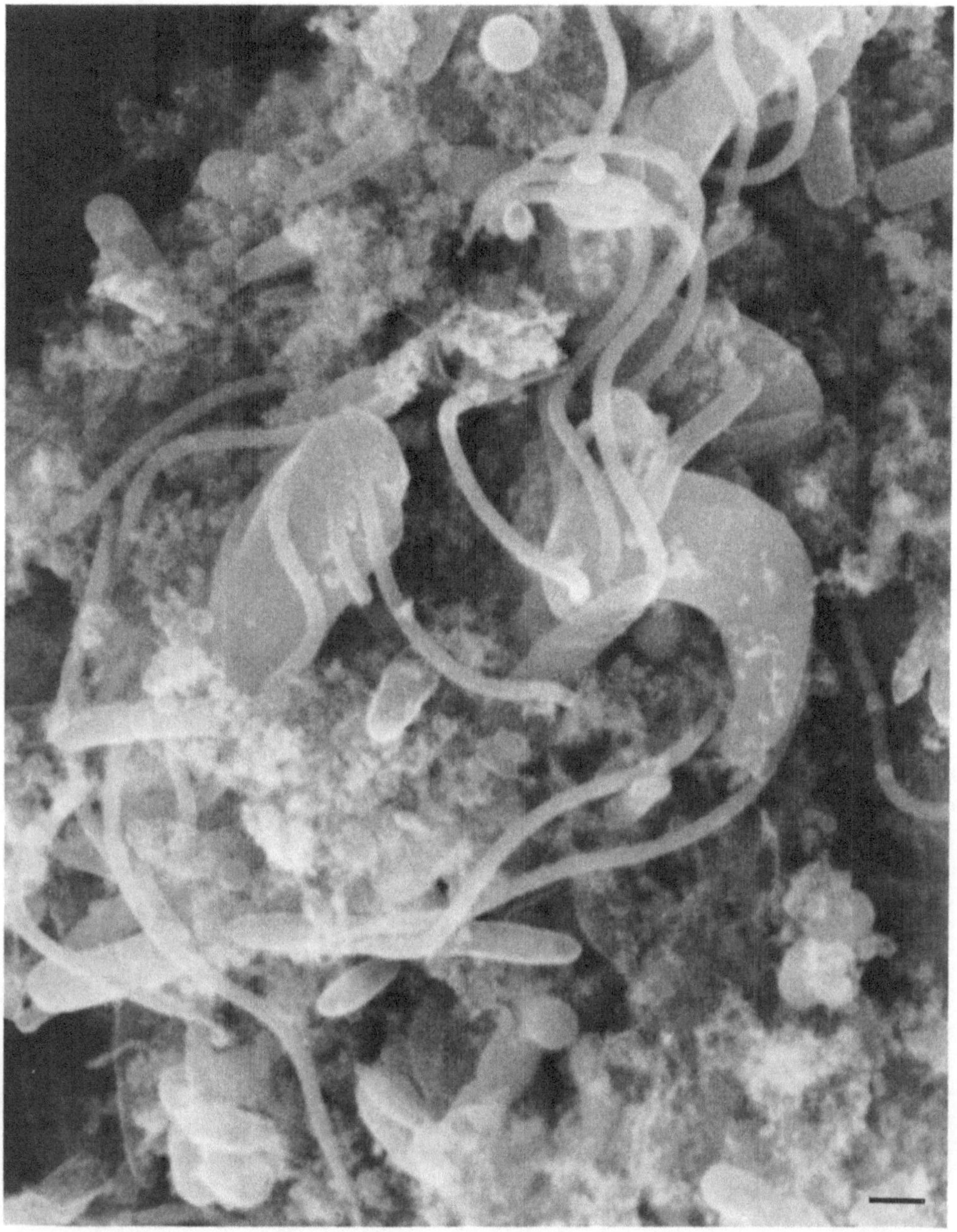

FIGURE 12. High magnification SEM of thiocarbohydrazide-metallized critical-point-dried antibody-stabilized preparation of the ileum of a 24-day-old mouse showing the morphology and flagellar insertion pattern of the predominent protozoan morphotype and the structures of several bacterial cells that are embedded in the amorphous matrix of condensed glycocalyces and mucous. The bar represents 1.0 μm.

long stationary filaments may actually traverse the full thickness of the mucous blanket and project at its lumenal aspect.

IV. DISCUSSION

The polysaccharides of the biological world have been ephemeral in the perceptions of the morphologist because of a complex of technical problems. These compounds do not

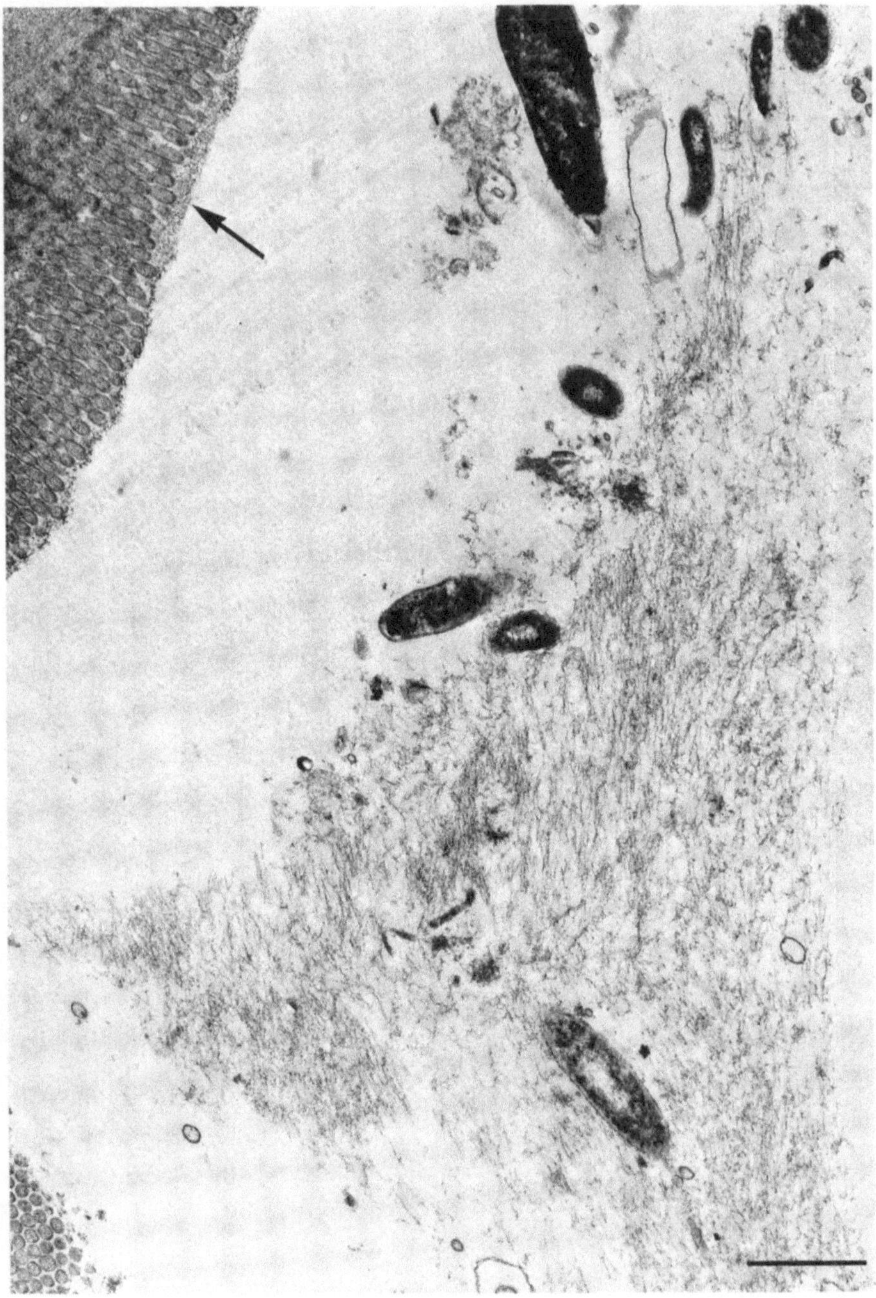

FIGURE 13. TEM of a ruthenium red-stained antibody-stabilized preparation of the ileum of a 24-day-old mouse. While the mucous blanket is not completely retained on the glycocalyx-covered surface (arrow) of the epithelial cells, its juxtaposition can be noted as can its fibrous nature and the morphological diversity of the bacteria that inhabit it. The bar represents 1.0 μm.

stain with the routine metal "stains" of electron microscopy, their degree of hydration (99%) makes dehydration a devastating artifact, and their lack of mechanical strength and tenuous adhesion to cells makes them easily lost during processing. For these reasons microbial morphologists have concentrated on the fascinating adherent bacterial populations of nonsecretory epithelia,[5,14,19,22-24] and on organisms whose avid adhesion (or insertion) mechanisms anchor them firmly to secretory epithelia.[26-29] We have projected our long-standing interest in polysaccharide matrices[9,10] into our morphological studies of the gastrointestinal tract, and we have retained the mucous blanket of the secretory abomasum of the cow, and of the ileum of the mouse, by a combination of antibody stabilization and very careful handling. When this is done, a concept emerges in which a small number of very adherent bacteria and protozoa remain immobile on the tissue surface while a continuous viscous stream of mucous flows past them and acts as the preferred menstruum[34] for a huge and varied population of autochthonous gut bacteria. Within this coherent mucous blanket, which is probably approximately 400 μm in thickness, bacteria and protozoa that have flagellar mechanisms specially adapted for viscous environments[38] move in response to chemotactic stimuli[37] and set up the consortia within which they actually live and function.

The fast-moving (approximately 12 mm/min) mucous blanket of the pulmonary airways is known to be a protective mechanism[39] that carries potential colonizing bacteria away from the epithelial tissues and prevents their colonization by all but the most avidly adherent pathogenic Mycoplasma.[40] The much slower-moving mucous blanket of the intestinal tract constitutes a similar protective mechanism in that it carries all but the most adherent autochthonous bacteria and protozoa past the tissue surface, and it represents a balanced microbial ecosystem into which most pathogenic bacteria must integrate themselves and compete to produce the "critical mass" of cells required to initiate an infection. In the absence of this colonized mucous blanket[34] newborn animals are exquisitely sensitive to pathogenic colonization by enteropathogenic *Escherichia coli*[33] and it is of interest to note that Sprunt and her colleagues[41] have prevented diarrhea in premature newborn humans by feeding autochthonous strains of Streptococci, just as we have protected newborn lambs from "scours" by feeding a mixture of 29 autochthonous gut organisms.[49] It seems logical to explore the feasibility of manipulating the microbial population of the human gut, following clindamycin therapy, to prevent the predominance of *Clostridium difficile* and the development of pseudomembranous colitis.[42]

When bacteria live in a matrix composed of charged polymers, as when intestinal bacteria grow in their own anionic exopolysaccharide glycocalyces within an extensive mass of similarly charged mucous glycoproteins, they are protected, in large measure, from many of the vicissitudes of life in an animal body. Unequivocal evidence shows that glycocalyx-enclosed cells are protected from surfactants,[43] from bactericidal serum factors,[44] from opsonizing antibodies,[45] and from phagocytosis by activated macrophages.[46] The additional ion-exchange effect of the mucous glycoproteins would be expected to protect bacteria embedded within it from antibiotics and biocides, just as growth in a thick (approximately 400 μm) biofilm protects biofilm bacteria from commercial biocides[11] and from antibiotics.[47] Thus, both autochthonous and pathogenic bacteria growing within the mucous blanket may have antibiotic sensitivities that differ radically from those of isolates growing in vitro, without an enveloping matrix of glycocalyx and mucous.

As we mentioned at the outset, our developing perception of how bacteria really grow in association with the nonsecretory and secretory epithelial tissues of the gastrointestinal tract should produce grave misgivings about our collective dependence on extrapolation from studies of "in vitro" cultures of isolates from this system. Improved methods are now available for the study of the structure and function of bacteria *in situ*[48] on a colonized surface. We must follow the lead of the aquatic microbial ecologists and begin to study the surface structure, the physiology, and the sensitivity to antibacterial agents, of bacteria actually growing within a biofilm (the mucous blanket) on a series of colonized surfaces (the gastrointestinal epithelia).

REFERENCES

1. **Bae, H. C., Cota-Robles, E. H., and Casida, L. E.,** Microflora of soil as viewed by transmission electron microscopy, *Appl. Microbiol.*, 23, 637, 1972.
2. **Geesey, G. G., Mutch, R., Costerton, J. W., and Green, R. B.,** Sessile bacteria: an important component of the microbial population in small mountain streams, *Limnol. Oceanogr.*, 23, 1214, 1978.
3. **Fletcher, M. and Floodgate, G. D.,** Electron microscopic demonstration of an acidic polysaccharide involved in the adhesion of a marine bacterium to solid surfaces, *J. Gen. Microbiol.*, 72, 325, 1973.
4. **McCoy, W. F., Bryers, J. D., Robbins, J., and Costerton, J. W.,** Observations on fouling biofilm production, *Can. J. Microbiol.*, 27, 910, 1981.
5. **Marrie, T. J., Lam, J., and Costerton, J. W.,** Bacterial adhesion to uroepithelial cells: a morphological study, *J. Infect. Dis.*, 142, 239, 1980.
6. **Lam, J., Chan, R., Lam, K., and Costerton, J. W.,** Production of mucoid microcolonies by *Pseudomonas aeruginosa* within infected lungs in cystic fibrosis, *Infect. Immun.*, 28, 546, 1980.
7. **Marrie, T. J., Harding, G. K. M., Ronald, A. R., Dikkema, J., Lam, J., Hoban, S., and Costerton, J. W.,** Influence of antibody coating of *Pseudomonas aeruginosa*, *J. Infect. Dis.*, 19, 357, 1979.
8. **Marrie, T. J. and Costerton, J. W.,** Colonization of a cardiac pacemaker by Staphylococci of bacteremic origin, *Circulation*, in press.
9. **Costerton, J. W., Irvin, R. T., and Cheng, K.-J.,** The bacterial glycocalyx in nature and disease, *Annu. Rev. Microbiol.*, 35, 299, 1981.
10. **Costerton, J. W., Irvin, R. T., and Cheng, K.-J.,** The role of bacterial surface structures in pathogenesis, CRC *Crit. Rev. Microbiol.*, 8, 303, 1981.
11. **Ruseska, I., Robbins, J., and Costerton, J. W.,** Biocide testing against corrosion-causing oilfield bacteria helps control plugging, *Oil Gas J.*, March 8, 253, 1982.
12. **Whittenburg, R. and Dow, C. S.,** Morphogenesis in bacteria, in *Companion to Microbiology*, Bull, A. T. and Meadow, P. M., Eds., Longman, London, 1978, 221.
13. **McInerney, M. J., Bryant, M. P., Hespell, R. B., and Costerton, J. W.,** *Syntrophomonas wolfei* gen. nov. sp. nov., an anaerobic, syntrophic, fatty acid oxidizing bacterium, *Appl. Environ. Microbiol.*, 41, 1020, 1981.
14. **Cheng, K.-J. and Costerton, J. W.,** Adherent rumen bacteria: their role in the digestion of plant material, urea and epithelial cells, in *Digestive Physiology and Metabolism in Ruminants*, Ruckebush, Y. and Thivend, P., Eds., MTP Press, Lancaster, UK, 1980, 227.
15. **Cheng, K.-J., Irvin, R. T., and Costerton, J. W.,** Autochthonous and pathogenic colonization of animal tissues by bacteria, *Can. J. Microbiol.*, 27, 461, 1981.
16. **Singer, S. J.,** The molecular organization of membrane, *Annu. Rev. Biochem.*, 43, 805, 1974.
17. **Costerton, J. W., Geesey, G. G., and Cheng, K.-J.,** How bacteria stick, *Sci. Am.*, 238, 86, 1978.
18. **Cheng, K.-J., McCowan, R. P., and Costerton, J. W.,** Adherent epithelial bacteria in ruminants and their roles in digestive tract function, *Am. J. Clin. Nutr.*, 32, 139, 1979.
19. **McCowan, R. P., Cheng, K.-J., Bailey, C. B. M., and Costerton, J. W.,** Adhesion of bacteria to epithelial cell surfaces within the reticulorumen of cattle, *Appl. Environ. Microbiol.*, 35, 149, 1978.
20. **Sutherland, I. W.,** Bacterial polysaccharides: their nature and production, in *Surface Carbohydrates of the Prokaryotic Cell*, Sutherland, I. W., Eds., Academic Press, New York, 1977, 27.
21. **Luft, J. H.,** Ruthenium red and ruthenium violet. I. Chemistry, purification, methods of use for electron microscopy and mechanism of action, *Anat. Rec.*, 171, 347, 1971.
22. **Foglesong, M. A., Walker, D. H., Jr., Puffe, J. S., and Markovetz, A. J.,** Ultrastructural morphology of some prokaryotic microorganisms associated with the hindgut of cockroaches, *J. Bacteriol.*, 123, 336, 1975.
23. **Savage, D. C.,** Microbial ecology of the gastrointestinal tract, *Annu. Rev. Microbiol.*, 31, 107, 1977.
24. **Fuller, R. and Brooker, E.,** Lactobacilli which attach to the crop epithelium of the fowl, *Am. J. Clin. Nutr.*, 27, 1305, 1974.
25. **Levin, R. J.,** Fundamental concepts of the structure and function of the gastrointestinal epithelium, in *Scientific Basis of Gastroenterology*, Duthie, H. L., Ed., Churchill Livingston, New York, 1979, 307.
26. **Davis, C. P. and Savage, D. C.,** Habitat, succession, attachment, and morphology of segmented, filamentous microbes indigenous to the murine gastrointestinal tract, *Infect. Immun.* 10, 948, 1974.

27. **Chase, D. G. and Erlandsen, S. C.,** Evidence for a complex life cycle and endospore formation in the attached, filamentous, segmented bacterium from murine ileum, *J. Bacteriol.*, 127, 572, 1976.
28. **Lee, A.,** Normal flora of intestinal surfaces, in *Adsorbtion of Microorganisms to Surfaces,* Bitton, G. and Marshall, K. C., Eds., John Wiley & Sons, New York, 1980, 145.
29. **Takeuchi, A.,** Spiral shaped organisms on the surface of the colonic epithelium of monkey and man, *Am. J. Clin. Nutr.*, 27, 1287, 1974.
30. **Birdsell, D. C., Doyle, R. J., and Morgenstern, M.,** Organization of teichoic acid in the cell wall of *Bacillus subtilis, J. Bacteriol.*, 121, 726, 1975.
31. **Bayer, M. E. and Thurow, H.,** Polysaccharide capsule of *Escherichia coli:* microscopic study of its size, structure, and sites of synthesis, *J. Bacteriol.*, 130, 911, 1977.
32. **Mackie, E. B., Brown, K. N., Lam, J., and Costerton, J. W.,** Morphological stabilization of capsules of group B streptococci, types Ia, Ib, II, and III, with specific antibody, *J. Bacteriol.*, 138, 609, 1979.
33. **Chan, R., Acers, S. D., and Costerton, J. W.,** The use of specific antibody to demonstrate glycocalyx, k99 pili, and the spatial relationships of k99+ enterotoxigenic *Escherichia coli* in the ileum of colostrum-fed calves, *Infect. Immun.*, in press.
34. **Rozee, K. R., Cooper, D., Lam, K., and Costerton, J. W.,** Microbial flora of the mouse ileum mucous layer and epithelial surface, *Appl. Environ. Microbiol.*, 43, 1451, 1982.
35. **Davis, C. P., McAllister, J. S., and Savage, D. C.,** Microbial colonization of the intestinal epithelium in suckling mice, *Infect. Immun.*, 7, 666, 1973.
36. **Savage, D. C.,** Indigenous microorganisms associated with mucosal epithelia in the gastrointestinal system, in *Microbiology 1975*, Schlessinger, D., Ed., American Society of Microbiology, Washington, D.C., 1975, 120.
37. **Freter, R., O'Brien, P. C. M., and Macsai, M. S.,** Role of chemotaxis in the association of motile bacteria with intestinal mucosa: in vivo studies, *Infect. Immun.*, 34, 234, 1981.
38. **Berg, H. C.,** How spirochetes may swim, *J. Theor. Biol.*, 56, 269, 1976.
39. **Costerton, J. W., Brown, M. R. W., and Sturgess, J. M.,** The cell envelope: its role in infection, in *Pseudomonas aeruginosa*, Doggett, R. G., Ed., Academic Press, New York, 1979, 41.
40. **Collier, A. M.,** Attachment of *Mycoplasma pneumoniae* to respiratory epithelium, in *Bacterial Adherence*, Beachey, E. H., Ed., Chapman and Hall, London, 1980, 159.
41. **Sprunt, K., Leidy, G., and Redman, W.,** Abnormal colonization of neonates in an ICU: conversion to normal colonization by pharyngeal implantation of alpha hemolytic Streptococcus strain 215, *Pediatr. Res.*, 14, 308, 1980.
42. **Chang, T. W., Bartlett, J. G., Gorbach, S. L., and Onderdonk, A. B.,** Clindamycin-induced enterocolitis in hamsters as a model of pseudomembrane colitis in patients, *Infect. Immun.*, 20, 526, 1978.
43. **Govan, J. R. W.,** Mucoid strains of *Pseudomonas aeruginosa:* the influence of culture medium on the stability of mucus production, *J. Med. Microbiol.*, 8, 513, 1975.
44. **Reynolds, B. L. and Pruul, H.,** Protective role of smooth lipopolysaccharide in the serum bactericidal reaction, *Infect. Immun.*, 4, 764, 1971.
45. **Baltimore, R. S. and Mitchell, M.,** Immunological investigation of mucoid strains of *Pseudomonas aeruginosa:* comparison of susceptibility by opsonic antibody in mucoid and nonmucoid strains, *J. Infect. Dis.*, 141, 238, 1980.
46. **Schwarzmann, S. and Boring, J. R.,** Antiphagocytic effect of slime from a mucoid stain of *Pseudomonas aeruginosa, Infect. Immun.*, 3, 762, 1971.
47. **Govan, J. R. W. and Fyfe, J. A. M.,** Mucoid *Pseudomonas aeruginosa* and cystic fibrosis: resistance of the mucoid form to carbenicillin, flucloxacillin, and tobramycin and the isolation of mucoid variants *in vitro, J. Antimicrob. Chemother.*, 4, 233, 1978.
48. **Costerton, J. W. and Colwell, R. R.,** *Native Aquatic Bacteria: Enumeration, Activity and Ecology*, ASTM Press, Philadelphia, 1979.
49. **Cheng, K.-J. and Costerton, J. W.,** unpublished data.

Chapter 19

THE ROLE OF NEGATIVELY CHARGED SURFACE POLYSACCHARIDE IN ADHERENCE OF *ESCHERICHIA COLI* STRAIN RDEC-1 TO GUT EPITHELIAL CELLS IN THE RABBIT

J. Robert Cantey and Lindsey R. Inman

TABLE OF CONTENTS

I. INTRODUCTION

The diarrheal disease caused by the RDEC-1 bacterium has been described in detail elsewhere in this volume. A striking feature of the disease is the manner in which the bacterium adheres to mucosal epithelial cells. We have been interested in determining what role the major RDEC-1 surface structures, pili, negatively charged (non-K) polysaccharide and O antigens, might play in adherence. Cheney et al. have produced evidence, discussed elsewhere in this volume that RDEC-1 pili mediate adherence of the bacterium to partially purified brush borders of rabbit ileal mucosal epithelial cells.[1,2] We have used electron microscopic techniques to examine the role of the surface polysaccharide. The results of these studies have been published[3] and will be briefly reviewed herein. At the time of publication of this material the RDEC-1 bacteria was felt to possess a K antigen. We have recently determined[16] that this was in error. Although it does not possess a K antigen it does have abundant negatively charged surface polysaccharides.

II. METHODS OF STUDY

Polysaccharides such as those found on bacterial surfaces and mucosal epithelial cell glycocalyx are not usually visible in sections of tissues prepared by routine techniques for electron microscopy. Not only are they not well stained but when the water is removed from the hydrated polysaccharides during dehydration for embedment, the polysaccharides collapse.[4,5]

Ruthenium red will stain negatively charged polysaccharides for visualization by electron microscopy,[6] but it will not prevent collapse of the polysaccharide during the dehydration process. For example, the negatively charged surface polysaccharide of RDEC-1 bacteria in the gut lumen or in culture, is stained as a dark to black matrix, closely applied to the bacterial cell wall. However, if the polysaccharide is stretched between two points, as between a bacterium and a microvillus, it will not collapse and will appear as a thin, black strand. Ruthenium red has disadvantages which one must be cognizant of. It will stain any negatively charged material, no matter what its composition. Fortunately, in the gut lumen, there is little of consequence that would stain, other than polysaccharides. Another, more serious shortcoming is the poor diffuseability of ruthenium red. It does not penetrate membranes, tight junctions, etc., unless they are damaged. It does not penetrate well into villus crypts and even relatively accessible areas are sometimes poorly stained. If one is aware of these shortcomings, and chooses the areas one wishes to examine carefully, none of these shortcomings need be serious.

The collapse of bacterial surface polysaccharides can be prevented by reacting these with specific antibody.[3-5] The antibody also enhances the polysaccharide for visualization, as the protein of the antibody in the antibody-polysaccharide matrix is well stained by the usual electron microscopic stains. It is only by this technique that one can accurately gauge the thickness of the bacterial surface polysaccharides.

Preliminary to the electron microscopic studies, we determined the specificity of the two major reagents, ruthenium red and antisurface polysaccharide antibody, for surface polysaccharides of the RDEC-1 bacterium. The polysaccharides or glycocalyx, consisting of O antigen and the negatively charged (non-K) polysaccharide were partially purified[7,8] and electrophoresed against ruthenium red.[3] Both antigens were precipitated by the ruthenium red. The precipitation of the O antigen was probably due to reaction of its fatty acid moiety with the ruthenium red. The fatty acid portion of the molecule would be inaccessible to the ruthenium red in the intact bacterium.

The antisurface polysaccharide serum was raised in rabbits by i.v. injection of live tryptic soy broth grown bacteria.[3] We determined by electron microscopic examination of negatively stained bacteria that tryptic soy broth grown bacteria lacked pili. The purified antigens were precipitated when electrophoresed against the antiserum.[3] To eliminate the possibility that the antiserum might, in spite of having been raised against bacteria lacking pili, have antipilus activity, it was checked in double diffusion against purified RDEC-1 pili. A precipitin line did not form, indicating that the antiserum lacked activity against strain RDEC-1 pili.

In summary, we used routinely stained tissues to demonstrate the lack of surface polysaccharide staining by conventional electron microscopic stains, antisurface polysaccharide serum stained tissues to gauge the true extent of the RDEC-1 surface polysaccharides and ruthenium red stained tissues to allow simultaneous visualization of both the microvillar and bacterial glycocalyx. The tissues were obtained from noninfected rabbits and rabbits with diarrhea due to the RDEC-1 bacterium.

Sections of the tissues 0.05 μm thick were examined by routine electron microscopy. Thick sections, 0.5 μm thick, were examined with the high voltage electron microscope. The examination of the thick secretions was particularly important to our study. The high voltage electron microscope located at the University of Colorado, in Boulder, Colorado, produces an accelerating voltage of 1 MeV.[3] The great section thickness that one can view with this instrument increases the probability of encountering previously unrecognized or tenuous aspects of the interaction between bacteria and mucosal epithelial cell glycocalyx, microvilli, etc. It also reduces the probability of errors of interpretation and sampling inherent in the examination of thin (0.05 μm) sections. Stereoscopic views of thick sections are dramatic and they allow three-dimensional conceptualization and interpretation otherwise possible only by reconstruction of serial thin sections.

III. EVIDENCE FOR A ROLE FOR SURFACE POLYSACCHARIDE IN ADHERENCE

Before discussing the evidence for a role of surface polysaccharide, it is necessary to define the concepts of early and late stage adherence that have evolved during our studies. The terms refer to spatial relationships between bacteria and mucosal cells. They are not based on examination of tissues at timed intervals postinoculation, as the effort required for such examination would have been prohibitive. Early stage attachment refers to adherence of bacteria to intact microvilli whereas late stage attachment refers to adherence to the surfaces of cells lacking microvilli.[3]

It is useful to begin by examining a micrograph of tissue from a rabbit with RDEC-1 diarrhea which was not stained with any special reagents (see Figure 5 in Chapter 5). Of interest to the present discussion is the fact that where the bacteria were closely associated with microvilli no stained material could be seen in the space between the bacterium and the microvilli. In subsequent micrographs it can be seen that this space contains ruthenium red and antisurface polysaccharide antibody stained material.

Tissues of rabbits with RDEC-1 diarrhea, stained with ruthenium red, confirmed the specificity of ruthenium red for microvillar glycocalyx of mucosal epithelial cells and bacterial negatively charged surface polysaccharide (glycocalyx) (Figures 1 and 2). The tissues in Figure 1 are not double stained with uranyl acetate and lead citrate, therefore ruthenium red is responsible for the dense staining of surface structures. In areas where the bacterial glycocalyx is not supported by an interaction with bacterial or microvillar glycocalyx it has fallen back onto the surface of the bacterium, giving the surface of the bacterium a black appearance (Figure 1). When uranyl acetate and lead citrate stains are added the polysaccharides appear significantly more dense (darker). The bacterial glycocalyx material has a fibrillar, amorphous, sometimes branched appearance in the ruthenium red stained tissues. Figures 3, 4, and 5 demonstrate the presence of ruthenium red stained materials extending between bacteria and epithelial cell microvilli (early stage adherence). Figures 5 and 6 show

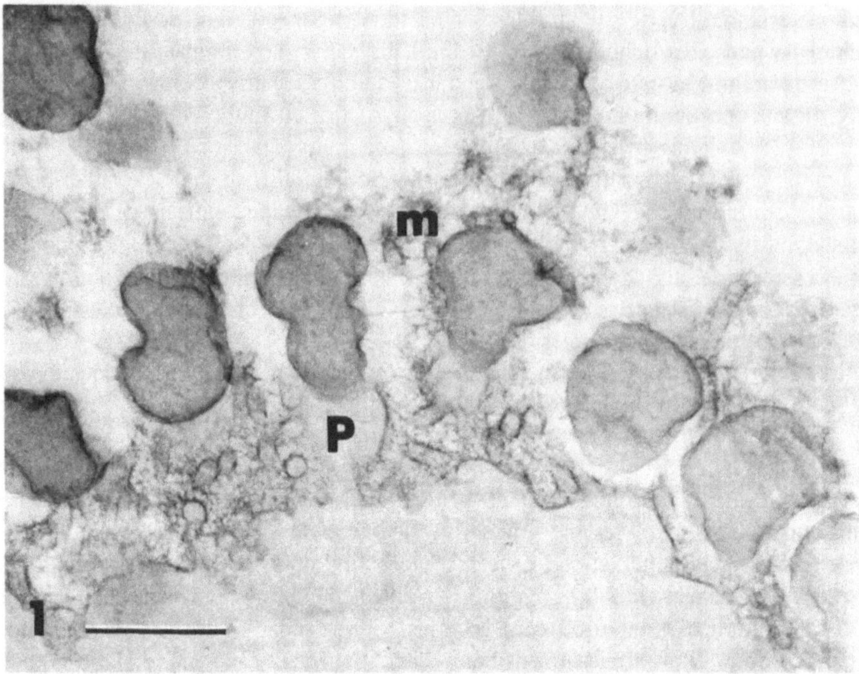

FIGURE 1. High voltage electron micrograph of 0.5 μm section of cecal mucosa from a rabbit with diarrhea due to strain RDEC-1 *Escherichia coli*. Tissue stained *en bloc* with ruthenium red and uranyl acetate. Bacteria are adherent to microvilli (m) and to pedestals (P) of epithelial cell plasmalemma. The ruthenium red has stained negatively charged polysaccharides (bacterial and epithelial cell glycocalyx). Fibrillar strands of stained material extend between bacterial surfaces and epithelial cell microvilli. Bar = 1 μm.

a bacteria closely adherent to the surfaces of mucosal epithelial cells lacking microvilli (late stage adherence) where fibrillar, ruthenium red stained material can be clearly seen spanning the distance between the bacteria and epithelial cell plasmalemma. Strands of ruthenium red stained material can also be seen stretching between bacteria (Figures 1 to 6). This is evidence that the strands of fibrillar material extending between the bacterial and epithelial cell surfaces are not due simply to the presence of epithelial cell glycocalyx.

Figure 7 shows in a vivid fashion the interaction of the bacterial glycocalyx and the glycocalyx mesh of the adjacent microvilli. Adherence pedestals are also prominent. The bacterial glycocalyx is not well stained, probably as a result of the poor diffuseability of ruthenium red.

The argument could be made that the fibrillar strands of ruthenium red stained material, which sometimes stretch more than half the length of the bacteria, are pili onto whose surface ruthenium red stained glycocalyx material has collapsed. Pili, which are not ordinarily visible in sections of tissues examined by electron microscopy, would become visible by virtue of the condensed ruthenium red stained bacterial glycocalyx. That argument cannot be refuted beyond a shadow of a doubt with this type of morphologic study. However, the variable, sometimes considerable length, the thickness, plus the amorphous occasionally branched appearance of the fibrillar strands do not seem compatible with the regular structure one might expect a pilus to have.

The limits of microvillar and bacterial glycocalyx are not defined in the ruthenium red stained tissues. However, the sections of the antiserum treated tissues that follow clearly define the limits of the bacterial capsule.

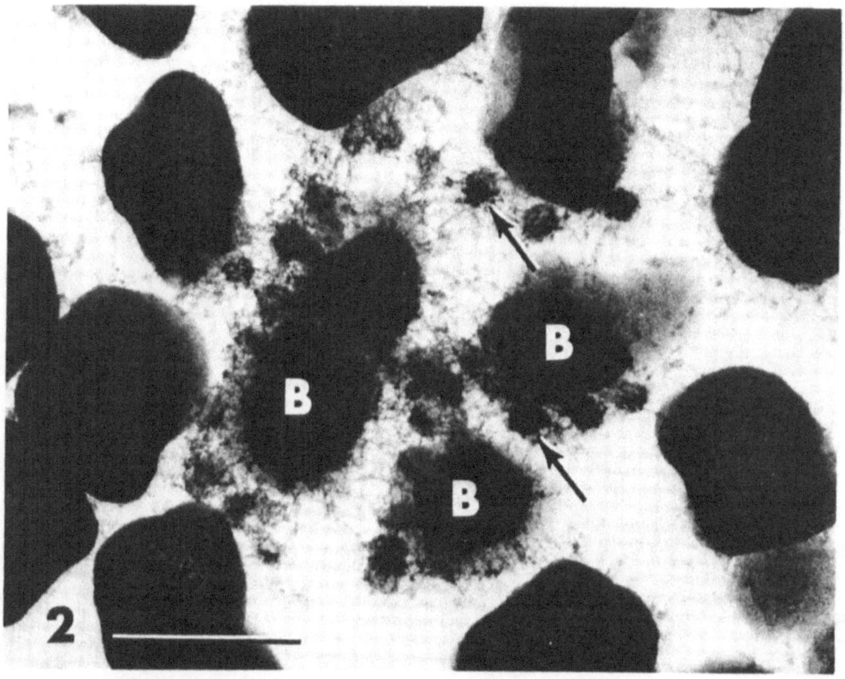

FIGURE 2. As for Figure 1 except that section double stained on the grid with uranyl acetate and lead citrate. The additional staining further darkens the ruthenium red stained materials. This and subsequent micrographs are printed at high density and low contrast to emphasize tenuous components of the bacterial and microvillar glycocalyx. This technique causes epithelial organelles to appear abnormally dark. The bacteria (B) in this section are connected by fibrillar, amorphous, sometimes branching ruthenium red stained strands to each other and to microvilli (arrows) in cross section. The presence of strands between bacteria is evidence that the ruthenium red stained material which extends between bacteria and microvilli is at least partially bacterial in origin. Bar = 1 μm.

The bacterial glycocalyx was visualized as a dense, flocculent, or fuzzy material of amorphous structure in tissues treated with antisurface polysaccharide serum (Figure 8). The surface polysaccharide has a greater density and is more evenly distributed around the bacterium than in the ruthenium red treated tissues, probably because the antibody cross-linking of the surface polysaccharides has prevented their collapse. Recall that ruthenium red does not prevent the collapse of surface polysaccharides. The antibody-coated material extended between bacteria in the lumen, between bacteria adherent to mucosal epithelial cells and bacteria in the lumen and between bacteria and mucosal epithelial cells (Figures 8 to 10). An early stage of adherence to intact microvilli can be seen in Figure 8. The bacterial polysaccharides clearly fill the space between the bacterium and the microvilli and extends to two adjacent bacteria as well. This space was filled with ruthenium red in earlier micrographs.

The role of antibody-coated material in late stage adherence is more difficult to evaluate as the density of the bacterial surface material is such that it is difficult to visualize the 11 nm space between the bacteria and the surfaces of mucosal epithelial cells. However, antibody-coated material can be seen to extend between closely adherent bacteria and adjacent bacteria or bacteria in the lumen (Figures 9 and 10).

IV. CONCLUSIONS

The prominence of ruthenium red and antibody-coated bacterial material at sites of bacterial

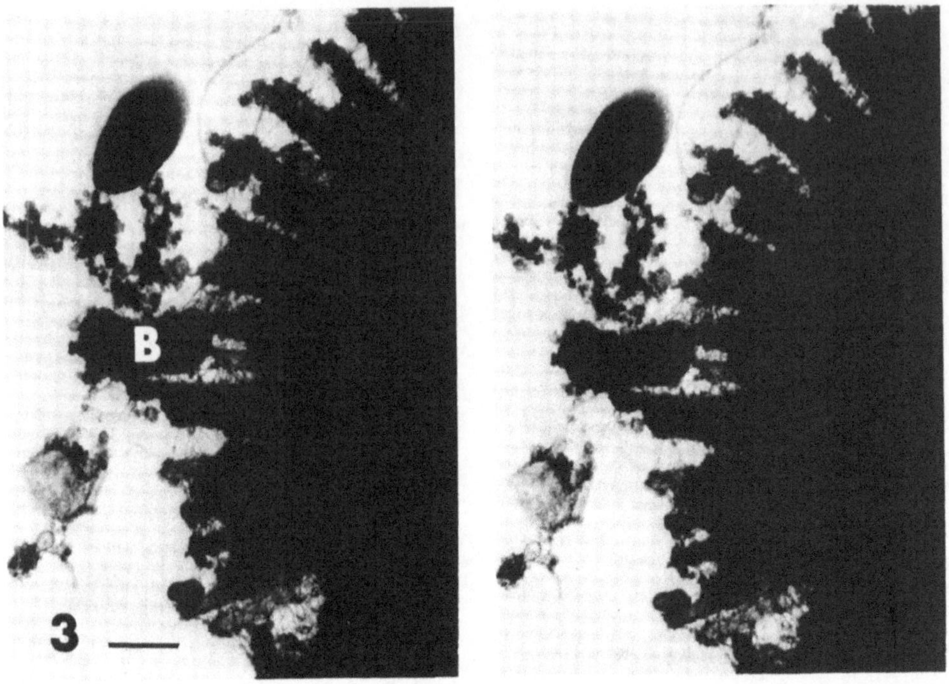

FIGURE 3. Stereopair high voltage electron micrographs. Tissues and staining as for Figure 2. The figures may be viewed in depth with a magnifying stereoviewer or by crossing the eyes until the two figures appear as three. The center figure will be seen in depth. A bacterium (B) is adhering to intact microvilli (early stage adherence) by strands of ruthenium red stained material. Strands and globs of ruthenium red stained material connect the adherent bacterium to two luminal bacteria (one of the bacteria was cut in a grazing section). Bar = 0.5 μm.

adherence, the structure of the material, the long distances it was capable of stretching to anchor bacteria to each other and to mucosal epithelial cells, etc., are all good reasons to suspect that the surface polysaccharides of the RDEC-1 bacterium play a role in adherence. Morphologic studies such as these cannot provide irrefutable evidence but they certainly are a sound basis for further exploration of the role of bacterial glycocalyx in adherence.

How could both pili and bacterial surface polysaccharides play a role in adherence? The RDEC-1 bacterium adheres first to the Peyer's patch lymphoid follicle epithelium and then to the absortive epithelium of the ileum, cecum, and colon. Perhaps pili and negatively charged surface polysaccharides mediate adherence to different epithelia. It is also possible that one surface structure mediates adherence to intact microvilli (early stage adherence) and the other mediates adherence to epithelial cells lacking microvilli (late stage adherence).

There is considerable morphologic evidence for a role of bacterial surface polysaccharides in bacterial adherence to mucosal epithelia.[9,10] Additional evidence has been gathered in infection experiments using mutants lacking negatively charged surface polysaccharides and in certain in vitro experiments.[11-15] However, a great deal more and varied proof is required before a substantial argument can be made. If one accepts the possibility of a role for bacterial glycocalyx in bacterial adherence, what mechanisms exist for its interaction with mucosal epithelial surfaces? It has been hypothesized that surface polysaccharides could be cross-linked with epithelial cell glycocalyx via divalent cations (glycocalyx is anionic), lectins, or enzymes.[9,10] As with the possible adherence role of bacterial surface polysaccharides, the possible mechanism of bacterial glycocalyx-epithelial cell interaction also requires a great deal of investigation before definite conclusions can be drawn. These questions will be the subject of future investigation of the RDEC-1 model of enteroadherent bacterial diarrhea.

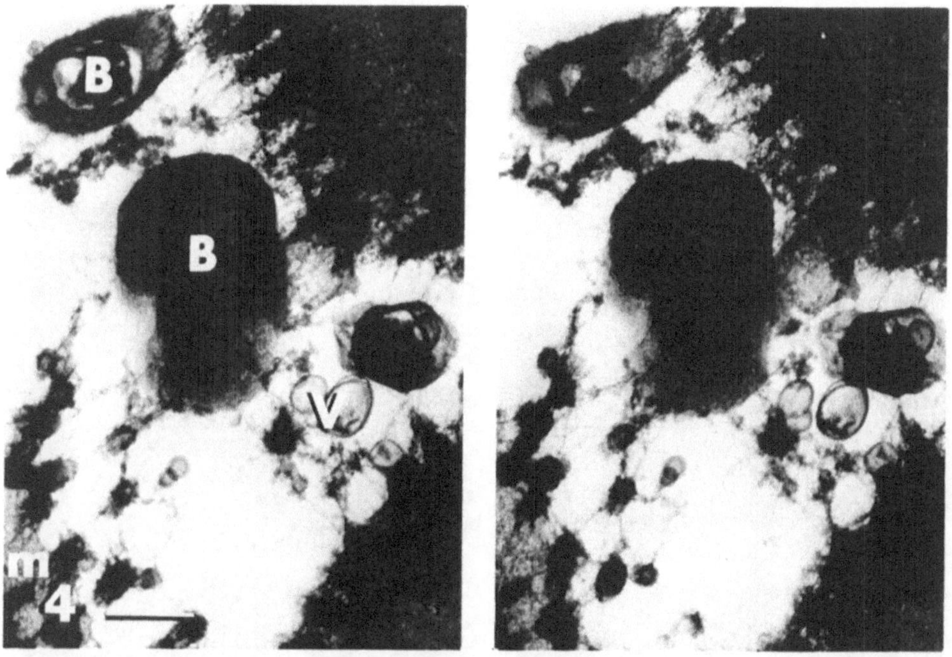

FIGURE 4. Stereopair high voltage electron micrographs. Tissues and staining as for Figure 2. Several bacteria or portions of bacteria (B) are adhering to microvilli [early stage adherence, microvilli cut in longitudinal and cross-section (m)] by fibrillar strands of ruthenium red stained material. Several vesicles (V) are present in the lumen. Bar = 0.5 μm.

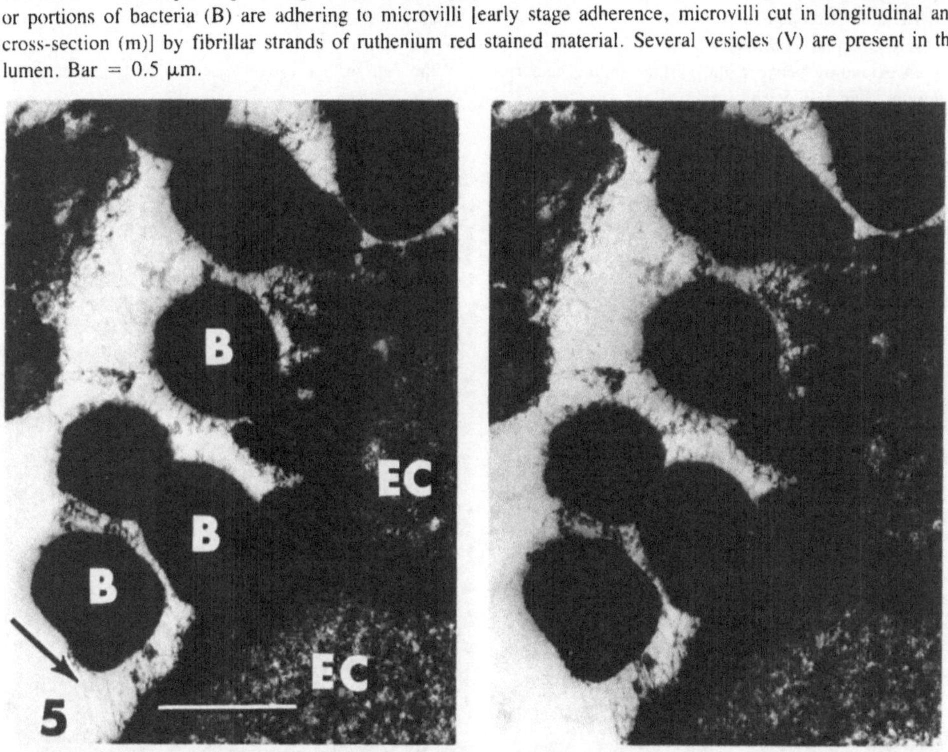

FIGURE 5. Stereopair of high voltage electron micrographs. Tissues and staining as for Figure 2. Bacteria (B) can be seen adhering to the surface of epithelial cells (EC) which have lost most of their microvilli (late stage adherence). The fibrillar, sometimes branching, sometimes amorphous strands of ruthenium red positive material can be seen extending a considerable distance between bacteria, between bacteria and microvilli and between a bacterium and the surface of the epithelial cell (arrow). Note the lack of strands on the luminal surface of the latter bacterium. This is probably due to the collapse of the unattached ruthenium red stained material onto the surface of the bacterium. Bar = 1 μm.

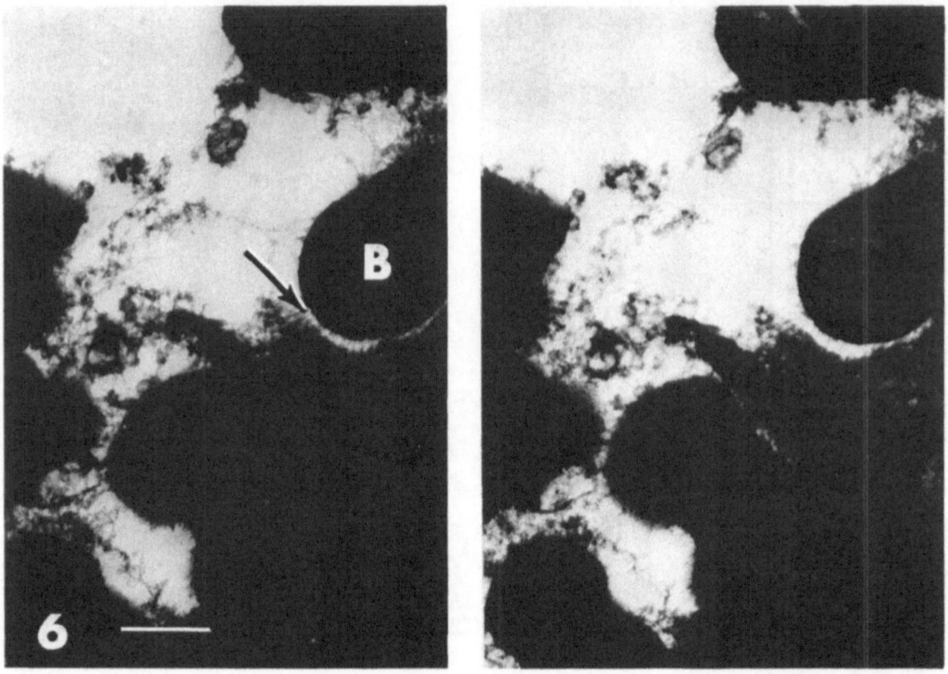

FIGURE 6. Stereopair high voltage electron micrograph. Tissues and staining as for Figure 2. A bacterium (B) is closely adherent to a mucosal epithelial cell lacking microvilli. Strands of ruthenium red stained material can be seen extending between the surface of the bacterium and the cell surface (arrow) as well as to an adjacent bacterium. Bar = 0.5 μm.

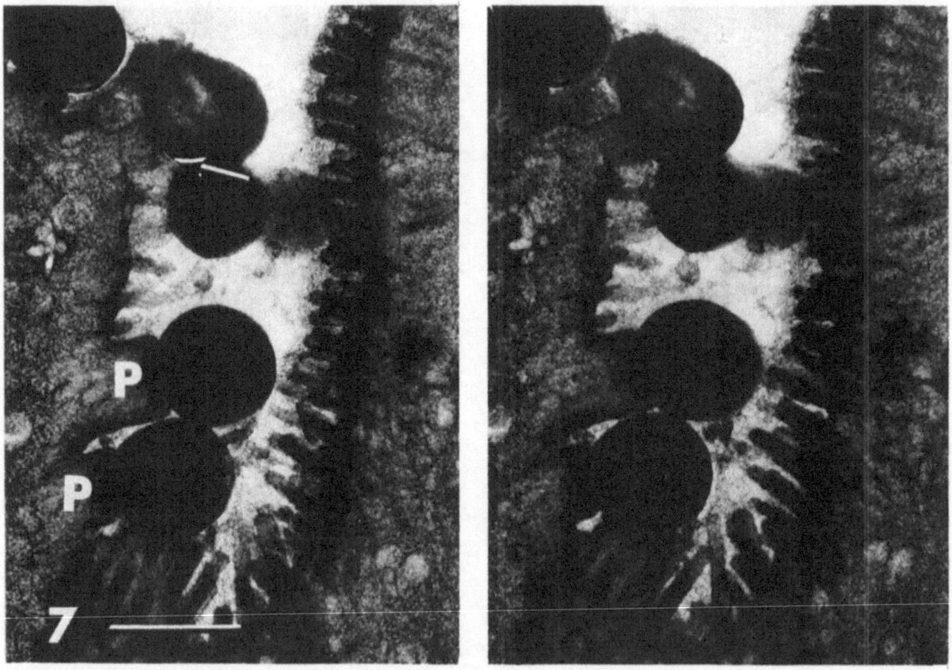

FIGURE 7. Stereopair high voltage electron micrographs. Tissues and staining as for Figure 2. Several bacteria are adhering to pedestals (P). The 11 nm space between a closely adherent bacterium and a pedestal is clearly visible (arrow). The ruthenium red positive negatively charged polysaccharide of the bacterial luminal surface is visible as fibrillar strands which are surrounded by the glycocalyx mesh of adjacent microvilli. Bar = 1 μm.

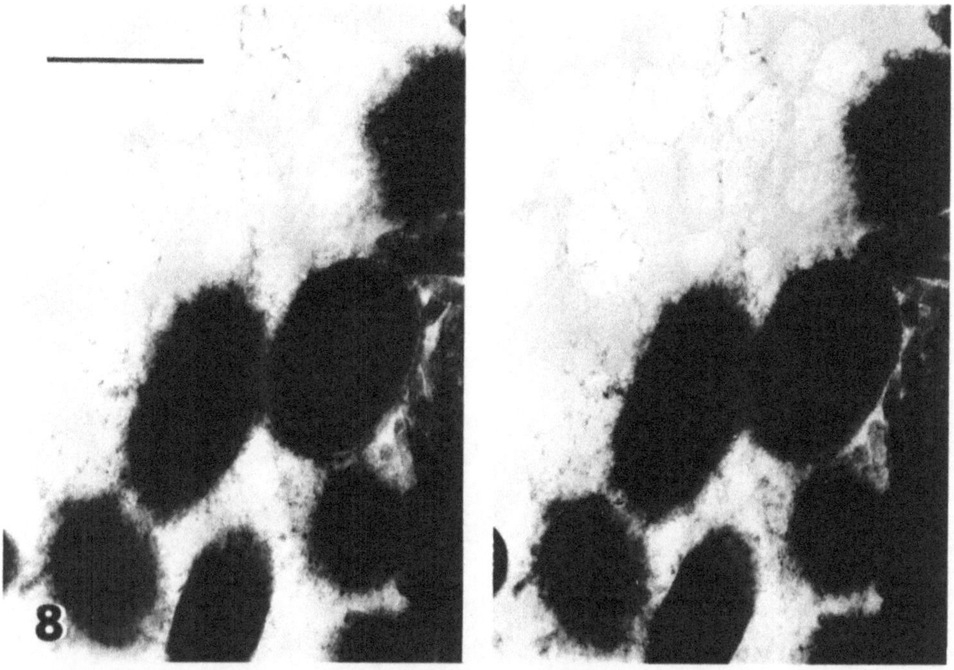

FIGURE 8. Stereopair high voltage electron micrograph of ileal mucosal tissue from a rabbit with strain RDEC-1 *Escherichia coli* diarrhea. Tissues treated with antiserum to strain RDEC-1 surface polysaccharide. The dense fuzzy coating of the bacteria is due to the reaction of the antiserum with the bacterial surface polysaccharides. The bacteria are adherent to each other and to intact microvilli of an epithelial cell. The antibody-coated material fills the space that is empty in conventionally stained tissues or that contains ruthenium red stained material in ruthenium red stained tissues. Bar = 1 μm.

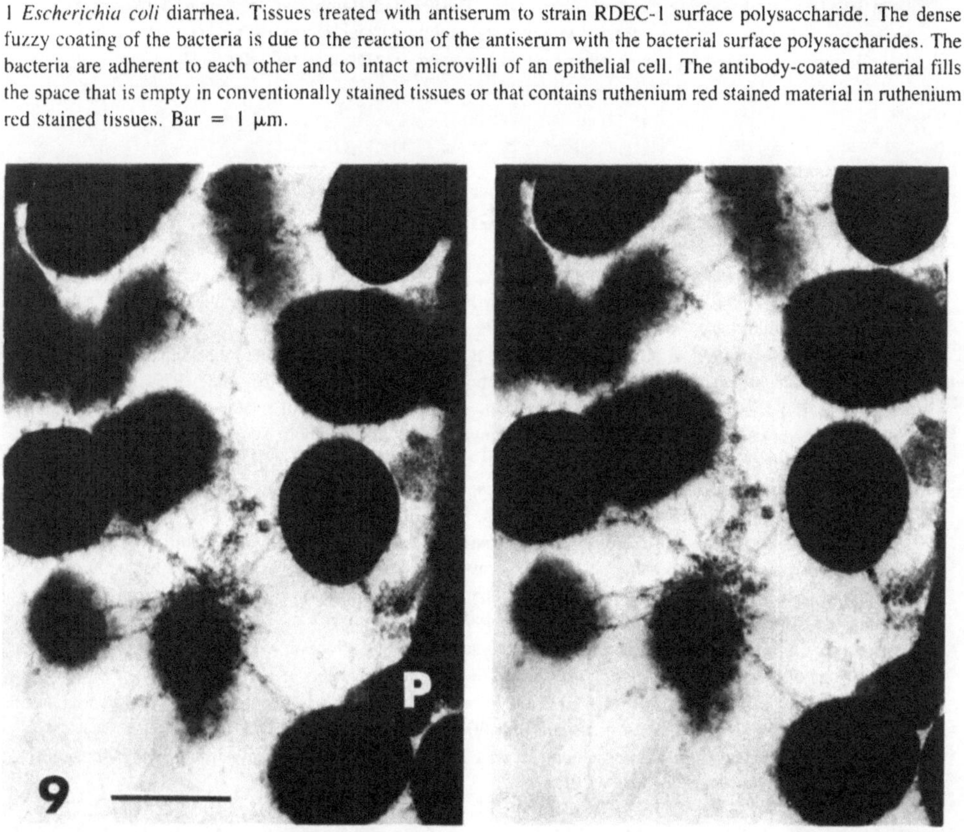

FIGURE 9. Stereopair high voltage electron micrographs. Tissues and staining as for Figure 8. Bacteria on pedestals (P) and in the lumen are connected to each other and to the plasmalemma of the epithelial cell by fuzzy strands of antibody-coated material. Bar = 0.5 μm.

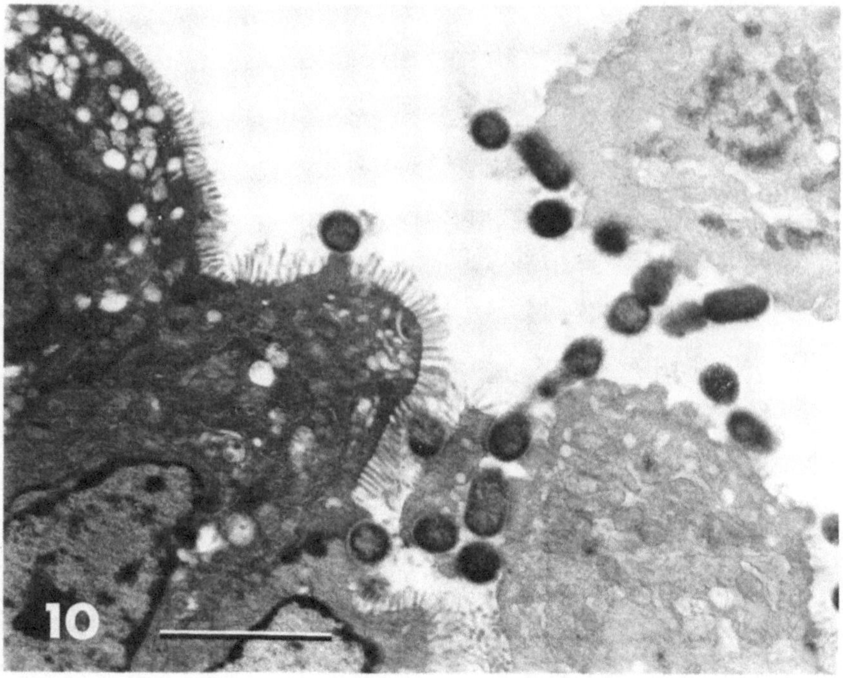

FIGURE 10. High voltage electron micrograph. Tissues and staining as for Figure 8. A single bacterium with a fuzzy coat is adherent to a pedestal among normal microvilli (top, center). Several bacteria are cross-linked with each other by fuzzy antibody-coated material. Bar = 2.5 μm.

REFERENCES

1. **Cheney, C. P., Schad, P. A., Boedeker, E. C., and Formal, S. B.,** The role of pili in adherence of enteropathogenic *Escherichia coli* to rabbit brush borders, Abstracts of the Annual Meeting, American Society for Microbiology, Washington, D.C., 1979, 19.
2. **Cheney, C. P., Boedeker, E. C., and Formal, S. B.,** Quantitation of the adherence of an enteropathogenic *Escherichia coli* to isolated rabbit intestinal brush borders, *Infect. Immun.*, 26, 736, 1979.
3. **Cantey, J. R., Lushbaugh, W. B., and Inman, L. R.,** Attachment of bacteria to intestinal epithelial cells in diarrhea caused by *Escherichia coli* strain RDEC-1 in the rabbit: stages and role of capsule, *J. Infect. Dis.*, 143, 219, 1981.
4. **Bayer, M. E. and Thurow, H.,** Polysaccharide capsule of *Escherichia coli:* microscope study of its size, structure, and sites of synthesis, *J. Bacteriol.*, 130, 911, 1977.
5. **Mackie, E. B., Brown, K. N., Lam, J., and Costerton, J. W.,** Morphological stabilization of capsules in group B streptococci, Types Ia, Ib, II, and III, with specific antibody, *J. Bacteriol.*, 138, 609, 1979.
6. **Luft, J. H.,** Ruthenium red and violet. I. Chemistry, purification, methods of use for electron microscopy and mechanism of action, *Anat. Rec.*, 171, 347, 1971.
7. **Westphal, O. and Jann, K.,** Bacterial lipopolysaccharides: extraction with phenol-water and further applications of the procedure, *Meth. Carbohydrate Chem.*, 5, 83, 1965.
8. **Hungerer, D., Jann, K., Jann, B., Ørskov, F., and Ørskov, I.,** Immunochemistry of K antigens of *Escherichia coli* 4. The K antigen of *E. coli* 09:K30:H12, *Eur. J. Biochem.*, 2, 115, 1967.
9. **Costerton, J. W., Irvin, R. T., and Cheng, K.-J.,** The role of bacterial surface-structures in pathogenesis, *CRC Crit. Rev. Microbiol.*, 8, 303, 1981.
10. **Costerton, J. W., Geesey, G. G., and Cheng, K.-J.,** How bacteria stick, *Sci. Am.*, 238, 86, 1978.
11. **Smith, H. W. and Huggins, M. B.,** The influence of plasmid-determined and other characteristics of enteropathogenic *Escherichia coli* on their ability to proliferate in the alimentary tracts of piglets, calves, and lambs, *J. Med. Microbiol.*, 11, 471, 1978.

12. **Carruthers, M. M.,** *In vitro* adherence of Kanagawa-positive *Vibrio parahaemolyticus* to epithelial cells, *J. Infect. Dis.*, 136, 588, 1977.
13. **Carruthers, M. M. and Anderson, B.,** Inhibition by polyanions of adherence by Kanagawa-positive *Vibrio parahaemolyticus:* a physicochemical effect, *J. Infect. Dis.*, 140, 119, 1979.
14. **Chan, R., Costerton, J. W., and Acres, S. D.,** Experimental infection of a colostrum-deprived calf by a pilus-minus strain of enteropathogenic *E. coli,* Abstracts of the Annual Meeting, American Society for Microbiology, Dallas, TX, 1981, p. 23.
15. **Hadad, J. J. and Gyles, C. L.,** The role of *Escherichia coli* polysaccharide K antigen in colonization of the intestine. Abstracts of the Annual Meeting, American Society for Microbiology, Dallas, TX, 1981, p. 32.
16. **Ørskov, F., Ørskov, I., and Cantey, J. R.,** unpublished observations

Chapter 20

OBSERVATIONS ON REOVIRUS TYPE 1: SPECIFIC ADHERENCE TO AND TRANSPORT THROUGH THE INTESTINAL MEMBRANOUS EPITHELIAL (M) CELL IN MICE

Jacqueline L. Wolf

Using the mammalian reoviruses as a model system, we have been studying how viruses that cause systemic disease penetrate the gastrointestinal tract and enter the host. We initially examined the penetration of the gastrointestinal mucosa by reovirus type 1 which causes ependymitis in suckling mice. For this purpose, closed ileal loops were prepared in suckling mice and injected with approximately 10^{11} particles of purified reovirus type 1. Loops were removed between 30 min and 4 hr after inoculation and processed for electron microscopy.

No viral particles were detected adherent to or within intestinal absorptive cells, or goblet cells lining Peyer's patches or adjacent villi. In contrast, many viral particles were applied to the stunted microvilli and microfolds of M cells. Figure 1 shows the events occurring 1 hr after inoculation of a closed ileal loop with reovirus type 1. An absorptive cell (A) is present on the far left with no adherent viruses on the surface. An intestinal M cell (M) is located to the right of the absorptive cell. On the luminal surface of the M cell are many adherent reoviruses (arrow) on and between the microvilli and microfolds. Specific M cell adherent virus was found at 30 min, the earliest time the loops were examined. The reovirus is transported through the M cell within vesicles and extruded into the extracellular (E) space (large arrow) in close proximity to the mononuclear cells. By 4 hr viruses can be found below the basement membrane within the Peyer's patch. In confirmation of these results, Kauffman et al. (*Virology*, 124, 403, 1983) have found, using plaque assays, that after oral inoculation of mice with type 1 reovirus, viable reovirus is found early within Peyer's patches followed by subsequent spread to the mesenteric nodes and finally to the spleen. Our studies suggest that the intestinal M cells are the site where reoviruses and possibly other viruses that cause systemic infection penetrate the intestinal epithelium.

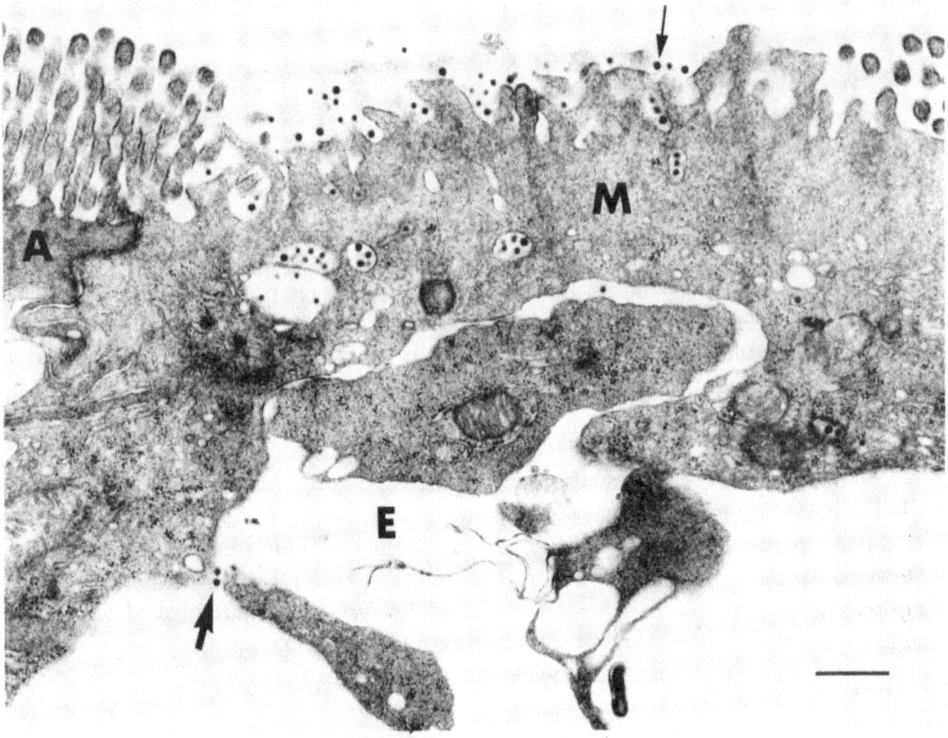

FIGURE 1. Events occurring 1 hr after inoculation with reovirus type 1 of a closed ileal loop made in a 10-day-old mouse. The electron micrograph shows a membranous epithelial cell (M) with reovirus type 1 (small arrow) adhering to its luminal surface being transported within vesicles across the cell and lying free (large arrow) within the extracellular space (E). The adjacent absorptive cell (A) has no adherent reovirus. Bar represents 500 nm. (Reprinted with permission from Fields, B., Molecular basis of reovirus virulence, *Arch. Virol.*, 71, 95, 1982.)

INDEX

A

B

C

D

E

F

G

H

I

J

K

L

M

N

O

P

R

S